INTRODUCTION TO NUCLEAR ENGINEERING

Introduction to Nuclear Engineering

INAM-UR-RAHMAN and PAULINUS S. SHIEH

Robert E. Krieger Publishing Company, Inc.
1981

Original edition 1981

Printed and Published by
ROBERT E. KRIEGER PUBLISHING CO., INC.
645 NEW YORK AVENUE
HUNTINGTON, NEW YORK 11743

Printed in the United States of America

Library of Congress Cataloging in Publication Data

Rahman, Inam-ur, 1937-
Introduction to nuclear engineering.

Bibliography: p.
Includes index.
1. Nuclear engineering. I. Shieh, Paulinus S., 1931- joint author. II. Title.
TK9145.R33 621.48 80-13160
ISBN 0-88275-972-8

TO ASMA AND GRACE

TABLE OF CONTENTS

PART III: RADIATION ENGINEERING

PREFACE

Since the early nineteen fifties, it has been difficult to visualize a vivid picture of the essential subject of studies in nuclear engineering. The rapid extension of the domain of nuclear engineering is making the picture more and more blurred. Presently, in most of the nuclear engineering programs major emphasis is directed towards the activities and courses related to power reactors alone. Though this certainly forms an essential part of nuclear engineering, we must not overlook other important aspects like radioisotope applications which are becoming increasingly significant in industry, medicine, agriculture and numerous other fields. Moreover, for imparting a balanced concept of the entire domain of nuclear engineering, it is imperative to have a well-organized and coherent curriculum which should not be amenable to drastic changes due to some transitory sociological problems such as the energy shortage and the environmental debate. A nuclear engineering curriculum offered by various institutions should therefore possess a sufficient degree of uniformity as well as stability. Such a conviction has inspired us to make an attempt to write this textbook for an introductory course of nuclear engineering to obtain a clear picture of subjects that should be included in a nuclear engineering curriculum.

The materials contained in this text are organized in three distinct parts: a) nuclear fundamentals, b) nuclear power engineering, and c) radiation engineering. It is intended to give a bird's eye view of the field of nuclear engineering in its entirety to the new entrants in this field. Most of the material in this text book has been derived from the lecture notes of the introductory course of nuclear engineering taught by us at the Pakistan Institute of Nuclear Science and Technology.

Since it is inevitable that some inaccuracies will creep into a work of this type, we shall welcome our readers to point out any mistakes whatsoever, and to forward their suggestions to make this book more beneficial.

The reader of this book is assumed to have one year each of college level general chemistry and general physics. He is also assumed to possess a basic knowledge of differential equations. The material in this book can be assimilated in a one semester course in about fifty contact hours. If extra material pertaining to detailed discussions of some chapters is utilized and examples and problems are studied, it can be used as a main reference text for a two semester course. This book is also intended as a general reference for those who have their basic training in other science or engineering fields and are working in radiation and nuclear power related fields.

No attempt is made to adopt a single unit system for this text. This is due to the fact that British, the cgs, and the MKS unit systems are widely used in nuclear engineering without any systematic effort exerted to streamline these usages into a single unit system. In fact, this is a challenge facing all nuclear radiation workers. The sooner we achieve a universal unit system (SI) the better for the nuclear development. In order to help the readers ofthis book to sort themselves out of the maze of units, we have compiled an exhaustive table of units and conversion factors as an appendix.

The authors would like to express their gratitude to all those who directly or indirectly helped in various ways. In particular the authors wish to gratefully acknowlege: Miss Atika Rabbani for proof reading the entire rough manuscript and making numerous valuable suggestions; Mr. Khalid M. Bukhari for his dedicated and very systematic work of compiling the appendices in the present shape; Mr. A. Hamid for typing the entire manuscript a couple of times with great skill and accuracy; Mr. Matloob Hussain for making all the figures and Mr. Javed Iqbal for typing most of the appendices. We would also like to thank the Pakistan Atomic Energy Commission for providing all the facilities without which it would not have been possible to bring out this book. Specifically, we would like to acknowledge the support and guidance given us by the Chairman of the Pakistan Atomic Energy Commission Mr. Munir Ahmad Khan, and our senior colleagues Dr. Ishfag Ahmed, Dr. Naeem Ahmad Khan, and Dr. S. A. Hasnain. The final form of the manuscript was reviewed by Dr. L. Wang Lau of TVA. We would like to express our deep appreciation for his comments and helpful suggestions.

I. U. R.
P. S. S.

PART I

NUCLEAR ENGINEERING FUNDAMENTALS

CHAPTER I

INTRODUCTION

Ever since the discovery of x-rays by Wilhelm Conrad Roentgen in 1895, the study of radiation and its applications has been carried out in varying degrees at different periods of time. This ultimately led to the formation of the branch of engineering known as Nuclear Engineering. With the passage of time, many subatomic "particles" were discovered to broaden the meaning of the term "radiation". When electrons were identified by John Joseph Thomson in 1897, the views of the structure of matter were drastically changed. Then, in 1899, alpha particles were found by Ernest Rutherford to be the nuclei of helium, as a result of his study of air conductivity due to the presence of uranium. Paulk Villard, in his paper entitled "On the Reflection and Refraction of Cathode Rays and Deviable Rays of Radium", published in 1900 in *Comptes Rendus,* discovered a highly penetrating ray. The propagation of this ray was not affected by a magnetic field. This, of course, marked the discovery of the gamma (γ)-ray. Perhaps the most significant events that effected the history of the world were the series of studies during the period extending from 1920 to 1932 on the highly penetrating radiation emitted from light atoms such as lithium and beryllium when they are bombarded with alpha particles. During the early part of this period many physicists such as Ernest Rutherford, O. Masson, and W.D. Harkins postulated the existence of "neutrons" as such a highly penetrating radiation. However, the characteristics of a neutron and the proof of its existence were not known until 1932 when James Chadwick calculated the neutron mass to be 1.008 amu. Since then, scores of subatomic particles have been discovered.

As early as 1939, Otto Hahn and Fritz Strassman utilized a radiochemical technique to identify barium as a fission fragment resulting from the neutron irradiation of uranium. Lise Meitner and Otton R. Frisch confirmed theoretically, by using the liquid-drop model of a nucleus, the possibility of fission reaction under suitable conditions. At about the same time, Niels Bohr established a firm theoretical foundation of nuclear fission. From that time onward nuclear chain reaction research was intensified to such an extent that many similar studies were carried out by scientists in numerous countries. It is uncertain as to who was the first person to succeed in achieving the first nuclear chain reaction. However, in the USA a group of scientists under the leadership of Enrico Fermi successfully constructed the first chain reacting pile in 1942. It is not an exaggeration to say that it marked the beginning of the end of the second world war. Fermi's first chain reacting system (CP-1) consisted of 23.9 tons of natural uranium in the form of cylindrical rods arranged in stacks of 14,000 graphite blocks. The system had no similarities in appearance with the nuclear explosives which were developed three years later under a crash program at Hanford Laboratories as part of the Manhattan project.

Since the end of World War II, the harnessing of the tremendous energy released from nuclear fission and fusion for peaceful applications has become a hotly pursued subject. This is prompted by many factors. First of all, the world population growth combined with the rising expectations make the future energy requirements seem staggering. Secondly, the availability of energy in a very concentrated form in nuclear fuels offers very attractive alternatives to the transportation industry, including space exploration. Even with new discoveries such as the Alaskan and numerous other oil-finds all over the world, presently available fossil fuel reserves, combined with the increased efficiency of utilization, will not suffice to fufill such huge energy demands in the future. It is very important, therefore, to develop new sources of energy production which can supplement the present fossil and hydro resources. In fact, it is a distinct possibility that in the near future we will be using nuclear powered air ships, trains, rockets as well as the presently available ocean vessels. The intellectual curiosity in the ever lasting search of better ways and means to provide service to mankind is yet another factor that is by no means negligible. Scientists and engineers from many disciplines have joined hands in carrying out many of the experiments centered around the design, construction, and utilization of nuclear reactors, thus marking the beginning of a new branch of engineering — Nuclear Engineering.

The present field of studies in nuclear engineering covers a wide range. These studies may be grouped into two categories: power engineering — the design, construction, and operation of nuclear reactors; and radiation engineering — the applications of the energy derived from radiation sources, including that extracted from an operating nuclear reactor. In a way they can be considered as parallel to the power and communication studies in electrical engineering with the exception that the range of radiation application is so diverse that it is practically unlimited. The American Nuclear Society (ANS) gives the following definition of Nuclear Engineering:

The Branch of Egineering concerned with the release, control, and utilization of energy from Atomic and Nuclear sources and systems

Figure 1.1 illustrates the nuclear engineering field and the typical activities associated with each of its two major functions. Although the modern nuclear reactor systems are much more sophisticated as compared with those of twenty years ago, new concepts and experiments are brought forth constantly to explore better methods of extracting enegry from nuclear reactions.

- Nuclear Engineering
  - Power Engineering
    - Reactor Engineering (fission)
      - New reactor concepts
      - Reactor design and analysis
      - Process and safety system engineering
      - Reactor materials and shielding
    - Thermonuclear Engineering (fusion)
  - Radiation Engineering
    - Radioisotope engineering
    - Geographical engineering
    - Environmental hazard analysis and control
    - Radiography and nondestructive testing
    - Radioactive tracer techniques
    - Neutron activation analysis
    - Radioisotope power
    - Radiogauging

Fig. 1.1: Nuclear engineering activities.

In contrast with the rapid production of nuclear explosives from 1939 to 1944, the development and improvement of reactor systems for power generation and other applications have been rather slow. This is mainly due to the lack of experience and the cost of carrying out experiments pertaining to new concepts and ideas on reactor design. In addition, the uncertainty involved in the values of many reactor design parameters makes the choice of an optimum design a rather difficult task. While these difficulties and uncertainties are inevitable, the basic process of reactor design is gradually developing. Today, a reactor design project is a sequence of standard events starting from a nuclear reactor concept to the maintenance and the operation of a nuclear reactor plant, The procedures one follows to carry out such a project are similiar to those for a conventional power plant with the only

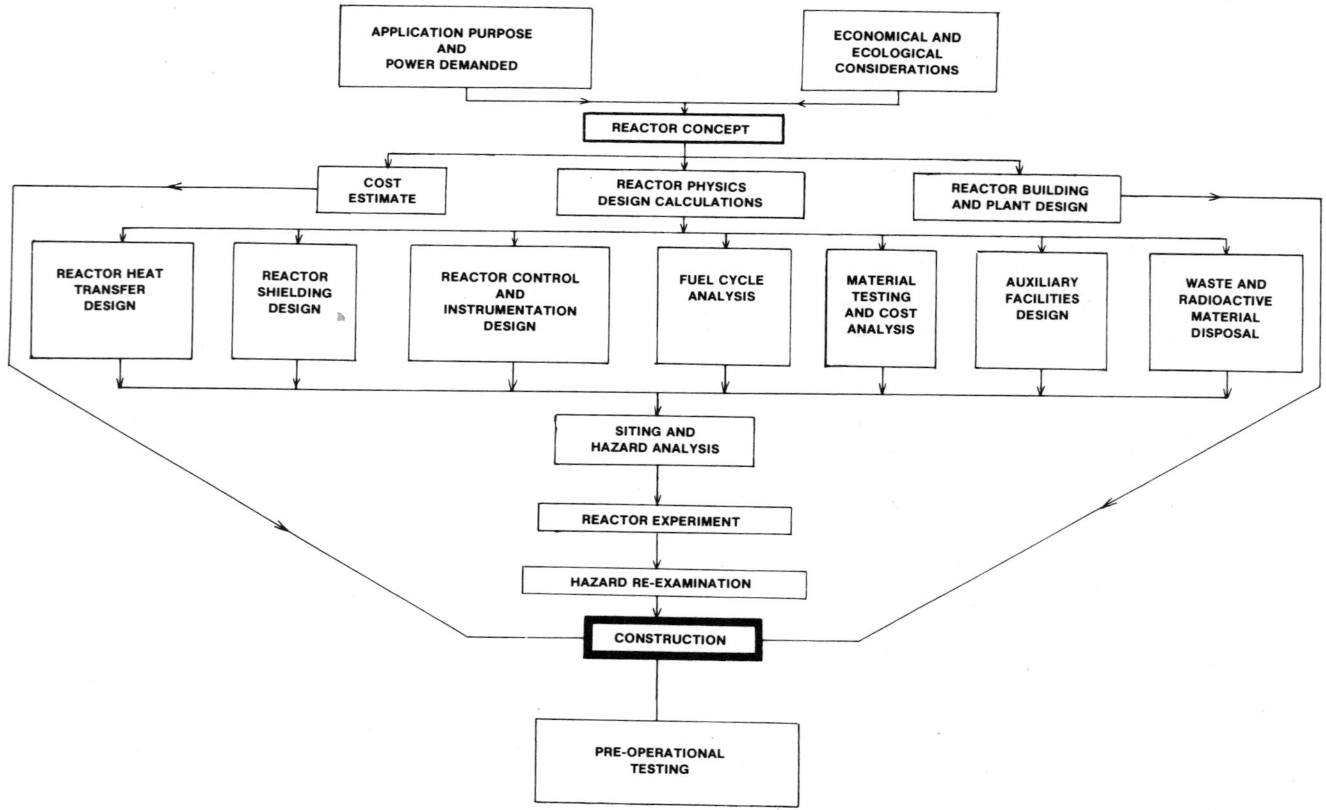

Fig. 1.2: Reactor system design flow chart

exception that radiation hazard (including decay heat removal) is the foremost consideration during every stage of planning. A typical nuclear reactor design flow chart (Figure 1.2) illustrates the various stages of planning and their order of execution.

The application of radiation and the associated enegry had its beginning when x-rays were discovered. Medical applications of x-rays had been in common use long before any nuclear reactor was built. The range of applications of nuclear energy in general has gradually broadened to such a state today that it is impossible to give a complete description of these activities. However, a few subjects of interest will be mentioned here to show the wide range of such applications (see Figure 1.1). Due to the wide applications of high intensity irradiations in industry and agriculture, the utilization of strong radiation sources ($\sim 10^4$ curies or above) and the design of such irradiation facilities have gradually become a field of specialization. This field is called radioisotope engineering. Neutron activation analysis is now utilized in a wide variety of studies such as crime detection, trace identification for biological specimens and geological samples, and, of course, radioisotope analysis for research, industry, the life sciences, medical diagnosis and therapy, and agricultural studies. The formation of health physics as a distinctive body of study is a recent development. Due to the close association between reactor operation and health physics activities, many nuclear engineers gain enough knowledge in hazard analysis and protection to carry out health physics activities in an industrial establishment or a research complex as safety officers. Lately, many new applications of nuclear energy have been brought to surface. The potentials in these directions are limited only by one's imagination. The applications of nuclear energy for geographical engineering is one of such examples.

CHAPTER 2

ENERGY PROBLEMS AND SOLUTIONS

2.1 Energy Requirements

The future of the world depends on new sources of energy since the present sources of fossil fuel will not meet future energy requirements. By the year 2000, the world population is expected to exceed six billion. While it took thousands of years for the world population to reach one billion in 1820, it is now taking only about ten years to add another billion. This is happening at a time when people all over the world are struggling for better living conditions. This projected population growth, combined with the pressing need to raise the living standards of a large portion of the world's population, makes the future energy requirements seem staggering. Figures 2.1 and 2.2 indicate the world population and total energy consumption based on the present predictions of population and energy growth. Presently available low cost fossil fuel reserves will not suffice to supply the huge energy demand in the future and therefore one of the most vital necessities for the continued progress of mankind is the increase of energy production. It is very important, therefore, to develop new sources of energy production which can supplement present fossil and hydro resources.

The fossil fuel resources of coal, oil and natural gas are economically

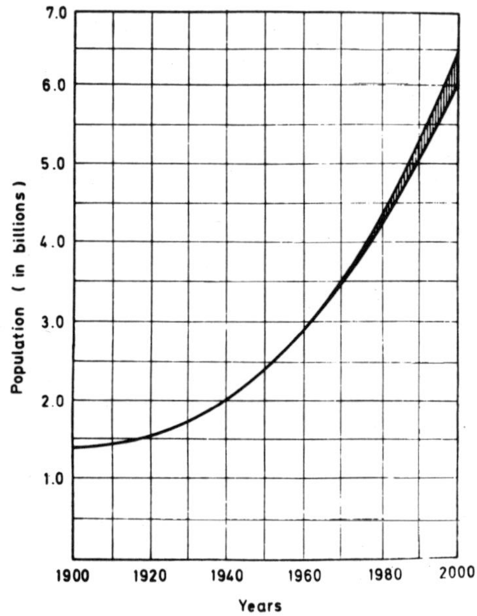

Fig. 2.1: World-wide growth of population.

finite, while hydro and other natural resources are fixed and have a limited exploitation potential even with impractical shifts of population. If the predictions regarding the increase in the world population and the reserves of various energy resources are even partially true, a tremendous increase in new levels of energy production will be required during the coming years. It is estimated that low cost and readily available fossil fuel will be exhausted in a few decades at the present rate of energy consumption. Conditions become more critical when one considers the increasing rate of energy consumption. Proper energy planning is therefore extremely important. Long before the existing energy resources are to be exhausted, long range plans should be drawn for the utilization of other energy sources. Such planning will not only keep the power costs low and the overall transition smooth and streamlined, but will also avoid a situation where the existing resources are all depleted and nothing is left to replace them. If the importance of energy is not realized well in time, the present civilization will soon approach the peak of its attainment, and a decline will follow in various aspects for life, including the living standards. This is especially frightening for the developing countries where the majority of the population is already on subsistence living! Also, some of the fossil fuels are very valuable for many nonpower producing purposes like

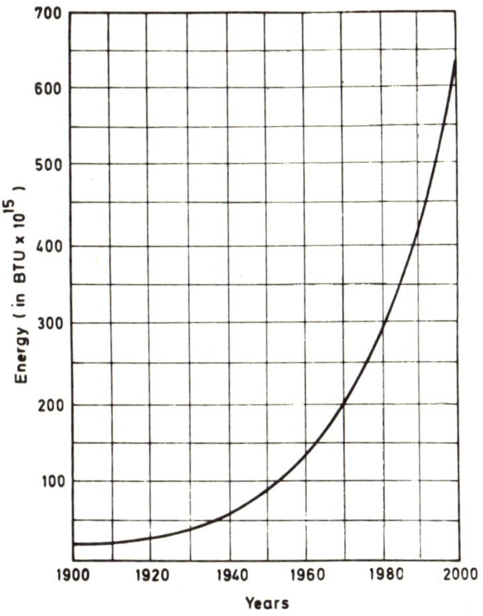

Fig. 2.2: World-wide growth of energy consumption.

transportation, small isolated heat installations, as sources of industrial chemicals, and should not be all burned for the production of electrical energy.

It is difficult to draw precise conclusions from a vast number of conflicting and incomplete predictions which confront anyone who attempts to analyze the adequacy of the world's fossil fuel resources. Various predictions are based upon different basic assumptions and methods of calculation. Sometimes personal prejudices and commercial interests enter these estimates. Besides, predictions regarding the future population increase and per capita energy demands may not be correct, for no one can speak about the future precisely. In spite of all these varying predictions, one thing which seems inevitable is that we must find new sources of energy, since the era during which we can rely completely on fossil fuels for energy production is fast approaching an end. Tables 2.1 and 2.2 are compiled from many sources (26,35,62,71) to show this possible trend. It becomes obvious from the numbers given in this table that coal supplies the bulk of the world's energy reserve. Out of an estimated 111 Q remaining to be produced, only 19.2 Q should be available at prices not over 25 percent above the 1958 prices, and only 6 Q at the 1958 prices. Medium cost increment refers to fossil fuel reserves available at cost increments from 25 to 50 percent, while high cost increment means 50 to 300 percent over the 1958 prices.

Table 2.2 depicts the fossil fuel reserves for various periods if all the low, medium, and high cost fossil fuel reserves are utilized, and if it is assumed that no nuclear power is introduced on a large scale as an alternate source of energy. It is evident that if no new sources of energy are introduced, all reserves of fossil fuel available at 1958 price levels will be exhausted before the year 2000, while others with low, medium, and high cost increments will last probably as long as the year 2052. The table has been prepared on the present projections of world population and about 2.3 percent compound rate of annual increase in the energy production through the year 2000. Beyond the year 2000 it is assumed that total energy use grows at four percent per annum and that all the increase is supplied by fossil fuels. If a lower rate of growth is assumed beyond the year 2000, the reserves of fossil fuels will last longer than those predicted in Table 2.2. For example, with a two percent increase beyond the year 2000 all fossil fuels would be consumed by the year 2072, and with a one percent increase, fossil fuel reserves will last as long as the year 2100.

Table 2.1: World fossil fuel reserves in Q (1Q = 10^{18} BTU)

Fuel type	Cost increment required for availability		
	Low	Medium	High
Coal and lignite	15.3	18.5	27.8
Liquid hydrocarbons	3.2	1.6	1.6
Natural gas	0.7	1.1	3.0
Oil shale and tar sands	0.0	13.9	24.4
Total	19.2	35.1	56.8

Table 2.2: World fossil fuel reserves at the end of various periods in Q (assuming that reserves towards the end of 1958 were equivalent to 111 Q).

Period	Production during period	Reserves at the end of period
1959-1980	3.8	107.3
1981-2000	7.9	99.4
2001-2020	18.4	81.0
2021-2040	41.0	40.0
2041-2052	40.0	0.0

The need for additional sources of energy has also been emphasized in the 1962 United States AEC Report to the President. In this report it is mentioned that the readily available low cost fossil fuels would be exhausted in 75 to 100 years. These are the predictions for a country where about six percent of the world's population enjoys approximately 30 percent of the world's reserves of fossil fuels. The availability of low cost fossil fuels in most of the developing countries is rather bleak. Rapid industrialization in some of these countries is putting an extra burden on the limited fossil fuel supplies. It is therefore obvious that except for a very few countries which possess great resources of coal, oil, or hydro-electric potential, there is a general need for the development of new sources of energy to meet the rising power needs during the coming years.

In the light of the above predictions, it is only natural to think and plan for alternate sources of energy. A constant replacement of one energy source by another, before the previous one was exhausted or became economically unattractive, is confirmed by past history. New and better fuels came into use long before resources of the then existing fuels were depleted. For example,wood used to be the world's most basic fuel till the late eighteenth century, but due to heavy industrialization and increased demands on larger sources of energy, patterns kept on changing. By the year 1910, coal accounted for approximately 88 percent of the fuel used throughout the world. In later years, fluid fuel replaced most of the diminishing coal reserves. By 1961, the world fuel balance for commercial fuel was approximately 51.1 percent coal, 32.5 percent oil, 14.4 percent natural gas, and about two percent hydro-electric power. Considering the fast diminishing fossil fuel resources and rapidly expanding energy requirements, it is apparent that new sources of energy must be introduced. In some of the countries the fossil fuel supply situation is so precarious that a stagnation in growth is expected within the coming decade unless these fuels are supplemented.

2.2 Alternate Energy Sources

It was in the light of the factors mentioned already that alternate sources of energy which can be technically satisfactory, economically feasible and available in large quantities were studied. Although there are numerous methods of energy production besides the use of fossil fuels, each one has its limitations. Nuclear energy seems to be the only source which can presently meet all the requirements for large scale energy production. Some alternate energy sources are briefly discussed here.

Hydro Energy

Energy from falling water is one of the earliest and still one of the most economic sources of power. However, most of the economical sites for hydro-electric power production are already in use. Sites which involve construction of dams are usually less attractive unless they serve several purposes such as irrigation, flood control and power production. Generally speaking, the capital outlay for such a multipurpose project is much larger than that of any other type having the same power output. Also, such plants are dependent on geographical characteristics and are strictly location bound. The cost of transmission lines to the industrial centers and the additional cost of plant construction due to the lack of transportation facilities should also be included as part of the capital investment of hydroplants. Many countries are using waterfalls as low as 1.5 meters with specially designed turbines and generators for the production of power on a small scale. Although such plants may not play any significant role in the overall power economy of a country, these can help in saving some fuel which can be utilized for other more useful purposes. Presently, in France about 375 hydroplants are below 100 kilowatts, while in Czechoslovakia more than 5,000 such stations of less than 200 kilowatt capacity are in operation. In the USSR, as in many other countries, hydro-electric plants of less than 1,000 kilowatt capacity run into thousands.

Wind Energy

The energy crisis has brought back the antique windmill and a new wave of interest has arisen to turn it into a reliable source of energy. Winds are created due to the uneven distribution of solar energy in different regions of the world, the rotation of the earth, and local geographical conditions. The magnitude of wind energy on the earth's surface is approximately 100 Q. The kinetic energy of wind is transformed into mechanical power with the help of a rotor or propeller which is then used for driving an electric generator. The basic elements determining the obtainable energy are the wind speed, the area swept by the rotor and the conversion efficiency of the plant. The sizes of wind plants are very flexible and may vary from a fraction of a kilowatt to as much as 2,000 kilowatts. One of the larger operating wind power plants was designed and built by the Electricity Authority of France and has a maximum output of 900 kilowatts at a wind speed of 20 meters per second. It has three propeller blades 30 meters in diameter. Great emphasis is being placed on the development of large industrial plants in many countries. In some areas where wind velocities are not steady, the wind plants may run in conjunction with diesel or other standby units.

Solar Energy

On the earth's surface, solar energy, in the form of electromagnetic radiation, greatly exceeds the total energy consumption density even in the most industrialized countries of the world. This means that from a purely theoretical point of view, solar energy can fulfill most of the world's energy requirements even at a very low conversion efficiency.

Work is being done to convert solar radiation into electrical energy using thermoelectric generators and photoelectric cells. It will, however, take quite some time before the results of these experiments will provide energy economically. As things appear today, solar energy is expected to fulfill small scale energy needs and it may take many years before it can be used on an industrial scale. One of the difficulties with solar energy is that it is not constantly available and is particularly unsuitable for those situations where energy is needed on a continuous basis. Presently, solar energy is being used on a small scale in different parts of the world for water heating, space heating, cooking, distillation of seawater, drying fruits, refrigeration, driving pumps and heat engines, etc.

Energy from Oceans

Energy from the oceans may be harnessed in various ways. Successful production of energy on a relatively large scale has been achieved from the ocean tides. The theory of tides is very complicated, but it might suffice here to say that tidal energy results from the combined effects of the earth, moon and sun. The major contribution is due to the rotation of the earth around its own axis. Industrial utilization of tidal energy consists in taking advantage of the difference in sea level produced by tides. The amount of energy derived from a given location depends on the height of the tides, which in some locations approaches as much as 20 meters! So far, France has done more work in this field that any other nation. One such plant, producing a peak output of 150,000 kilowatts, is in operation with moderate electrical cost. Some important sites for the utilization of tidal energy are the Bay of Fundy in Canada with expected peak daily output of two million kilowatts, St. Michel in France with three million kilowatts capacity, along with numerous other sites with relatively smaller capacities. The Passamaquoddy Tidal Power Project which is under construction in Canada will utilize tides produced in the Bay of Fundy where, in a 12-hour cycle, the tide brings in about two billion tons of water each time it rises and falls an average of 5.5 meters.

There are some other procedures which are still in the experimental stages. These are the energy derived from ocean currents and from the considerable temperature differences that exist between the surface and the deeper layers of the ocean. Work is also being done to harness the energy from the ocean waves.

Geothermal Energy

The interior of the earth is very hot — less than 64 kilometers under the earth's surface are molten rock and gases which boil at temperatures up to 1000°C. These very high temperatures are caused mainly by the decay of radioactive elements and very high subsurface pressures. Very little of the energy stored deep inside the earth has been exploited for useful purposes. It is estimated that the total accessible energy stored in the earth's crust may exceed all the fossil fuel and fissionable nuclear reserves.

Most of the energy contained in the earth's interior is dissipated by thermal conduction, while a very small fraction comes out in the form of volcanic eruptions or hot springs and natural steam. No suitable methods have been devised for making use of the volcanic energy; however, energy contained in hot springs and natural steam has been utilized in many places. A geothermal plant of 389 Megawatts capacity has been in successful operation in Lorderello in Italy since 1904 and similar power plants are being used in several other countries including the USA, New Zealand, Japan and the USSR.

Hydrogen

The predictions that all oil reserves will be exhausted in a few decades have accelerated the development of a substitute for oil. Of all the alternatives suggested so far, hydrogen appears to be the most fascinating. It is believed that it will not only replace oil for transportation purposes but will also prove to be a good substitute for natural gas. The cost of producing hydrogen is high as it requires huge amounts of electricity to obtain hydrogen from the electrolysis of water. However, if electricity from nuclear sources becomes available on a large scale and oxygen is available as the by-product for a number of industrial uses, the cost of hydrogen might not appear very unreasonable. The advantages of hydrogen are so many that it may become one of the most popular energy sources of the future next to the nuclear sources. It is an extremely clean burning fuel with an energy content 2.5 times higher than conventional fuels. Research in various countries has indicated that hydrogen can be used to power internal combustion engines. Efforts are being made to utilize it as a fuel for road transport, railways, ships and airplanes in place of oil.

Nuclear Energy

After Albert Einstein proved the equivalence of mass and energy, it became obvious that energy must be released in huge quantities in any process associated with a net decrease in mass. There are two types of nuclear reactions in which energy is released in huge quantities. The first is the fusion or thermonuclear reaction and is due to the combination of some of the lighter nuclei like the isotopes of hydrogen. The second is nuclear fission where heavier nuclei such as uranium and plutonium split apart thereby releasing huge amounts of energy.

The development of controlled fusion reaction for large scale power production is one of the greatest scientific challenges of the present time. Such a development would have profound consequences on energy costs and hence the living standards of the world. The basic fuel for the fusion reaction is deuterium. For every 6,500 atoms of light hydrogen in water there is one atom of deuterium and the cost of extracting deuterium is small. One can get a rough idea regarding the magnitude of fusion energy by noting that the nuclear energy in one gallon of water is equivalent to the chemical energy in 350 gallons of gasoline. The same has been put more dramatically by Seaborg, "The energy from deuterium nuclei in the oceans is equivalent to the energy of 500 Pacific Oceans filled with high grade fuel oil". This fuel can supply 1000 times the present power needs of the world for the next million years! Such a potential fuel source cannot be easily ignored. Although the bright future prospects combined with the scientific challenge are leading towards great advances in the understanding of the fusion reaction, the prospects of obtaining large scale commercial power from such reactors seem to be quite remote during the next few decades.

Nuclear reactors based on nuclear fission are now well out of their state of infancy. At the end of the second world war, there were two schools of thought regarding the civilian use of nuclear energy. Some were very certain about the production of very cheap energy, while others were almost certain that energy produced by nuclear fuels would be so expensive that it will have no future at all. However, within two decades this new source of energy became economically feasible in many high cost fossil fuel regions of the world. Presently, nuclear energy is playing a very important role in meeting the world energy requirements.

The need and importance of nuclear energy (fission and fusion) is expected to increase with time because of its availability in large quantities and relative independence of location. This mode of producing electricity on large scales is, and will be, the most economical method. The last factor weighs quite heavily in favor of nuclear energy since the demand for electric energy is increasing much faster than the total energy consumption. For example, in 1966 less than one-fifth of the world's energy consumption was used to produce electricity, but with present trends, it is expected to rise by 1980 to about one-third. This development is shown in Figure 2.3. Hydro resources, although very attractive for the production of electrical energy, are very limited and unevenly distributed throughout the world. In addition, their relative contribution to the world's electricity requirement is declining. This simply means that the future electricity sources will be either fossil or nuclear fuels. Since fossil fuel reserves are not enough for future requirements and these can be used as raw materials for nonpower producing industries, a major part of future electrical energy growth is expected to be supplied by the nuclear fuels. Figure 2.3 indicates the approximate trend of nuclear electric energy growth. According to this estimate, by the year 2000 nuclear electric energy consumption is expected to become almost one-half of the total electric energy consumption.

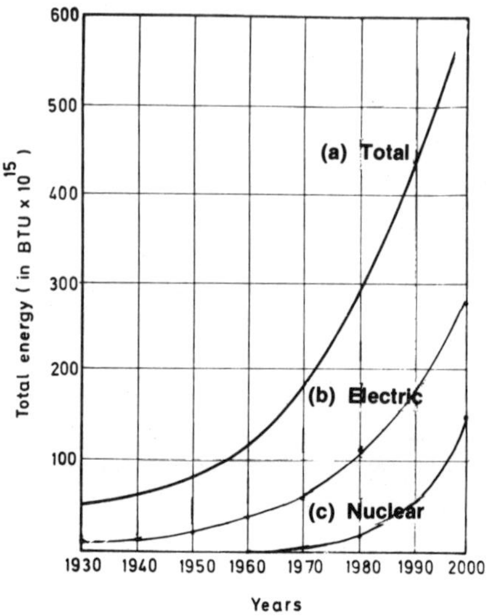

Fig. 2.3: World-wide growth of total energy consumption.

If nuclear energy is expected to be used in increasingly large quantities all over the world it should fulfill two major requirements. Firstly, it must be economical wherever it is used and secondly, sufficient reserves of nuclear fuels must be available so that a long range nuclear power program can be initiated in various energy deficient areas of the world. The first requirement is already met in many regions of the world, while the second is slightly controversial. It has been admitted by many authorities that if presently known reserves of uranium and thorium are fully utilized, with the help of breeders, nuclear fuels will have more than 20 times as much energy content as all the conventional fuels put together. However, if breeders are not developed the total energy content of nuclear fissile fuels will be reduced to a quantity much smaller than that of fossil fuels. Of course, if fusion reactors become commercial, there will be plenty of fuel available because the vast oceans are literally a huge pool of fuel.

Nuclear energy has many attractive features besides conservation of fossil fuels and the production of inexpensive electric power. Radioisotopes which are by-products of nuclear reactors are being used in increasing quantities in medicine, industry, the physical sciences, and space. The most important application of nuclear energy after the production of low cost energy seems to be conservation of food by irradiation. Large radiation sources may be used in specially designed irradiation facilities to preserve food after harvesting and before the crop ever reaches the final consumer.

Refrigeration of such a crop is very expensive in these countries and therefore specially designed irradiators to destroy insect pests in grain and vegetables can solve the food problem to a great extent. Desalination of sea water using large single- and dual-purpose nuclear power plants may be a great boon in selected regions of the world. There is also potential in the use of nuclear energy for geographical engineering. All these and other applications of nuclear energy can greatly help in solving many problems facing the human race today.

CHAPTER 3

ATOMIC AND NUCLEAR PHENOMENA

3.1 Atomic Structure and Radiation

Radiation is an all embracing word that means the propagation of subatomic particles or electromagnetic waves. It can be classified into three categories. The first group, electromagnetic radiation, includes in decreasing order of frequency gamma rays, x-rays, ultraviolet waves, broadcast radio waves, and long radio waves. These radiations are propagating in free space at the speed of light. The second group is the charged particulate radiation such as alpha particles (helium nuclei) or beta particles (electrons). The third category includes the uncharged particulate radiation such as neutrons. Frequently, in its narrower sense, radiation refers to alpha particles, beta particles, gamma rays, and neutrons. Radiation originates in the disruption of the atomic or nuclear structure or in the composition of matter.

It has been established that an atom has a heavy and positively charged nucleus which is surrounded by electrons. The number of electrons in an atom is indicated by the atomic number Z for a neutral atom. The nucleus contains N (neutron number) neutrons and Z (atomic number) protons. Both of these particles are called nucleons. The number of nucleons in a nucleus is given by the mass number $A = N + Z$. The behavior of an electron inside an atom may be examined by using a one-electron atom such as hydrogen.

Consider an atom with a positively charged nucleus and an electron moving in a circular orbit around the nucleus as shown in Figure 3.1. Based on Bohr's atomic theory, the electronic angular momentum is quantized. This is represented by

$$mvr = n\hbar, \quad n = 1, 2, 3, \ldots, \tag{3.1}$$

where m, v, r, are the electron mass, speed, and orbital radius, respectively, n is the principal quantum number, and $\hbar = h/2\pi$ with h being Planck's constant. The expressions for the electron energy and orbital radius are derived as follows:

In the electric field due to the positive charge Ze of the nucleus, the total electron energy is given by

$$E = \frac{1}{2} mv^2 - \frac{1}{4\pi\epsilon_0} \frac{Ze^2}{r}, \tag{3.2}$$

where Z is the atomic number and ϵ_0 is the electric permitivity of vacuum. Note that the second term in this equation represents the electric potential energy of the electron. A circular orbit of the electron about the nucleus is due to the centripetal force between the nucleus and the electron

$$\frac{mv^2}{r} = \frac{1}{4\pi\epsilon_0} \frac{Ze^2}{r}. \tag{3.3}$$

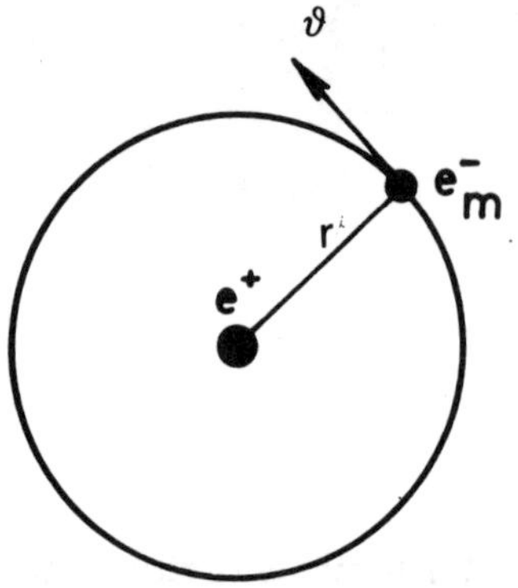

Fig. 3.1: The hydrogen atom.

Combining equations (3.1), (3.2), and (3.3), we obtain the total electron energy

$$E = -\frac{1}{4\pi\epsilon_0} \frac{Ze^2}{2r} = -\left(\frac{mZ^2e^4}{8\pi^2\epsilon_0^2 h^2}\right) \cdot \frac{1}{n^2} \tag{3.4}$$

and the radius

$$r = \frac{4\pi\epsilon_0\hbar^2}{mZe^2} n^2. \tag{3.5}$$

Equations (3.4) and (3.5) are very important for understanding the electronic structure of an atom. The principal quantum number n may be used to determine the orbital radius. In fact, corresponding to the K, L, M, N, O, and P electronic orbits we have n = 1, 2, 3, 4, 5, 6, etc. Thus one can estimate the first orbital radius for hydrogen (Z = 1), to be

$$r_1 = \frac{4\pi\epsilon_0 h^2}{mZe^2}(1)^2 = 5.3 \times 10^{-11} \text{ m.}$$

It is obvious that the atomic dimension is of the order of magnitude of 10^{-10}m. The total electron energy in allowable orbits increases progressively in discrete fashion from

$$E_1 = -\frac{mZ^2e^4}{2\epsilon_0^2 h^2} \cdot \frac{1}{(1)^2} = -13.5 \text{ eV for n} = 1$$

to $E_\infty = 0$ for $n = \infty$ for a hydrogen atom. Therefore one finds that the ionization potential of a typical hydrogen atom is approximately 13.5 volts.

When an electron goes from an orbit of lower n to a higher orbit, energy must be supplied to it. The amount of engery required is equal to the difference between the electron energies of these two orbits. When an electron "jumps" from an orbit designated by n' to n'', ($n'' > n'$) the energy supplied must be

$$E = E_{n''} - E_{n'} = \frac{mZ^2e^4}{2\epsilon_0^2 h^2}\left(\frac{1}{n'^2} - \frac{1}{n''^2}\right). \qquad (3.6)$$

Conversely if the electron drops from a higher orbit, i.e. from the orbit designated by n'' to the n' orbit, then the excess enegry ΔE is carried off as a photon (gamma ray) as illustrated by Figure 3.2. This situation is illustrated in an energy level diagram, Figure 3.3. The transition lines represent the gamma ray emission due to electron transition from higher orbits to lower orbits. These facts have been proved experimentally by observing the atomic emission spectra of excited atoms.

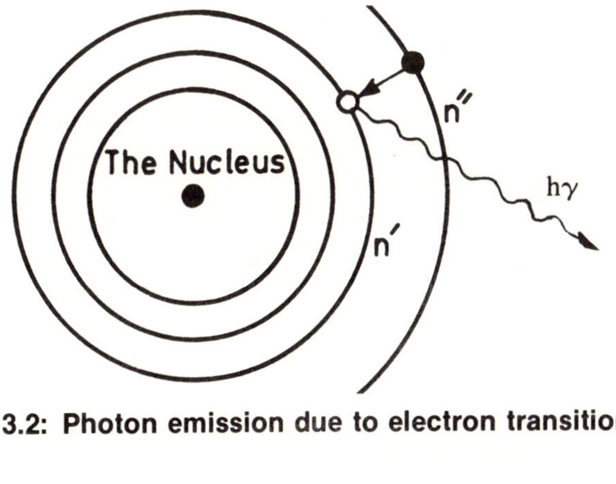

Fig. 3.2: Photon emission due to electron transition.

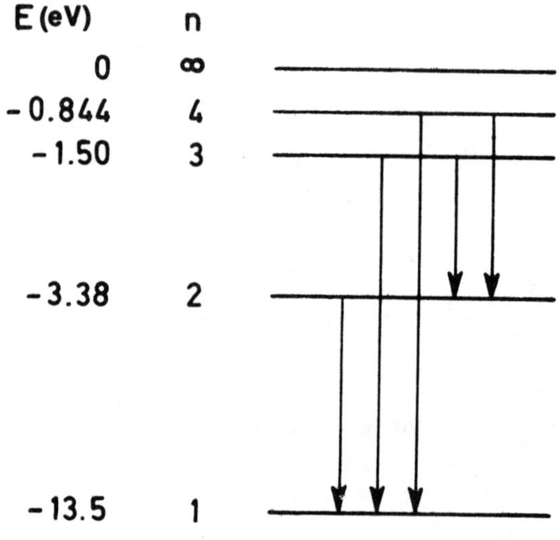

Fig. 3.3: Energy level diagram of hydrogen atoms

The term excited atom means that one or more of the electrons is in a higher orbit where it will "jump" to a lower orbit thereby emitting gamma rays. The frequency of the emitted gamma rays can then be calculated from $\Delta E = h\nu$, i.e.

$$\nu = \frac{\Delta E}{h}. \tag{3.7}$$

To what level will an electron move if it is originally at a higher level (or shell)? To answer this question the electronic structure must be examined in more detail. In quantum mechanics this problem is solved using the Schrodinger wave equation. The detailed calculations are beyond the scope of this book. The results, however, reveal that not just one but at least four quantum numbers are needed to fully explain the electronic structure of an atom and the radiation emission phenomena, mainly gamma and x-rays in this case. As described in the earlier treatment of a hydrogen atom, the principal quantum number n and its direct bearing on the electron orbital radius and energy applies to atoms in general. If the structure of the atom is postulated as a miniature solar system it is immediately apparent that the orbital and spinning motions of an electron must both be considered. These are represented by the orbital quantum number l and the spin quantum number s, respectively. Furthermore, due to the electric and magnetic interactions among the charged particles (the nucleus and the electrons) the orbital and the spin angular momenta of an electron always assume certain space orientations. This fact is represented by the magnetic quantum numbers m_l and m_s. The resultant angular momentum of an electron has a corresponding quantum number j. When $l = 0$, $j = \frac{1}{2}$. In general,

$$j = l \pm s, \; l = 1,2,3,... \text{ and } s = \tfrac{1}{2}. \tag{3.8}$$

The total angular momentum due to all electrons may be obtained in either of the following two ways. Designating the quantum numbers with capital letters, the sum of the respective quantum numbers of the individual electrons is

$$J = L + S \text{ where } L = l_1 + l_2 + ...; \; S = s_1 + s_2 + ...$$

or

$$J = j_1 + j_2 + ... \text{ where } j_i = l_i \pm s_i, \; i = 1,2,3, ... \tag{3.9}$$

Here i designates a particular electron.

The permissible values of these quantum numbers have direct bearings on the state of an atom. These are

$$n = 1, 2, 3, ...$$
$$l = 0, 1, 2, ..., n - 1,$$
$$m_l = -l, -l + 1, ... , -1, 0, 1, ..., l - 1, l,$$

and

$$m_s = + \tfrac{1}{2} \text{ or } - \tfrac{1}{2}.$$

These four quantum numbers together identify a particular electron in an atom. This is based on Pauli's exclusion principle which states that within one atom no two electrons can have the same quantum numbers. Due to the limitations of the allowed values of these quantum numbers and the exclusion principle we can easily see that in a shell designated by n, the total number of electrons is given by $2n^2$.

In detailed format the energy level diagram of an atom should also include other quantum numbers. A representative case is the sodium energy level shown in Figure 3.4. In this figure we see some of the electron transition lines. These are determined by the selection rules $\Delta j = \pm 1$ or 0 and $\Delta l = \pm 1$. The values of j must be positive so that there are only single levels for $l = 0$. Athough levels with the same n are drawn approximately at the same height, their corresponding energies are usually different.

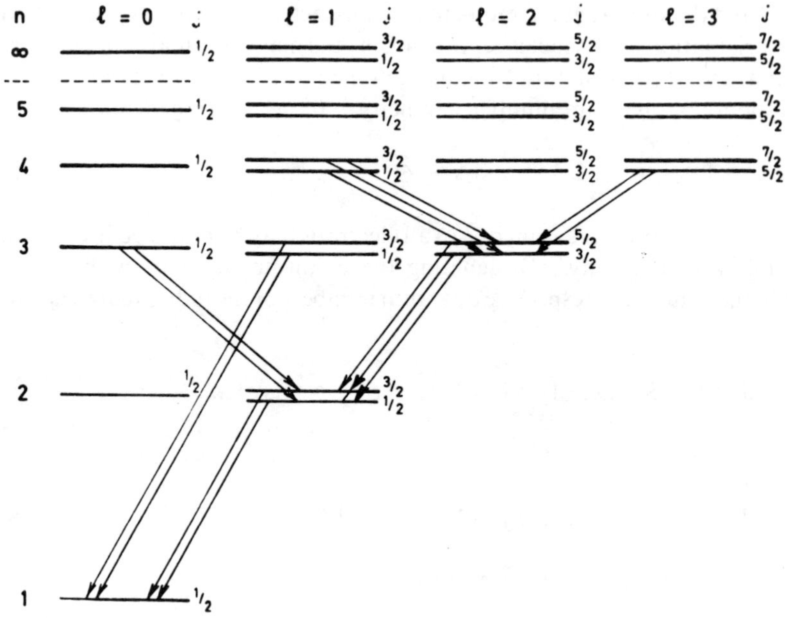

Fig. 3.4 : Sodium energy level diagram.

When all the electrons in an atom occupy their lowest possible orbit or shell, the atom has a minimum total energy and is said to be in its ground state. With some exceptions one may think of filling of the subshells in an increasing order of quantum numbers. That is, for a fixed set of n, l, m_l, and m_s, the first electron has $m_s = -\frac{1}{2}$ and the second one $m_s = +\frac{1}{2}$. Then m_l starts from -l. In each case there are two electrons, i.e. one with $m_s = -\frac{1}{2}$ and another with $m_s = +\frac{1}{2}$. When all m_l values (up to $m_l = l$) are used up, l increases up to n-1. In this way one can construct a periodic number. For example, ground state hydrogen has its electron designated by $n = 1$, $l = 0$, $m_l = 0$, $m_s = -\frac{1}{2}$. For helium at its ground state its electrons have $n = 1$, $l = 0$, $m_l = 0$, $m_s = -\frac{1}{2}$ and $n = 1$, $l = 0$, $m_l = 0$, $m_s = +\frac{1}{2}$. It is usual to use the letters s, p, d, f, g, to represent $l = 0, 1, 2, 3, 4$, respectively. Therefore, all the ground state hydrogen or helium electrons are s-electrons.

For an atom with more than one electron, it is conventional to designate the electrons by using the principal quantum number multiplied by the letter designation of the orbital quantum number. The superscript of the letters, s, p... etc. then gives the number of electrons in each level. For example, the copper atom is represented by

$$1s^2\ 2s^2\ 2p^6\ 3s^2\ 3p^6\ 3d^{10}\ 4s^1.$$

In other words it has 2 s-electrons in its K-orbit; 2 s-electrons in the L-orbit; 6 p-electrons occupying the L-orbit; 2 s-electrons in the M-orbit; 6 p-electrons in the M-orbit; 10 d-electrons in the M-orbit; and 1 s-electron in the N-orbit. Following the same rule, many electronic transitions can be represented by a notation such as

$$2p \rightarrow 1s.$$

This represents the jump of a p-electron ($l = 1$) in the L-orbit ($n = 2$) to the $n = 1$ and $l = 0$ level. The corresponding energy of the gamma or x-ray can then be quickly calculated.

3.2 Binding Energy and Nuclear Structure

According to the neutron-proton concept of a nucleus, the nuclear constituents are neutrons and protons. The fact that positively charged protons can be packed together seems puzzling. This can be explained only if we assume that a very strong nuclear force exists. For a given number of nucleons, the nuclear forces provide a definite amount of binding energy to hold the nucleons together. It has been actually found that, on the average, the binding energy per nucleon is about 8 MeV, except for very light nuclei. Figure 3.5 shows these values for different nuclei as a function of their mass numbers. The availability of the binding energy to hold nucleons together to form a nucleus can be easily explained from the difference between the sum of the masses of N neutrons and Z protons and the mass of the nucleus. Taking Mg-24 as an example, we have

$$\Delta E = 12\,(1.00989) + 12\,(1.00758) - (23.99267) - 12\,(0.00055) \text{ amu}$$

$$\therefore\ \frac{\Delta E}{A} = 8.25 \text{ MeV per nucleon.}$$

The last term in the equation for ΔE is the total electron mass of the Mg atom. This calculation reveals that it takes 8.25 MeV energy to hold a neutron or a proton within the nucleus. From Figure 3.5 one can see that nuclear fission (splitting of heavy nuclei) or nuclear fusion (fusing of light nuclei) yields product nuclei with higher binding energies per nucleon. Extra energy is therefore obtained from either type of nuclear reaction. This will be discussed later.

A rough estimate of the radius of a nucleus can be made by assuming that the binding energy per nucleon, $\frac{B.E.}{A} \cong 8\,\text{MeV}$, is equal to the Coulomb repulsive energy experienced by a proton due to the rest of the protons inside of the nucleus. In this way

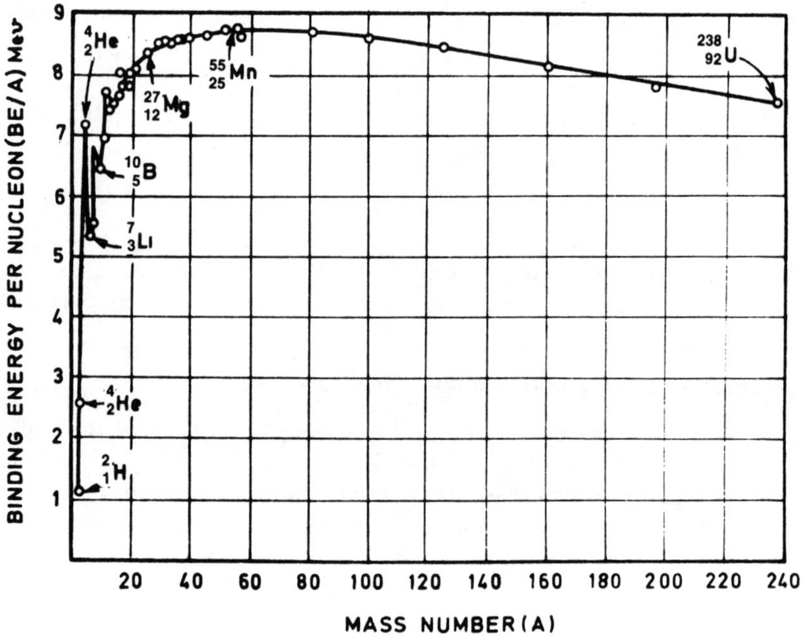

Fig. 3.5: Binding energy per nucleon.

$$\frac{B.E.}{A} = \frac{1}{4\pi\epsilon_0} \cdot \frac{(Z-1)e^2}{r}.$$

$$\therefore r = (9\times10^9 \frac{Nm^2}{C^2})(Z-1)\ (1.6\times10^{-19}C)^2/(8\times10^6\times1.6\times10^{-9}J)$$

$$= 1.8 \times 10^{-14}\ (Z-1)\ \text{cm}.$$

Many detailed experiments have been carried out to measure the nuclear radius. A semi-emprical formula commonly used for the purpose is

$$r = 1.2 \times 10^{-13}\ A^{1/3}\ \text{cm}, \qquad (3.10)$$

where A is the mass number of the nucleus. It is interesting to note that based on this formula and the masses of the nucleons the nuclear density may be estimated to be

$$\rho = \frac{\text{nuclear mass}}{\text{nuclear volume}} = \frac{A \times 1.675 \times 10^{-24}\text{g}}{4/3\ (1.2 \times 10^{-13}\ A^{1/3})^3\ \text{cm}^3}$$

$$\cong 2.4 \times 10^{15} \text{ g/cm}^3.$$

Out of many proposed nuclear structure models, the liquid-drop-model and the shell model are the most frequently used. Whereas the shell model is mor adaptable in explaining nuclear stability, the most valuable tools for nuclear calculations are derived from the liquid-drop-model of a nucleus. In this concept the nucleus is considered to be a homogeneous liquid-drop type of structure made up of condensed matter with the mass directly proportional to the mass number A. This leads to the assumption that the binding energy is proportional to the mass number. The nucleons near the surface are not interacting with so many nucleons as the ones deep inside the nucleus. This reduces the binding energy in proportion to the surface area, i.e. $A^{2/3}$. The Coulomb effect due to the electric charge carried by the protons contributes another negative term to the binding energy. This term is proportional to $Z^2/A^{-1/3}$. The symmetric effect that favors equal neutron and proton number for stability gives another negative contribution that is proportional to $(A - 2Z)^2/A$. Furthermore, depending upon whether the neutron and proton numbers, N and Z, are even or odd, the binding energy differs. The net effect can be summed up as

$$B(A,Z) = a_v A - a_s A^{2/3} - a_c \frac{Z^2}{A^{1/3}} - a_a \frac{(A - 2Z)^2}{A} + \delta, \qquad (3.11)$$

where

$$\delta = \begin{cases} \dfrac{33.5}{A^{3/4}} \text{ MeV for N, Z even} \\ 0 \quad \text{for odd A.} \\ \dfrac{-33.5}{A^{3/4}} \text{ MeV for N, Z odd.} \end{cases}$$

The values of the constants in this equation are $a_v = 14.0$ MeV, $a_s = 13.0$ MeV, $a_c = 0.6$ MeV, and $a_a = 19.1$ MeV. Thus we can write the mass of a nucleus to be

$$M(A,Z) = Zm_p + (A - Z)m_n - B(A, Z), \qquad (3.12)$$

where m_p and m_n are the proton and neutron masses. These equations (3.11 and 3.12) form a set of very useful formulas for calculating the Q-values of nuclear reactions.

3.3 Nuclear Reactions and Radioactivity

A nuclear reaction may be represented by

$$^{A_1}_{Z_1}a + ^{A_2}_{Z_2}X \rightarrow [^{A_1+A_2}_{Z_1+Z_2}C] \rightarrow ^{A_3}_{Z_3}b + ^{A_4}_{Z_4}Y, \tag{3.13}$$

where a and X are the parent nuclei, b and Y are the product nuclei, and C is the compound nucleus. According to Bohr, all nuclear reactions can be described in two steps.

Step I : The formation of a compound nucleus in a highly excited state.
Step II : The dissociation of the compound nucleus, the lifetime of which is about 10^{-16} s. The nature of this dissociation and the product nuclei formed are independent of the manner in which the compound nucleus is formed.

A frequently used short version of (3.13) has the form

$$X\ (a,b)\ Y, \tag{3.14}$$

where a and b are light nuclei and X and Y are heavy nuclei.

According to the types of particles a and b, nuclear reactions may be classified into:

a) scattering : $a = b$,
b) absorption : $a \neq b$, and
c) radioactive decay : there is no incident particle a.

The nuclear fission reaction is an exception in that there are two or more product nuclei. Therefore, the equation for this reaction must be written as

$$n + ^{A_2}_{Z_2}X \rightarrow ^{A_3}_{Z_3}Y_1 + ^{A_4}_{Z_4}Y_2 + (2 \text{ to } 3)\ n \tag{3.15}$$

or

$$X\ (n, \nu n)\ Y_1 + Y_2,\ \nu = 2 \text{ to } 3. \tag{3.16}$$

In a nuclear reaction represented by equation (3.13), the atomic number, the mass number, and the mass and energy of all the nuclides must be conserved, i.e.

$$Z_1 + Z_2 = Z_3 + Z_4, \tag{3.17}$$

$$A_1 + A_2 = A_3 + A_4, \tag{3.18}$$

and

$$E_1 + M(A_1,Z_1) + E_2 + M(A_2,Z_2) = E_3 + M(A_3,Z_3) + E_4 + M(A_4,Z_4), \tag{3.19}$$

where the E's and M's are the kinetic energies and masses of the respective nuclides.

In all nuclear reactions energy is either absorbed or released. this energy is commonly called the Q-value of the reaction and is defined by the following equation

$$Q \equiv M(A_1, Z_1) + M(A_2, Z_2) - M(A_3, Z_3) - M(A_4, Z_4). \tag{3.20}$$

From equations (3.19) and (3.20) one obtains

$$Q = E_3 + E_4 - E_1 - E_2. \tag{3.21}$$

In other words, the Q-value is a measure of the kinetic energies of the participants in the reaction.

Experimentally, the Q-value can be determined from a collision-type of experiment. Assume that particle a strikes the target nucleus X resulting in the production of b and Y as shown in Figure 3.6.

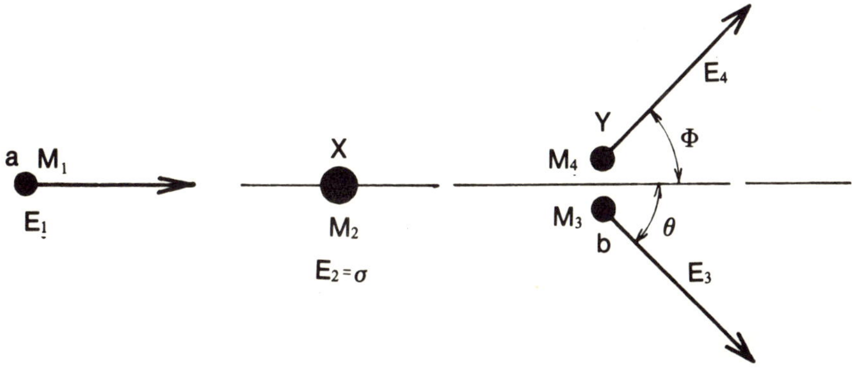

Fig. 3.6: A simple nuclear reaction.

From the conservation laws the following relationship can be derived

$$Q = E_3 \left(1 + \frac{M_3}{M_4}\right) - E_1\left(1 - \frac{M_1}{M_4}\right) - \frac{2(M_1E_1M_3E_3)^{1/2}}{M_4} \cos\theta. \tag{3.22}$$

Measuring the kinetic energies of the outogoing particles b and Y at specified angles with respect to the direction of motion of the incident particle a yields the Q-value of this nuclear reaction.

If the nucleus X disintegrates into Y and b without the incident particle a, the process is called radioactive decay. Radioactive materials containing many atoms will undergo decay in statistical manners. It is impossible to predict the exact time when any particular atom will disintegrate. Radioactive decay may be compared with human life. The average life of a population may be taken as 60 years, but it is impossible to say when a particular individual will die. Although one cannot predict the time of disintegration of a specific radio-active atom, the probability of radioactive decay can be predicted.

Suppose N radioactive nuclei are present at a certain time t, then the number of nuclei, dN, decaying in a time interval, dt, is proportional to the original number of nuclei. Thus

$$dN \sim - N\, dt, \tag{3.23}$$

the minus sign indicates that the number of nuclei present decreases with time. This relationship can be rewritten as

$$dN = -\lambda N\, dt, \tag{3.24}$$

where λ is called the disintergration (decay) constant and is characteristic of the radioactive sample.

Assuming that at $t = 0$, $N = N_0$ and at time t the number of atoms present is N, we have, on integrating equation (3.24),

$$\int_0^N \frac{dN}{N} = - \int_0^t \lambda dt \tag{3.25}$$

or

$$\ln \frac{N}{N_0} = - \lambda t.$$

$$\therefore N = N_0\, e^{-\lambda t}. \tag{3.26}$$

This is the exponential decay law.

In actual practice, the decay of a radioactive isotope is usually given in terms of its half-life (T). This is defined as the time required for one-half of the atoms originally present to decay. Substituting $N = N_0/2$ and $t = T$ in equation (3.26), we obtain

$$\frac{N_0}{2} = N_0\, e^{-\lambda T}$$

or

$$\lambda T = \ln 2.$$

$$\therefore\ T = \frac{0.693}{\lambda}. \qquad (3.27)$$

This is a basic equation; it relates the half-life and the radioactive decay constant of a radionuclide. In one half-life the number of radioactive atoms originally present will be reduced by a factor of 2, at the end of a second half-life by a factor of 4, and so on. Hence, theoretically, it requires an infinite time for the complete decay of a radioactive sample. Another characteristic time for a radioactive sample is the one-tenth life. This is given as

$$T_{1/10} = \frac{\ln 10}{\lambda} = \frac{2.303}{\lambda}. \qquad (3.28)$$

Since the radioactive disintegration is a statistical process, the actual life of any atom can have any value between zero and infinity. The mean life can be determined by summing up the time of decay of each radioactive atom, and dividing by the number of radioactive atoms originally present. Thus if τ denotes the mean life of a radioactive isotope, then it can be defined by

$$\tau \equiv \frac{\text{Total lifetime of all radioactive atoms}}{\text{Total number of radioactive atoms}}.$$

$$\therefore\ \tau \equiv \frac{\int_0^N t\,dN}{\int_0^N dN}. \qquad (3.29)$$

Substituting the value of dN from equation (3.24) in equation (3.29),

$$\tau = \frac{1}{N_0} \int_0^\infty t\lambda N dt.$$

The integration limits denote that it takes an infinitely long time for all the nuclei to disappear statistically. Therefore,

$$\tau = \lambda \int_0^\infty t e^{-\lambda t}\, dt.$$

Integrating and simplifying we obtain

$$\tau = \frac{1}{\lambda} \cdot \tag{3.30}$$

From equations (3.27) and (3.30),

$$T = 0.693\tau. \tag{3.31}$$

Radioactive decay and a relationship between the mean life and the half-life of radioactive elements is given in Figure 3.7.

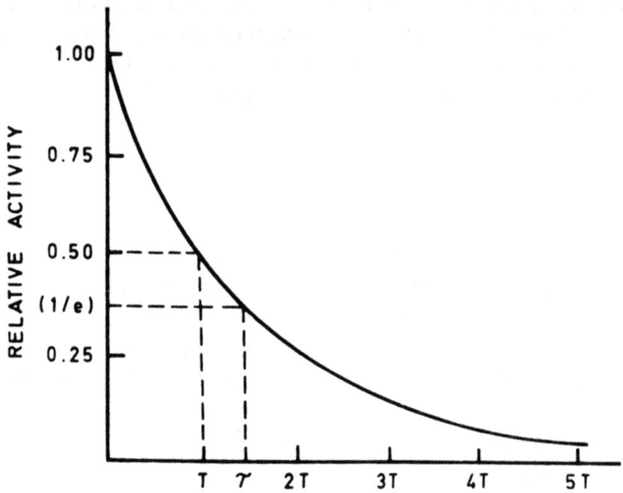

Fig. 3.7: Radioactive decay and the relationship between mean life and half-life.

The disintegration rate or the number of atoms decaying per unit time, $\frac{dN}{dt}$, is called the activity of the sample. The activity (A) is expressed by the equation

$$A = \lambda N. \tag{3.32}$$

The activity per gram of a sample is called its specific activity.

Standard units are used for the measurement of activity. The most common unit for activity is 'curie' (Ci). One curie is defined as 3.7×10^{10} disintegrations per second. Since the curie is quite a large unit, millicurie (1 mCi = 10^{-3}Ci) and microcurie (1 μCi = 10^{-6} Ci) generally are used. Other units used in the metric system for radioactivity are the Becquerel (1 dis./s) and the Rutherford (10^6 dis./s).

Up to now we have considered only those radioactive nuclei which decay radioactively to stable nuclei. Consider a naturally radioactive nuclide, 1, which decays into another radioactive nuclide, 2. The latter in turn decays into a stable product 3. We then get

$$\frac{dN_1}{dt} = -\lambda_1 N_1, \tag{3.33}$$

$$\frac{dN_2}{dt} = \lambda_1 N_1 - \lambda_2 N_2, \tag{3.34}$$

and

$$\frac{dN_3}{dt} = \lambda_2 N_2, \tag{3.35}$$

If N_1^0, N_2^0, N_3^0 are the number of atoms of nuclide 1, 2, 3, at time $t = 0$, respectively, expressions can be derived for values of N_1, N_2, N_3 as functions of time. The results are given below.

$$N_1 = N_1^0 e^{-\lambda_1 t}. \tag{3.36}$$

$$N_2 = \frac{\lambda_1}{\lambda_2 - \lambda_1} N_1^0 (e^{-\lambda_1 t} - e^{-\lambda_2 t}) + N_2^0 e^{-\lambda_2 t}. \tag{3.37}$$

$$N_3 = N_1^0 (1 + \frac{\lambda_1}{\lambda_2 - \lambda_1} e^{-\lambda_2 t} - \frac{\lambda_2}{\lambda_2 - \lambda_1} e^{-\lambda_1 t}) + N_3^0 e^{-\lambda_3 t}. \tag{3.38}$$

A typical case of such successive disintegrations is given in Figure 3.8. This treatment can be extended to a chain of any number of radioactive products.

Example 3.1

One of the successes of the Bohr atomic theory is the establishment of the electronic structure of a hydrogen atom. It confirms the experimental measurements of the wavelengths of the hydrogen emission spectrum. Use the concept of electron transitions in a hydrogen atom to calculate the wavelengths of the spectral lines due to electrons "jumping" from various higher orbits (n = 3, 4, 5, 6, ...) to the second orbit (n = 2).

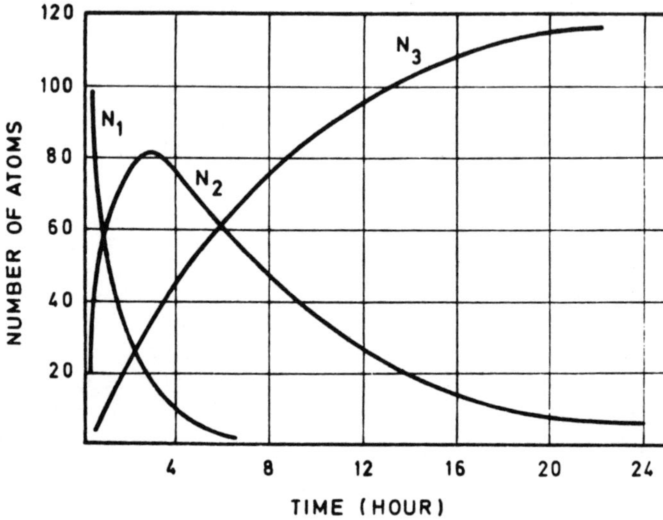

Fig. 3.8: A radioactive series with three members. Only the parent (T = 1 hr.) is present initially; the daughter has a half-life of 5 hours and the third member is stable.

Solution

The electron energies at orbits n = 2, 3, 4, 5, 6, ... etc. can be calculated from equation (3.4). They are approximately $E_2 = -3.38$ eV, $E_3 = -1.50$ eV, $E_4 = -0.844$ eV, $E_5 = -0.540$ eV, ... From these, the energe differences between each of the higher energies and E_2 can be obtained. The wavelengths are then calculated from these energy differences ΔE, using $\lambda = hc/\Delta E$. A representative calculation is

$$\lambda_1 = \frac{hc}{E_3 - E_2} = \frac{(6.63 \times 10^{-34} \text{Js})(3 \times 10^8 \text{m/s})}{(1.88 \text{ eV})(1.6 \times 10^{-19} \text{J/eV})}$$

$$= \frac{6.63 \times 3}{1.9 \times 1.6} \times 10^{-7} \text{m} = 6612 \text{ Å}.$$

The calculated results and experimentally observed data are listed in the following table.

n	E_n(eV)	$E = E_{n+1} - E_n$(eV)	λ_{n-1}(Å)	Observed λ (Å)
2	- 3.38	1.88	6612	6570
3	- 1.50	2.54	4894	4861
4	- 0.844	2.84	4377	4102
5	- 0.540	3.01	4130	
6	- 0.375	3.11	3997	
7	- 0.276	3.17	3922	
8	- 0.211	3.21	3873	
9	- 0.167	3.25		
10	- 0.135			
∞	0	3.38	3678	3650

The detailed calculations demonstrate excellent agreement with the experimentally observed wavelengths of a series of spectral lines. This series of spectral lines is the Balmer series.

Similar calculations with respect to orbits designated by n = 1 and n = 3 correspond to the Lyman and Paschan series of the hydrogen spectra.

Example 3.2

Estimate the radius of a gold atom. How does it compare with the radius of the gold nucleus?

Solution

$$\text{Volume of an atom} = \frac{A}{\rho N_0}$$

$$\text{Volume of a gold atom} = \frac{197 \text{ g/mole}}{(19.33 \text{ g/cm}^3)(6.02 \times 10^{23} \text{atom/mole})}$$

$$= 1.69 \times 10^{-23} \text{cm}^3 .$$

Assuming the atom to be spherical, the radius of the atom, r_a, can be estimated from

$$\frac{4}{3}(\pi r_a^3) = 1.2 \times 10^{-23} \text{cm}^3.$$

$$\therefore r_a = 1.59 \times 10^{-8} \text{cm}.$$

The nuclear radius of gold, $r_n = 1.2 \times 10^{-13}(197)^{1/3}$ or 6.98×10^{-13} cm. Hence

$$\frac{r_a}{r_n} = \frac{1.59 \times 10^{-8}}{6.98 \times 10^{-13}} = 2.28 \times 10^{4}$$

Example 3.3

Calculate the binding energy per nucleon in MeV of the uranium-238 nucleus, assuming it has a mass of 238.1250 amu.

Solution

Proton mass	=	1.00759 amu
Neutron mass	=	1.00898 amu
Mass of 92 protons	=	92.6983 amu
Mass of 146 neutrons	=	147.3111 amu
Total	=	240.0094 amu

$$\text{Total nucleon binding energy} = (240.0094 - 238.1250) \times 931 \text{ MeV}$$

$$= 1760 \text{ MeV.}$$

$$\therefore \text{ Binding energy per nucleon} = \frac{1760}{238} \text{ MeV} = 7.9 \text{ MeV.}$$

Example 3.4

In the reaction $^{10}B(\alpha, p)^{13}C$ a group structure exists in the energy distribution of the protons. When 4.77 MeV alpha particles bombarded on a B-10 target, four groups of protons were observed at 90° to the incident alpha beam. The energies of these protons, as determined by the magnetic deflection method, were found to be 6.84, 3.98, 3.43 and 3.23 MeV, respectively. Determine the Q-value of each proton group and draw an energy level diagram of the reaction.

Solution

The maximum proton energy is $E_3 = 6.84$ MeV. The Q-value of the reaction is therefore given by the following equation

$$Q_0 = 6.84(1 + \frac{1}{13}) - 4.77(1 - \frac{4}{13}) = 4.07 \text{ MeV.}$$

Similarly, the smaller values of E_3 give the following values of Q.
$E_3 = 3.98$ MeV, $Q_1 = 0.98$ MeV.

$E_3 = 3.43$ MeV, $\quad Q_2 = 0.39$ MeV.
$E_3 = 3.23$ MeV, $\quad Q_3 = 0.22$ MeV.

$Q_0 = 4.07$ MeV is called the Q-value of the reaction or the ground-level Q-value of the $^{10}B(\alpha, p)^{13}C$ reaction. All the Q-values calculated above are due to the formation of C-13 nuclei in excited states. The energies corresponding to the various excited levels are given by $(Q_0 - Q_1)$, $(Q_0 - Q_2)$, and $(Q_0 - Q_3)$. The gamma rays corresponding to these energies are emitted by the excited nuclei upon reaching the ground state. The energy level diagram of this reaction is shown in Figure 3.9.

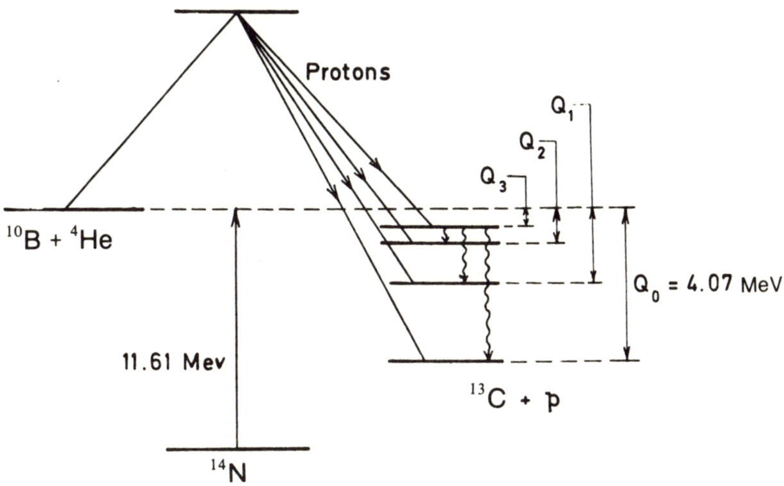

Fig. 3.9: Energy level diagram of $^{10}B(\alpha, p)^{13}C$.

Example 3.5

The earth was created due to a cosmological event some 5×10^9 years ago. Suppose two radioactive nuclei X and Y existed at that time in their natural forms. Knowing that presently for every atom of X there are 100 atoms of Y, calculate this ratio at the time of the origin of the earth. Assume the half-lives of X and Y to be 8×10^8 and 2×10^9 years, respectively.

Solution

$$\frac{N_X(0)}{N_Y(0)} = \frac{1}{100},$$

where $N_X(0)$ and $N_Y(0)$ are the current numbers of atoms of X and Y, respectively.

$\lambda_X = 8.69 \times 10^{-10}$ year^{-1}

$\lambda_Y = 3.47 \times 10^{-10}$ year^{-1}.

Suppose the earth came into being T years ago. The number of atoms $N_X(T)$ and $N_Y(T)$ present at that time would be related to the current numbers by the relation:

$$\frac{N_X(0)}{N_Y(0)} = \frac{N_X(T)e^{-\lambda_x T}}{N_Y(T)e^{-\lambda_y T}} = \frac{1}{100}.$$

$$\therefore \frac{N_X(T)}{N_Y(T)} = \frac{1}{100} \times \frac{e^{-3.47\times10^{-10}\times5\times10^9}}{e^{-8.69\times10^{-10}\times5\times10^9}} = 0.135.$$

Problems

3.1 Calculate the value of the first excitation potential for a hydrogen atom. Evaluate the frequency and wavelength of the emitted quanta.

3.2 Write down the electron configurations in the atoms of carbon and iron. Explain the letters and superscripts used.

3.3 Calculate the nuclear densities of aluminum-27, gold-197 and uranium-238. What conclusion can you draw from the results?

3.4 Show that the binding energy per nucleon can be written as

$$[(m_n - 1) - \frac{Z}{A}(m_n - m_p) - (\frac{M - A}{A})]c^2,$$

where m_n and m_p are the masses of a neutron and proton, M is the isotopic mass, A is the mass number, and c is the speed of light in a vacuum.

3.5 The binding energy of the last proton or neutron in a nucleus is different from the average binding energy per nucleon. Knowing the masses of $^{40}_{20}Ca$, $^{39}_{19}K$ and the proton, calculate the binding energy of the last proton in $^{40}_{20}Ca$. Will the binding energy of the last neutron in the nucleus be the same as that of the last proton?

3.6 It is desired to have a nuclear reaction represented by $^7Li(p, \alpha)^4He$. Suppose the rest masses of the interacting and the resulting particles are not known, but the binding energies per nucleon for lithium-7 and the alpha particles are given as 5.60 and 7.05 MeV, respectively. Calculate the Q-value of the reaction.

3.7 Protons were first produced by the bombardment of nitrogen atoms by alpha particles emitted by naturally occurring radioactive elements. Calculate the Q-value of the reaction. What is the minimum energy of the alpha particles for initiating such a reaction?

3.8 It is desired to produce protons and neutrons from the following photodisintegration reactions

$$^{8}Be + \gamma \rightarrow {}^{7}Li + {}^{1}H.$$
$$^{7}Li + \gamma \rightarrow {}^{6}Li + {}^{1}n.$$
$$^{9}Be + \gamma \rightarrow 2\,{}^{4}He + {}^{1}n.$$

Calculate the Q-values of these reactions and determine the minimum gamma energies to initiate such reactions.

3.9 Calculate the mass of cobalt-60 (in grams) that will produce an activity of 50 curies. What will be the strength of the cobalt source ten years later?

3.10 Xenon is formed during the fission of uranium-235 according to the following scheme

$$^{135}I \xrightarrow[6.7hr]{\beta^-} {}^{135}Xe \xrightarrow[9.2hr]{\beta^-} {}^{135}Cs \xrightarrow[2x10^6yr]{\beta^-} \cdots$$

Suppose the production of iodine-135 was stopped at t = 0, when the number of iodide-135 atoms was 5×10^{14}. Calculate the time when the number of xenon-135 atoms will be at a maximum. Draw a graph illustrating the change in quantity of xenon with time. See chapter 8.

3.11 The decay scheme of the nuclide N_1 is as follows

$$N_1(t) \xrightarrow{\lambda_1} N_2(t) \xrightarrow{\lambda_2} N_3(t) \xrightarrow{\lambda_3} N(t) \qquad \text{(Stable)}.$$

Assume that initially only N_1 nuclides are present and $N_1(0) = N_1^0$. Show that $N_3(t)$ can be written as

$$N_3(t) = \lambda_1\lambda_2 N_1^0 \left\{ \frac{e^{-\lambda_1 t}}{(\lambda_2 - \lambda_1)(\lambda_3 - \lambda_1)} + \frac{e^{-\lambda_2 t}}{(\lambda_1 - \lambda_2)(\lambda_3 - \lambda_2)} + \frac{e^{-\lambda_3 t}}{(\lambda_1 - \lambda_3)(\lambda_2 - \lambda_3)} \right\}.$$

CHAPTER 4

RADIATION MEASUREMENT

4.1 Basic Radiation Detection Methods

The detection of radiation is based on the effects of the interaction between radiation and matter. The nature of such interactions varies widely with the types of radiation and materials. In general, for charged particulate radiation, ionization and excitation of the absorbing atoms are the major reasons for radiation energy loss. Ion pairs produced as a result of the interaction of radiation with atoms can be collected by electrodes, creating an electric current or voltage pulse to be registered by a meter or a recording system. For uncharged particulate or electromagnetic radiation, besides ionization and excitation of the absorbing medium, other processes also occur to cause the loss of energy of the incident radiation. These are elastic or inelastic scattering, Compton scattering, pair production and the photoelectric effect. Conventional methods such as a nuclear emulsion and calorimeters are also used for radiation detection. Recently, solid state detectors have been commonly used because of their high sensitivities.

A typical radiation detection system contains three major components: detector, amplifier, discriminator and counting meter (scaler). This is shown in Figure 4.1. Radiation impinges upon the detector from the left. The incident radiation causes ionization in the detector, resulting in an electric pulse. After amplification this pulse is counted by a meter or a scaler. Depending on the type of system, the data obtained must be corrected for the

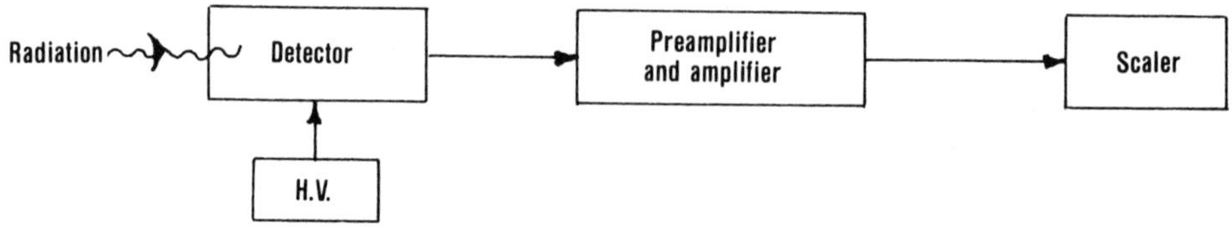

Fig. 4.1: Radiation detection system.

efficiency of the counting system, attenuation, and statistical distribution. Therefore, it is neccessary to calibrate the detection system using a known radiation source or by comparing with another system. An absorption experiment is usually carried out to evaluate the radiation attenuation by the absorbing materials such as air and/or detector window or wall between the source and the active volume of the detector. Excellent references on these discussions can be found in *Nuclear Radiation Detection* by W. J. Price, McGraw Hill Book Co., and *Radioisotope Techniques* by R. T. Overman and H. M. Clark, McGraw Hill Book Co. In the following, some detection systems for radiations other than neutrons are briefly described.

A cloud chamber operates on the principle that supersaturated (either by expansion or diffusion) condensable vapor in air will condense on dust or ionic particles. When radiation enters a chamber containing supersaturated vapor, ionization of the air creates condensation sites for the vapor. As a result, the path of the incident radiation can be traced along these sites or droplets. This path can be photographed. The structure of a cloud chamber is illustrated in Figure 4.2. For charged particles, an external electric or magnetic field may be applied to the chamber to deflect the path and measure the speed of radiation propagation.

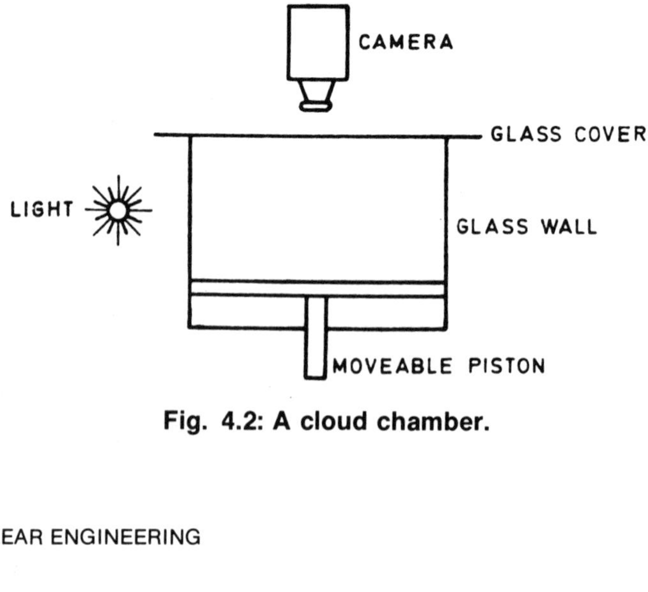

Fig. 4.2: A cloud chamber.

An ionization chamber is, in essence, a capacitor. Usually it has a central wire as the positive electrode with a cylindrical wall serving as the negative electrode as shown in Figure 4.3.

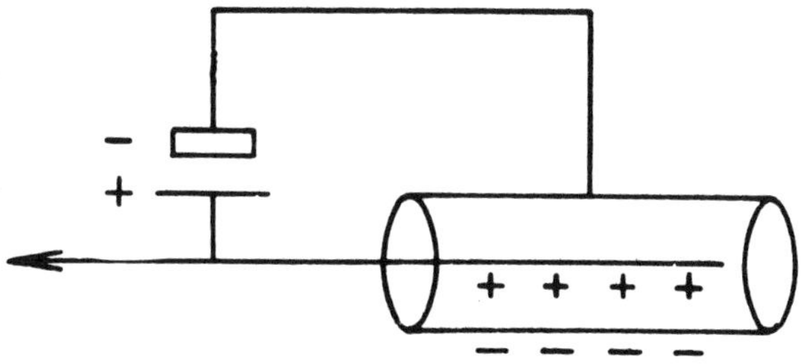

Fig. 4.3: An ionization chamber.

The chamber contains gas at various pressures depending on the specific design and the purpose of the detector. When the radiation enters the chamber, ionization of the gas creates charges which, under the influence of the electric field, move to the respective electrodes. The amount of charges created directly or indirectly by the incident radiation and the charges deposited on the electrodes increase with the increasing voltage applied between the electrodes. The rate of increase of the charge collection is, however, not uniform. In general, a plot of the charges collected by the electrodes (the electric pulse height sent out from the detector against the applied voltage between the electrodes) can be divided into six distinctive regions (Figure 4.4): 1. recombination, 2. ionization, 3. proportional, 4. limited proportional, 5. Geiger-Mueller, and 6. continuous discharge. In region 1, the ions recombine before they reach the electrode at a decreasing rate as the applied voltage is raised. In region 2, the collection of charges is essentially constant because all ions have gained enough kinetic energy to reach the respective electrode without recombination. It is important to realize that in this region the pulse size is approximately proportional to the incident radiation intensity. The electroscope is an example of an ionization chamber operating in this mode. In region 3, the electrons gain enough kinetic energy to cause secondary ionization. The degree of secondary ionization, i.e. the total ion pairs produced by an original ion pair can be expressed in terms of the gas amplification factor (note that in the ionization region this factor is unity). At a fixed operating voltage the gas amplification factor is constant. Therefore the output pulse size is proportional to the original amount of ionization. With increasing voltage, however, the gas amplification factor increases rapidly until a limit is reached due to the buildup of space charges. When this limit is reached, the pulse size will no longer be proportional to the original ionization. Thus region 4 is called the limited proportional region. When the applied voltage is

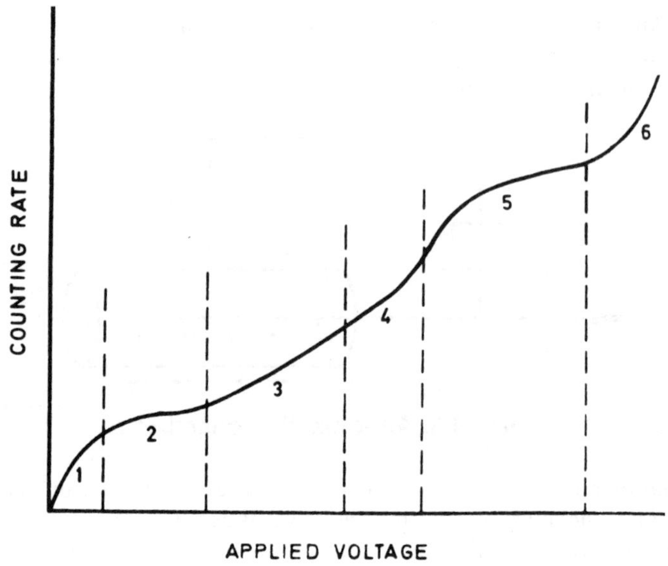

Fig. 4.4: Ionization chamber counting rate vs. applied voltage.

raised further, one reaches the so-called Geiger region. In this region the entire tube is flushed with an avalanche of ion pairs triggered by a single primary ion pair. In this way, the output pulse size does not depend on the nature or energy of the radiation that produces the initial ion pair. In region 6 the voltages are so high as to cause continuous discharge to occur in the chamber. Needless to say, when operating an ionization chamber we must avoid such a high operating voltage.

In an electroscope, when the electrodes are charged, a quartz fiber indicator is repelled from one of the electrodes to read zero on a scale of the viewing scope. When ionizing radiation enters the region between the electrodes ionization takes place, resulting in the production of ion pairs. These electrons and ions migrate to the oppositely charged electrodes. The resulting current causes a reduction in potential until there exists no potential difference between the electrodes. Such a deflection of an electroscope is a measure of the potential difference between the electrodes. The average electric field along any line of force is proportional to the voltage drop across the electrodes. The charges on the electrodes of the electroscope are proportional to the electric field. Thus the potential difference across the electrodes is a direct measure of the collected charges and hence the radiation intensity. An electroscope can be used to measure various types of radiation. Since the ionizing effect of alpha radiation is much greater than that of either beta or gamma radiation, the use of an unshielded radioactive sample will give a result primarily due to alpha emission. If beta emission measurement is desired, the sample should be shielded with .01 cm of Al. If gamma emission alone is to be measured, the shielding should be 2 or 3 mm of Pb. Once the background radiation has been determined by allowing the fiber to drift over the scale, a source of known strength and composition can be used to calibrate the fiber movement across the scale. If an uncalibrated source of the same composition and energy is then placed in the chamber in the same geometrical configuration as the standard, the apparent activity of the sample can be determined.

The Geiger-Mueller detector is the most frequently used radiation detector. It operates in the Geiger region. The counting rate versus applied voltage curve is shown in Figure 4.4. The operating high voltage is usually chosen to be one-third of the voltage range of the plateau. The plateau is considered to be that region of the voltage where the overall slope is less than ten percent per 100 volts.

The gas flow proportional counter has different operating voltages for alpha and beta particles due to the difference in their ionizing characteristics. The pulses due to the alpha radiation are large enough to be recorded by the meter at lower voltages while those due to the beta radiation will be registered only if higher voltages are applied to the detector.

Dosimeters and survey meters are portable ionization chambers designed for personal and background radiation monitoring. The most frequently used pocket dosimeters, self-indicating or indirect reading, are essentially capacitors with a quartz fiber electrode. Before use, the dosimeter is charged so that the initial position of the fiber is read from a calibrated scale. The discharging due to the ionization caused by incident radiation moves the fiber along the calibrated scale to indicate the radiation dosage. For indirect reading pocket dosimeters, a charger-reader is used to charge the dosimeter and to read the radiation dosage. Figure 4.5 shows the workings of a pocket chamber. The initial charged position is usually labelled as zero. There are several kinds of pocket chambers for measuring gamma ray and neutron doses. A portable survey meter is basically the same as a G-M detector system with a count rate meter. It is powered by batteries which operate the counter and supply a high voltage to the detector.

A scintillation detector is a crystal scintillator connected to a photomultiplier tube. Radiation incident on the scintillator causes the emission of photons. Some of these photons strike the cathode of the photomultiplier tube. The photoelectrons emitted at the cathode are accelerated by the electric potential difference established between the cathode and the first dynode. When the accelerated photoelectrons strike the first dynode, secondary electrons are emitted. The number of secondary electrons per photoelectron (between 3 and 5) is a function of the potential. Several stages later, the

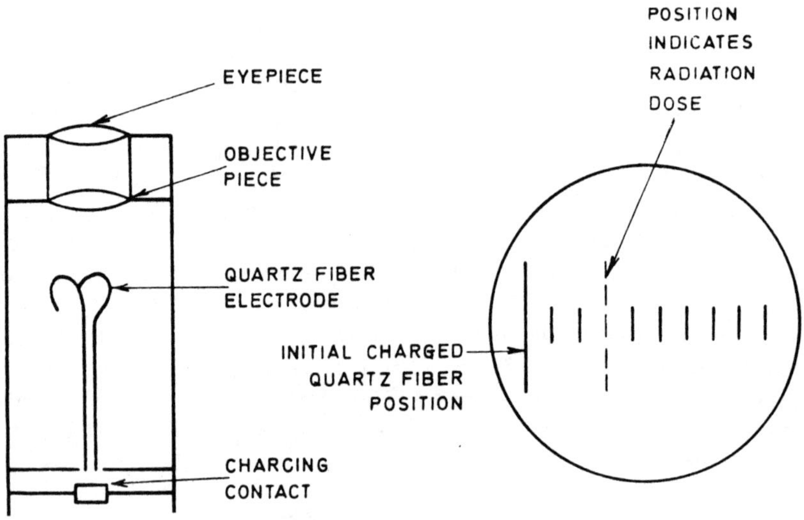

Fig. 4.5: A pocket chamber.

number of secondary electrons is increased tremendously. This causes a detectable pulse at the anode of the photoamplifier tube as shown in Figure 4.6. Since the number of photons created in the scintillator is proportional to

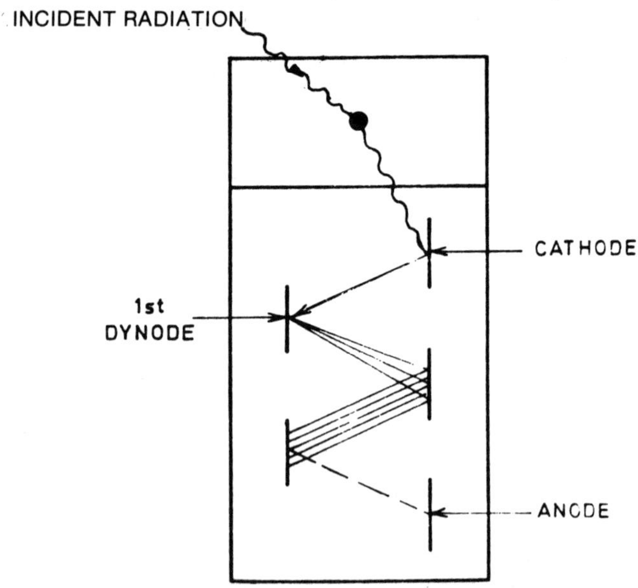

Fig. 4.6: Photomultiplier tube.

the energy and intensity of the incident radiation, the same is also true for the electric pulses at the anode. Because of this feature a scintillation detection system is frequently used for measuring the energy of the incident radiation. The electric pulses from the anode of the photomultiplier tube are amplified through a preamplifier, amplifier, and discriminator, and recorded by a meter or scaler. Such a system is called a scintillation spectrometer.

There are two modes of operation of a scintillation spectrometer: integral and differential. In the integral mode an electronic discriminator is used to reject all voltage pulses that have energies below a chosen value. Since the magnitude of the voltage pulses is directly proportional to the incident gamma ray energy, the discriminator setting represents this energy. In the differential mode, one more discriminator is used to reject pulses with higher amplitudes.

If an isotope which emits only one characteristic gamma ray is placed on the detector and counts are taken at various discriminator settings, a curve similar to Figure 4.7 can be obtained from the data. Note that at a discriminator setting of 50 there is a sharp decrease in the counting rate. This break point represents the energy of the characteristic gamma ray of the sample being counted.

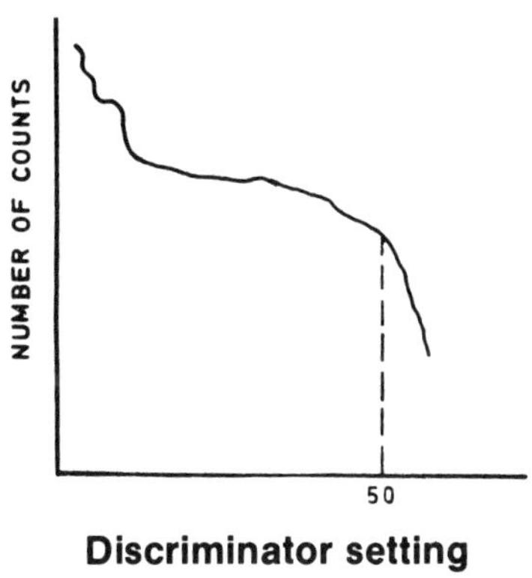

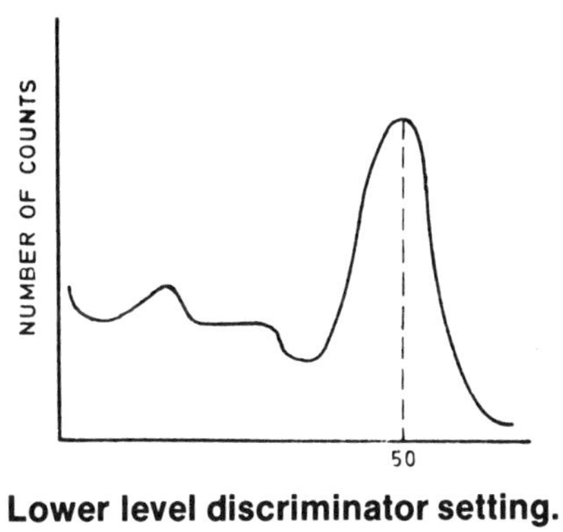

Fig. 4.7: Integral spectrum

Fig. 4.8: Differential spectrum

On most scintillation spectrometers there is an adjustable discriminator setting above which all voltage (energy) signals are rejected. This upper level discriminator, ULD, together with the lower level discriminator, LLD, may be adjusted so that only those voltage (energy) pulses falling within a very small range, called window width, are counted. If the window width is kept constant and counts are taken at various LLD settings, data for drawing a differential spectrum can be obtained (Figure 4.8). A multichannel analyzer (MCA), containing this discriminator, the preamplifier and amplifier, a memory bank, a logic module, and a displaying oscilloscope, is usually used to show such a spectrum when all channels are exhibiting simultaneously. When the operating voltage is lowered, the spectrum is compressed towards the y-axis. If the operating voltage is raised, the spectrum is extended away from the y-axis. Lowering or raising the gain of the linear amplifier has the same effect.

The term resolution, used to denote the degree of broadening, is defined as the full width of the photopeak of the differential spectrum as measured at half its maximum divided by the pulse height at the peak. Resolution of a system is a measure of its ability to separate or resolve photopeaks with energies close to each other.

The passage of radiation through matter may cause chemical changes in it. Some plastics undergo marked changes in their structure when exposed to radiation. Both phenomena pertain to the production of ions in a chemical reaction and their subsequent combination with other ions stimulated by the radiation energy loss. The quantity G is used to see if a specific chemical reaction can be utilized for radiation dosimetry. The G unit is defined as the number of reacting molecules produced per 100 eV of absorbed radiation energy. It is a measure of the chemical yield. In general, the value of G depends upon the type of radiation. Some examples of chemical dosimeters are the Fricke, ferrous sulphate, and ceric sulphate dosimeters. One specific type of chemical indicator is of particular interest to those who desire continuous monitoring of radiation. This involves the gas evolution from aqueous solutions caused by radiation. In this case one milliliter of potassium iodide solution is used. The volume of gas released is proportional to the total radiation dose.

A calorimeter is an instrument used to measure heat quantities. The problem in utilizing this type of instrument for measuring radiation is that only small amounts of heat from radiation energy losses are involved. This is why, in radiation dosimetry work, the microcalorimeter is used. Since calorimeter measurement does not involve the usual parts of a radiation detection system, it has the advantage of inherent accuracy in activity measurements. However, due to the low heat quantities involved, it has the disadvantage of low sensitivity.

4.2 Neutron Measurements

The detection of neutrons is based on their interaction with matter. The detection process must necessarily be indirect in nature because neutrons are not charged particles and hence are unable to ionize and excite the material atoms. Some methods of measuring and detecting neutrons are

a) neutron scattering with charged particles such as protons; b) neutron induced transmutations yielding radioisotopes; c) neutron induced transmutations yielding alpha particles or gamma rays. A neutron detection system is very similar to an ionization chamber counting system. In fact, the only major difference is the detector used, (Figure 4.9).

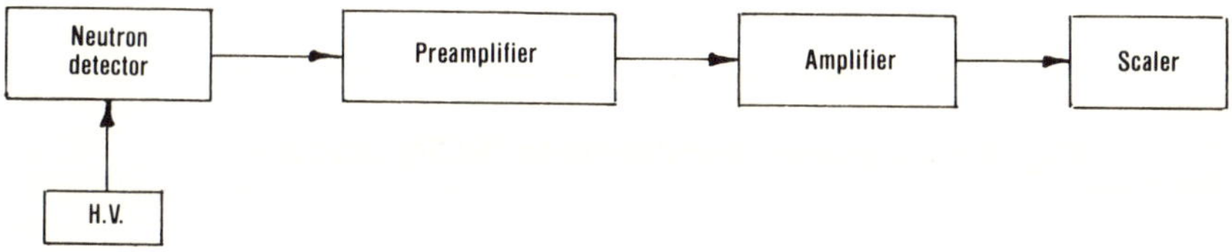

Fig. 4.9: Neutron detector system.

Neutron interaction with matter is greatly influenced by the energy of the neutron. Therefore the detectors for low energy neutrons (thermal neutrons in the eV or lower range) and those for high energy neutrons (MeV range) must be considered separately.

For thermal neutrons, some of the frequently used detectors are the BF_3-detector, the parallel-circular-plate (PCP) neutron detector, the compensated ion chamber (CIC), the scintillation detector, the foil detector, the fission chamber, and the thermopile. Fast neutrons may be detected by a threshold detector, a long counter, a scintillation detector, a Szilard-Chalmer detector or a proton recoil detector. Also, if one needs to measure the neutron energy spectrum or select a specific energy group of neutrons from a neutron flux of mixed energies, a neutron chopper or a neutron crystal diffraction spectrometer may be used.

The BF_3-proportional counter is an ionization chamber filled with BF_3 gas operating in the proportional region. The basic reaction is $^{10}B(n, \alpha)^7Li$. The produced alpha particles are easy to detect because of their high specific ionizing power. A parallel-circular-plate detector is a special design of a BF_3-counter in which B-10 coated circular plates are the capacitive electrodes (Figure 4.10).

The compensated ion chamber (CIC) is an ionization chamber partitioned into two chambers. One chamber is gamma sensitive and the other is lined with boron-10 so that it responds to both gamma rays and neutrons. By subtracting the two outputs, the detector can be made to respond to neutrons only. Figure 4.11 illustrates the compensated ion chamber.

The fission chamber has a U-235 coated electrode. The fission products lose their kinetic energy by producing ion pairs in the chamber.

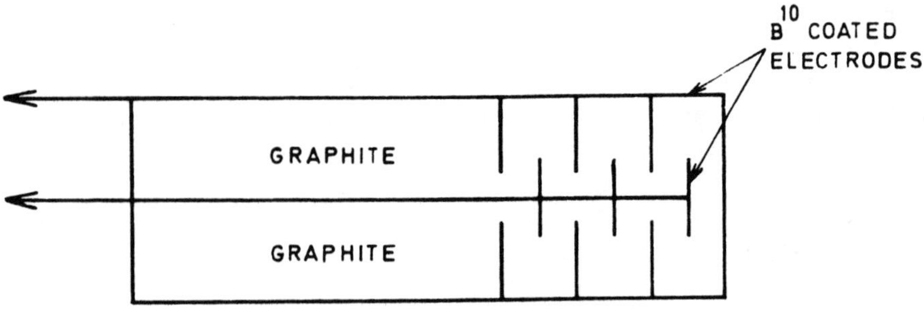

Fig. 4.10: Parallel-circular-plate (PCP) detector.

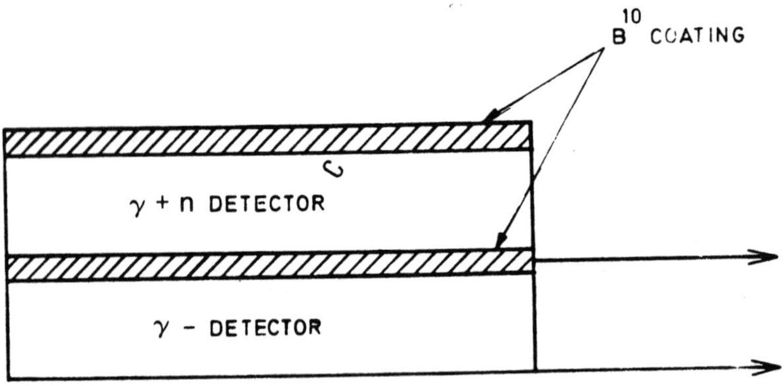

Fig. 4.11: Compensated ion chamber (CIC).

These charges are then collected by the electrodes.

For an operating nuclear power reactor, particularly during its startup and shutdown operations, the BF_3-counter may be used for its counter range of power (10^{-11} to 10^{-6} of its full power), a fission chamber may be used for the upper portion of the counter range and the lower portion of the period; a CIC or a PCP is normally used for the power range (10^{-2}–1).

Foil detectors are thin foils that can be activated by neutrons. The induced activities are then used to calculate the irradiating neutron flux. The detailed procedure of calculating the neutron flux concerning the neutron activation analysis is described in Chapter 20.

A neutron thermopile is simply a thermal couple with its hot junction coated with B-10 or U-235. Some heat is produced when neutrons interact with the coating material. A potential difference is established in the thermal couple. Many of these thermal couples are usually used simultaneously to form a thermopile to obtain a proportionate height reading on the voltmeter.

Some of the materials that can be used for thermal neutron detection are 2, 5-diphenyloxazole (PPO), Ag, Eu, and Xe. For fast neutrons, PPO or Ag can be used as scintillators.

A long counter is a BF_3-detector or a B-10 coated ion chamber surrounded by a thick (5-10 inches) paraffin cylinder as the neutron moderator. It is frequently used for fast neutron detection.

A threshold detector is a foil whose neutron absorption cross section has a known threshold energy. Above the threshold energy, the absorption cross section is essentially constant. For example, P-31 has a threshold energy at 2.5 MeV and a neutron capture cross-section of 0.075 barns.

The Szilard-Chalmer detection method is based on (n, β) or (n, γ) reactions. For example, in an aqueous solution of ethyl-iodide (C_2H_5I), fast neutrons from an immersed source are slowed down by water. The recoil energy of the excited ethyl-iodide which is the product of the reaction

$$C_2H_5I + n \rightarrow C_2H_5I^* + \gamma \tag{4.1}$$

causes radioactive iodine I* to be separated from the radical C_2H_5. By adding $AgNO_3$ to the solution, AgI, which is easily detectable, is formed. This method is frequently used for neutron source calibration. With care, a source of as low as 10 fast neutrons per second may be detected.

Proton recoil neutron detectors are widely used for detecting fast neutrons. The recoil protons serve as the ionizing and exciting agents. Hydrogenous materials such as gaseous hydrogen or solid paraffin may be enclosed in an ionization chamber or a compensated ionization chamber for neutron slowing down. The other components and the operation of the proton recoil detector are similar to the ionization chamber radiation detection system.

In many neutron studies, the energy distribution of a beam of neutrons must be known. Sometimes one needs to select neutrons of the same energy from such a neutron beam. Two frequently used methods to achieve these objectives are the neutron chopper (time-of-flight) and the neutron diffraction spectrometer.

The time-of-flight method of neutron measurement is usually used in neutron spectroscopy. A beam of intense neutrons is first passed through a set of rotating discs made of materials of high neutron absorption cross section. Cadmium discs are used for slow neutron beams. In each of the discs a slot is made for the neutrons to pass. They rotate in such a way that the slots of the first and the subsequent discs remain one revolution apart so that they can be synchronized to allow the passage of neutrons that have passed through the first disc. This arrangement is sometimes called a monochromator because the neutrons passing throught the slot of the last disc have essentially the same energy. The neutron energy is then measured by measuring the time it takes to

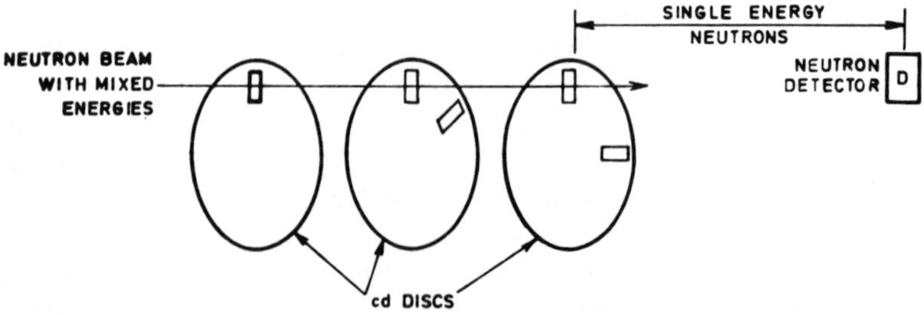

Fig. 4.12: Time-of-flight experimental setup.

travel from the last disc to the detector D. If the neutron energy is E (in eV) and the time per unit distance travelled by the neutron (in microseconds per meter) is t, they are related by

$$\sqrt{E} = 72/t. \qquad (4.2)$$

High speed electronic counting systems must be used for fast neutron energy measurement. For example, if the neutron energy is 14 MeV,

$$t = \frac{72}{\sqrt{14 \times 10^6}} = \frac{72}{3.75 \times 10^3} = 2.0 \times 10^{-2} \quad \mu s/m.$$

Usually, because of its high efficiency and sensitivity, scintillation detectors are used as neutron detectors in a time-of-flight experiment.

A neutron diffraction spectrometer operates on the same principle as x-ray diffraction. The wavelength, λ, of a thermal neutron at room temperature (0.025 eV) is approximately 2 Å which is in the range of a soft x-ray. To correlate the neutron wavelength, λ, to energy, we have

$$\lambda = \frac{h}{mv} = \frac{h}{\sqrt{2mE}}, \qquad (4.3)$$

where m, v, and E are the neutron mass, speed, and energy, respectively. The diffraction of neutrons from a crystal such as LiF is governed by Bragg's equation, i.e.

$$n\lambda = 2d\sin\theta, \tag{4.4}$$

where n is the order of the spectrum, d is the lattice spacing and θ is the incident glancing angle. Combining the last two equations, we can correlate the neutron energy to the diffraction angle θ,

$$E = \frac{n^2h^2}{8md^2}\csc^2\theta. \tag{4.5}$$

Thus the angular measurement of the diffracted neutrons measures the incident neutron energies. Figure 4.13 ullustrates a neutron diffraction spectrometer. Both the time-of-flight and neutron crystal diffraction techniques are suitable methods for accurately measuring the neutron cross section, see Chapter 6.

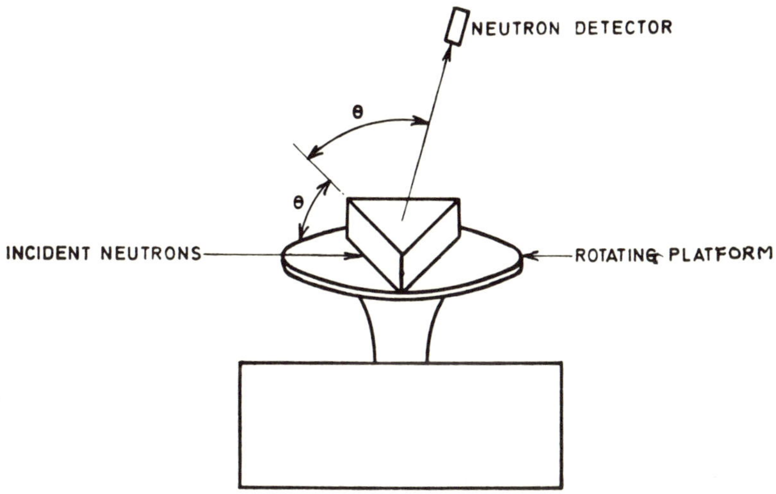

Fig. 4.13: Neutron diffraction experiment.

Example 4.1

A G-M counter requires some time to restore the potential difference across its electrodes immediately after a discharge caused by the incidence of radiation. This period can be called the resolving time. When a counting rate n_1 is obtained, due to this inactive time of the counter, the actual counting rate is $N_1 = \frac{n_1}{1 - n_1\tau}$ with τ as the resolving time. A split source method can be used to determine the resolving time of a counting system. Two radioactive sources of approximately equal strengths are counted separately (n_1 and n_2) and jointly (n_{12}). Assuming negligible background we have

$$\frac{n_1}{1 - n_1\tau} + \frac{n_2}{1 - n_2\tau} = \frac{n_{12}}{1 - n_{12}\tau} .$$

From this equation we can obtain the following approximation

$$\tau \cong \frac{n_1 + n_2 - n_{12}}{n_{12}^2 - n_1^2 - n_2^2} .$$

Suppose, in such an experiment, $n_1 = 5132$ c/min, $n_2 = 5493$ c/min and $n_{12} =$ 10531 c/min, what is the resolving time of the counting system?

Solution

$$\tau = \frac{(5132 + 5493 - 10531)}{(10531)^2 - (5132)^2 - (5493)^2} \text{ min} = \frac{94.0}{1.11 \times 10^8 - 2.63 \times 10^7 - 3.02 \times 10^7} \text{ min}$$

$$\therefore \tau = \frac{94.0 \times 60}{0.54 \times 10^8} \text{ s} = 104 \mu\text{s}.$$

Example 4.2

The efficiency of a G-M counting system may be represented by

$$\epsilon = 1 - e^{\frac{-l}{I_m}\frac{dE}{dx}} ,$$

where l is the track length of the particle in the counter, I_m is the average energy loss per ion, and $-\frac{dE}{dx}$ is the stopping power of the gas filling the conter. Calculate the counter efficiency if it is filled with argon.

Solution

If the gas is argon the stopping power

$$\frac{dE}{dx} = - 350 \text{ eV/cm} .$$

$l = 0.2$ cm.

$I_m = 26.4$ eV

$$\therefore \epsilon = 1 - e^{-(0.2 \text{ cm}/26.4 \text{ eV})(350 \text{ eV/cm})} \cong 0.93$$

Problems

4.1 Calculate the voltage decrease of a pocket chamber (to be considered as a capacitor) due to a gamma ray radiation dose of 20 mrem. The capacity of the chamber is 40×10^{-11} farads. The enclosed air has a volume of 12 cm^3. The chamber is to be operated under STP. One mrem (millirem) is equivalent to the amount of energy required to produce about 4×10^{-11} C of charge of either sign in the chamber.

4.2 A thermal neutron is detected by a fission chamber. If the capacitance of the chamber is 40 micromicrofarads, what is the current output due to fission caused by this thermal neutron? Assume that all fission energies are utilized and use an ionization energy of 35 eV/ion pair.

4.3 A gamma radiation source of unknown energy has a photopeak at a LLd setting at midway between its entire range. The maximum LLD setting is calibrated to the 1.33 MeV of Co-60. What is the energy of the unknown source? What might be this radioisotope?

4.4 In radiation measurement, statistical and error analysis must be used for data analysis. Explain the following terms:
a) Standard deviation.
b) Gaussian distribution.
c) Least square method of plotting a straight line.
d) Chi-square test.

4.5 A radiation counter subtends a solid angle of 0.2 steradian with respect to a Cs-137 source. The registered counting rate is 12 disintegrations per second. How many Cs-137 nuclei are present in the source? The half-life of Cs-137 is 30 years.

4.6 A radiation counter has a resolving time of 200 microseconds. The observed counting rate is 5000 c/min. What is the correct counting rate?

4.7 Neutron detection by a foil detector is based on neutron activation. Examine the properties of indium and gold to explain why they are the most frequently used foil detectors.

4.8 A G-M counter is used to determine the half-life of a radioactive sample. The following set of data is obtained:

Time after the beginning of the first counting (min)	**Counts per minute**
0.5	9335
2.5	8409
4.5	7317
6.5	6371
8.5	5531
10.5	4831
12.5	4205
14.5	3661
16.5	3187
18.5	2771
20.5	2410

a) Plot this set of data on semi-log graph paper to obtain a curve of counting rate versus time.

b) Graphically determine the half-life of the sample.

4.9 The kinetic energy of a neutron can be calculated from the usual formula $E = \frac{1}{2}mv^2$. From this expression, using appropriate units, derive equation (4.2).

4.10 Show that the thermal neutron wavelength is approximately 2 Å.

CHAPTER 5

RADIATION AND THE ENVIRONMENT

In dealing with the many beneficial applications of radiation one must constantly remember the possible harmful effects to living things and the environment. In the absence of sufficient experimental evidence, one must adopt the attitude that, if it is not absolutely necessary, a radiation dose of any kind should be avoided. The basic methods of avoiding or reducing radiation exposure are (1) time — as short as possible; (2) distance — as far from the source as one can manage; (3) shielding — as massive as the circumstances permit.

In the natural environment, the background radiation due to cosmic rays, terrestrial radioactivities, and naturally existing radioisotopes amounts to about 100 mrem per year. This natural radiation is hardly avoidable. It follows that the background or other radiation dose rate of about 100 mrem per year is not damaging and is acceptable.

There are many radiation sources other than natural background radiation. Some of these are unavoidable in our everyday life. These are (a) occupational exposure, (b) medical exposure, (c) weapons fallout, (d) radiation accidents, and (e) others. The radiation exposure received by a radiation worker is discussed later in this chapter. Routine x-ray examination does pose some radiation hazard if it is not administered carefully. For example, if many of the blood making cells in the bone marrow are destroyed due to x-rays, leukemia might result. The main concern, however, is on the long term genetic effect. If reproductive cells in the male and femal genitals are damaged by x-rays, there is an increased chance of genetic defects in the offspring. The x-ray exposure of a mother before and during her pregnancy could also result in infant leukemia and cancer. Due to these factors the International Commission on Radiological Protection recommends that all radiological examinations of the lower abdomen and pelvis of women of reproductive capacity be limited to the ten-day period following the onset of menstruation. In order to reduce the possible x-ray hazard, the patient may be protected in two ways when x-ray examination is necessary for medical reasons. First, a collimator can be used to narrow the x-ray beam size so that only the part of the body that needs examination will be exposed. Secondly, shielding such as lead aprons or bricks can be used to protect critical areas such as the brain, eyes, and genitals.

The world-wide radiation background has been, and still is, increasing due to nuclear weapons tests. Radioactive debris from such tests rises into the troposphere and returns to the earth's surface in a few months. The fall-out material includes the long lived radioisotopes such as Sr-90, Cs-137, and Ba-140. The increase of radiation levels due to this cause varies. In one estimate, the United Nations Scientific Committee has reported that in 1958 the radiation level was increased by 0.5 mrem/year.

Other radiation sources such as color television, watches, etc. also contribute to possible radiation hazard. Appendix 2 lists some of these sources and the approximate exposure rates.

Radiation biological effects can be divided into three types. They are acute, chronic, and genetic. The first classification pertains to the effects of large radiation doses delivered in short time intervals. The chronic effects are due to exposure to low level radiation for a long period of time. This accumulated radiation biological effect is a very important consideration for radiation workers. It is because of this reason that all nuclear energy establishments as well as agencies dealing with radiation and the use of radioisotopes require all employers to keep and preserve radiation dose records of all radiation workers. The genetic effect is a chronic effect and refers to the irreversible injury caused by radiation doses, the effects of which might be cumulative through generations to come. Special care is required on account of the uncertainties.

The biological effects of radiation depend on the amount and manner of radiation energy transfer to the receiver. The second factor is usually measured in terms of the linear energy transfer (LET) or specific ionization, i.e. the number of ion pairs formed in the receiver's body per unit length of radiation path. The linear energy transfer is defined as the energy dissipation per unit distance.

Various types of radiations interact differently with matter. To examine the radiation effects due to all possible radiation sources some form of unifying factors must be used. To this end, the International Commission on Radiological Units and Measurements (ICRU) recommended the following terminologies:

rad — radiation unit for radiation absorbed dose
roentgen (R) — unit for radiation exposure
curie — unit for measuring radioactivity
distribution factor (DF) — for non-uniform irradiation
dose equivalent —

$$DE = (D)(QF)(DF), \tag{5.1}$$

where DE is in rem (roentgen equivalent man) and D is the absorbed radiation dose in rad. The quality factor QF is referred to as the relative biological effectiveness (RBE) in radiobiology work.

The roentgen is defined as the quantity of x- or gamma radiation that produces ions carrying 3.3×10^{-10} C of electricity of either sign per 0.001293 g (1 cm^3) of air at STP. The radiation absorbed dose is defined as the amount of energy imparted to matter by ionizing radiation per unit mass of irradiated material. In this way the radiation absorbed dose (rad) is simply set to be 100 erg/g of matter.

The quality factor is a measure of the degree of injuries or effects of an equal absorbed dose incurred by a standard man due to different types of radiation (see table 5.2). A standard man is assumed to be over 18 years of age and has a mass of approximately 70 kg. The mass distribution of the standard man is as follows:

O	65%
C	18%
H	10%
N	3%
Others	4%

Daily intake of water and air by a standard man are set to be 200 cc and 2×10^7cc, respectively. If the distribution factor is unity,

$$QF = \frac{DE}{D}. \tag{5.2}$$

The quality factors for neutrons depend on the neutron kinetic energy. For thermal neutrons the quality factor is 3 which increases to approximately 11 for neutrons with kinetic energies in the vicinity of 1 MeV, it decreases to around 6 for 30 MeV. Detailed information may be found in Handbook 63,

National Bureau of Standards, U.S. Department of Commerce. For gamma rays and x-rays, the quality factor is usually chosen to be unity.

Different media absorb different amounts of energy from a given amount of radiation. This absorption is proportional to the mass absorption coefficient μ_ρ (cm^2/g) so that

$$DE_{\text{(medium)}} = DE_{\text{(air)}} \frac{\mu_{\rho\text{ (medium)}}}{\mu_{\rho\text{ (air)}}}. \tag{5.3}$$

Since by definition

$$DE \text{ (air)} = 0.877\ R_0 \text{ rad/hr}, \tag{5.4}$$

with R_0 being the exposure dose rate in roentgen per hour in air, equation (5.3) may be rewritten as:

$$DE_{\text{(medium)}} = 0.877\ R_0 \frac{\mu_{\rho\text{ (medium)}}}{\mu_{\rho\text{ (air)}}}. \tag{5.5}$$

This equation gives the relationship between the exposure dose rate in air and the absorbed dose rate in the medium. A frequently used standard reference is the one-curie Co-60 source. At one meter from this source the exposure rate in air is 1.28 R/hr. Using the inverse square law, the exposure rate is 0.32 R/hr at two meters and 5.12 R/hr at 50 cm.

A rough rule of thumb for radiation workers in judging whether or not a whole body radiation dose might be considered as safe is

$$DE = 5\ (A - 18), \tag{5.6}$$

where the dose equivalent in rem is the total accumulated maximum permissible dose up to the time when the age is A years. This formula pertains to the dose due to constant radiation exposure. In exceptional cases due to unavoidable elements, a one-time acute dose of up to 100 rem may be tolerable. For nonradiation workers and the public it is recommended that the maximum permissible dose be reduced by a factor of 10. Thus the safety level is 0.5 rem per year. For this reason, the U.S. AEC has set up a maximum permissible level for granting licenses for the possession and utilization of radioactive material and radiation sources as 0.5 rem. The exceptions can go to an ultimate maximum level of about 5 rem per year in one of the following ways

i) a 2 rem dose per hour, or
ii) a continuous dose of 100 mrem per week.

In order to avoid the maximum permissible radiation dose, an individual must utilize the time, shielding, and distance factors. However, the ultimate protection rests in observing the maximum permissible concentration (MPC) of radioactive materials in the environment. The federal code of regulations (FCR), 10 energy, part 20, lists the procedures of granting licenses for using, possessing, transporting, or disposing radiation source materials. Some representative data from this FCR are listed in table 5.1.

Table 5.1: Concentrations in air and water above natural background.

Isotope *	**Concentration Levels**** (μCi/cc) (for 40 hr/wk)		**MPC** (μCi/cc) (for 168 hr/wk)	
	in air	in water	in air	in water
Ac 227	3×10^{-11}	9×10^{-3}	9×10^{-13}	3×10^{-4}
Sb 124	2×10^{-8}	7×10^{-4}	7×10^{-10}	2×10^{-5}
A 37	6×10^{-3}	—	1×10^{-4}	—
As 76	1×10^{-7}	6×10^{-4}	4×10^{-9}	2×10^{-8}
Ba 140	4×10^{-8}	7×10^{-4}	1×10^{-9}	2×10^{-8}
Br 82	2×10^{-7}	1×10^{-3}	6×10^{-9}	4×10^{-8}
C 14	4×10^{-6}	2×10^{-2}	1×10^{-7}	8×10^{-4}
Cs 137	1×10^{-8}	1×10^{-3}	5×10^{-10}	4×10^{-8}
Co 60	9×10^{-8}	1×10^{-3}	3×10^{-10}	3×10^{-8}
F 18	3×10^{-8}	1×10^{-2}	9×10^{-8}	5×10^{-4}
I 131	3×10^{-7}	2×10^{-3}	1×10^{-8}	6×10^{-8}
Fe 59	5×10^{-8}	2×10^{-3}	2×10^{-9}	5×10^{-8}
Mn 56	5×10^{-7}	3×10^{-3}	2×10^{-8}	1×10^{-4}
Hg 197	3×10^{-6}	1×10^{-2}	9×10^{-8}	5×10^{-4}
Hg 203	1×10^{-7}	3×10^{-3}	4×10^{-9}	1×10^{-6}
P 32	8×10^{-8}	7×10^{-4}	3×10^{-9}	2×10^{-5}
Pu 239	4×10^{-11}	8×10^{-4}	1×10^{-12}	3×10^{-5}
Pu 240	4×10^{-11}	8×10^{-4}	1×10^{-12}	3×10^{-5}
Pu 241	4×10^{-8}	4×10^{-2}	1×10^{-9}	1×10^{-3}
Rn 220	3×10^{-7}	—	1×10^{-8}	—
Ag 11	3×10^{-7}	1×10^{-3}	8×10^{-9}	4×10^{-5}
Na 24	1×10^{-7}	8×10^{-4}	5×10^{-9}	3×10^{-5}
Sr 90	5×10^{-9}	1×10^{-3}	2×10^{-10}	4×10^{-8}
S 35	3×10^{-7}	8×10^{-3}	9×10^{-9}	3×10^{-4}

*The values given here are for insoluble isotopes. For solubles the concentration values are in general doubled, except for G-14, Rn-220, and Xe-135 which are solubles.
**Any area that has 25 percent or more radioisotope concentration above these levels designated as high radiation area.

Isotope	Concentration Levels (μC_i/cc) (for 40 hr/wk)		MPC (μC_i/cc) (for 168 hr/wk)	
	in air	in water	in air	in water
U 233	1×10^{-10}	9×10^{-4}	4×10^{-12}	3×10^{-5}
U 235	1×10^{-10}	8×10^{-4}	4×10^{-12}	3×10^{-5}
U 238	1×10^{-10}	1×10^{-3}	5×10^{-12}	4×10^{-5}
Xe 135	4×10^{-6}	—	1×10^{-7}	—

If there is more than one radioisotope present in air or water for the purpose of determining the MPC, the sum of the ratios of the concentration to the MPC for each radioisotope must be less than or equal to unity, i.e.

$$\frac{C_A}{(MPC)_A} + \frac{C_B}{(MPC)_B} + \cdots \leqslant 1, \tag{5.7}$$

where C's are the actual concentrations of radioisotopes A, B, ... etc.

In Appendix 2, the biological effects of radiation dose on a standard man and the aprroximate radiation doses from various sources are listed. Rough practical guides for radiation safety are given in table 5.2.

Table 5.2: Maximum permissible average flux densities for a 40 hr/week exposure.

Radiation	Energy (MeV)	Flux density (#/cm²s)
Beta particles	1	100
Fast neutrons	10	30
Fast neutrons	2	40
Thermal neutrons	0.025 eV	1000
Gamma rays	1	4000

For accuracy in neutron absorbed dose rate calculations, apply the appropriate conversion factor to each energy group and then add the total dose rate. Table 5.3 gives these conversion factors.

Table 5.3: Approximate neutron flux density dose equivalents.

Neutron energy (MeV)	Neutron flux density equivalent to 1 rem
Thermal	970
0.001	720
0.01	460
0.1	120
1.0	29
2.0 - 10.00	25
10.0 - 30.0	15

A simpler but more conservative version of this is to group the neutrons into two energy groups: fast and thermal. The conversion factors are

$$40 \text{ fast neutrons}/\text{cm}^2\text{s} = 1 \text{ mrem}/\text{hr}$$
$$\text{and } 110 \text{ thermal neutrons}/\text{cm}^2\text{s} = 1 \text{ mrem}/\text{hr}.$$

Example 5.1

The energy transferred from one roentgen of x- or gamma rays to air can be calculated from the definition of a roentgen. Since one electronic charge is 4.8×10^{-10} esu and the ionization energy of air is 34 eV per ion pair,

$$1 \text{ roentgen} = \left(\frac{1 \text{ esu}}{0.001293 \text{ g}}\right)\left(\frac{34 \text{ eV}/\text{ion pair}}{4.8 \times 10^{-10} \text{ esu}/\text{ion pair}}\right)$$
$$= 5.48 \times 10^{10} \text{ eV}/\text{g} = 87.7 \text{ erg}/\text{g of air}.$$

Example 5.2

Calculate the allowable working time for a radiation worker in a radiation field for which the exposure rate is 60 mrad/hr due to gamma rays, 20 mrad/hr due to fast neutrons (average energy 2 MeV) and 16 mrad/hr due to thermal neutrons.

Solution

Using equations (5.6), obtain the total accumulated maximum permissible radiation dose in a year to be

$$(DE)_{year} = 5\ (19 - 18) \text{ rem} = 5 \text{ rem}.$$

Therefore, assuming 50 working weeks per year, the weekly

permissible dose will be

$$(DE)_{week} = 5 \text{ rem}/50 = 100 \text{ mrem}.$$

The dose rate (mrem/hr) due to the radiation field is

$$(DE) = (60 \text{ mrad/hr}) \times 1 + (20 \text{ mrad/hr}) \times 10 + (16 \text{ mrad/hr}) \times 3$$

$$= 308 \text{ mrem/hr}.$$

$$\therefore \text{Allowable working time} = \frac{100 \text{ mrem/week}}{308 \text{ mrem/hr}} \cong 19.5 \text{ min.}$$

Problems

5.1 A patient is given a local therapeutic exposure of gamma radiation of 100 roentgens. If the mass absorption coefficient for air is 0.10 cm^2/g, calculate the amount of energy per square centimeter that passes into the patient. The density of air is 0.001293 g/cm^3.

5.2 A person at age 25 has received an accumulated radiation dose prescribed by $DE = 5\,(A - 18)$. In the following 13 weeks he receives 2.5 rem. What radiation dose may he receive in the next 13 week period?

5.3 Calculate the radiation dose delivered by carbon-14, the source strength of which is one millimicrocurie in five days at a distance of one meter.

5.4 Show that all the following statements are equivalents:
a) one roentgen of x- or gamma rays
b) 2.1×10^2 ion pairs produced per cm^3 of air at STP
c) 5.5×10^7 MeV absorbed per gram of air at STP
d) 1.6×10^{12} ion pairs produced per gram of air at STP
e) 7.1×10^4 MeV absorbed per cm^3 of air at STP.

5.5 Calculate the safe working distance from a one gram radiation source of Ra-226, assuming that the disintegration emits 2 gammas at an average energy of 0.8 MeV.

5.6 A unit frequently used to indicate the strength of a radiation source is the roentgen per hour at one meter (rhm).
a) Calculate the radiation energy flux density (MeV/cm^2s) one meter from a one Curie Cs-137 source.
b) If the gamma ray linear absorption coefficient is 3.5×10^{-5} cm^{-1}, calculate the absorption rate (MeV/cm^2s).
c) Calculate the corresponding rhm. (See problem 5.4).

5.7 The following are present in an air sample: A-37 (0.1×10^{-4}

μCi/cc), Co-60 (0.3×10^{-10} μCi/cc), and Hg-203 (2.8×10^{-9} μCi/cc). It is discovered that the same air sample also contains Na-24. What is the largest permissible amount of Na-24 concentration in this sample?

Ans: 0.5×10^{-9} μCi/cc.

5.8 A Co-60 source has a mass of $5 \times 10^{\circ 5}$ g. When it is held at one meter from a survey meter the reading is 0.7 mr/hr. What is the efficiency of the survey meter?

5.9 A radiation worker is accidentally exposed to a whole body radiation dose of 150 rads of gamma rays and 150 rads of thermal neutrons. Describe the possible biological effects and chronic symptoms.

5.10 The radiation absorbed dose (rad) from a gamma source in air is 17.54 rad/hr. What is the corresponding rad/hr in water, assuming that the distance between the source and the exposure point remains unchanged.

5.11 A person, age 40, works in a laboratory 40 hours per week. If his working area is a radiation area having thermal and fast neutrons in a two to one ratio, calculate the maximum neutron fluxes permissible in this area for him to maintain a 40-hour per week schedule.

PART II

NUCLEAR POWER ENGINEERING

CHAPTER 6

NEUTRON AND NUCLEAR FISSION REACTION

6.1 Neutrons

The neutron, a highly penetrating particle resulting from the incidence of protons on beryllium, was discovered by James Chadwick in 1932. To study the nature of the "new" particle, Chadwick performed experiments where the particle was brought to a head-on collision with a proton (the hydrogen nucleus in a paraffin sheet) and the velocity of the recoil proton was inferred from the range measurement of the aluminum foil absorption (Figure 6.1) When the same experiment was repeated replacing hydrogen with a thin sheet containing nitrogen, the recoil nitrogen nucleus had a velocity approximately one seventh of the velocity of the recoil proton. These results were used to calculate the mass of the new particle which was approximately 1.16 times the mass of a proton. When sheets up to a few cm in thickness containing lead were placed between the $(B + \alpha)$ source and the detector which contained nitrogen gas, no noticeable counting rate decrease was observed. This confirmed the high penetrating power of the new particle. Such high penetrating power could only have been achieved if the new

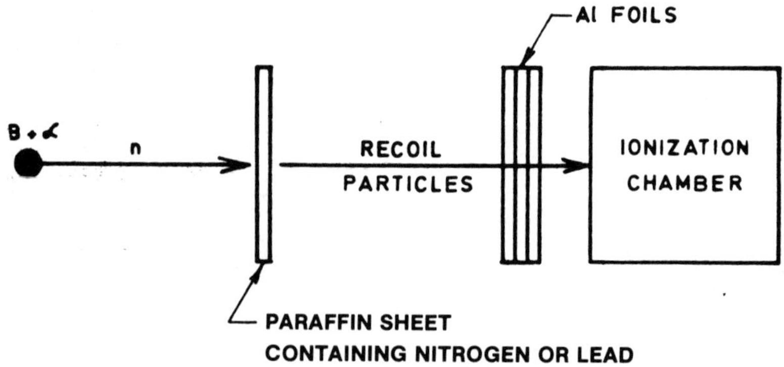

Fig. 6.1: Discovery of neutron.

particle was electrically neutral. Chadwick therefore called it the NEUTRON.

The size of a neutron in its free state is usually referred to in terms of its wavelength calculated from the dualistic property of matter (the deBroglie wavelength). For a 0.1 eV neutron, its wavelength is

$$\lambda = \frac{h}{\sqrt{2mE}} = \frac{6.63 \times 10^{-34}\ \text{Js}}{\sqrt{2(1.675 \times 10^{-27}\text{g})(0.1 \times 1.6 \times 10^{-19}\text{J})}}$$

$$= 1.66 \times 10^{-10}\text{m}.$$

This wavelength is of the same order of magnitude as that of an atom and is actually not contradictory to the fact that the neutron is a nucleon. In a nucleus, a neutron is moving very fast and hence the corresponding wavelength is reduced. The kinetic energy of a neutron inside a nucleus is approximately 50 MeV. The wavelength corresponding to this energy is about 10^{-14}m, which is roughly of nuclear dimension.

A free neutron disintegrates into a proton, an electron, and a neutrino. Enrico Fermi named this particle the neutrino because it behaves like a neutron having negligible mass. The half-life of a free neutron is approximately thirteen minutes. This does not mean that a neutron is a combination of a proton and an electron. In fact, compared with the time intervals involved in nuclear reactions or the lifetimes of reactor neutrons, a neutron can be considered as a stable elementary particle. The atmosphere, which contains approximately 10 neutrons per cubic centimeter, is the only environment in which free state neutrons exist in equilibrium.

What is the neutron structure if a neutron decays into a proton-electron pair? Due to the fact that the overall net charge of a neutron is zero and it exhibits a magnetic moment, one may think of a neutron as a negatively charged rotating 'cloud' surrounding a positively charged (proton) core as shown in Figure 6.2. This is purely a hypothetical concept. Its validity remains to be proved.

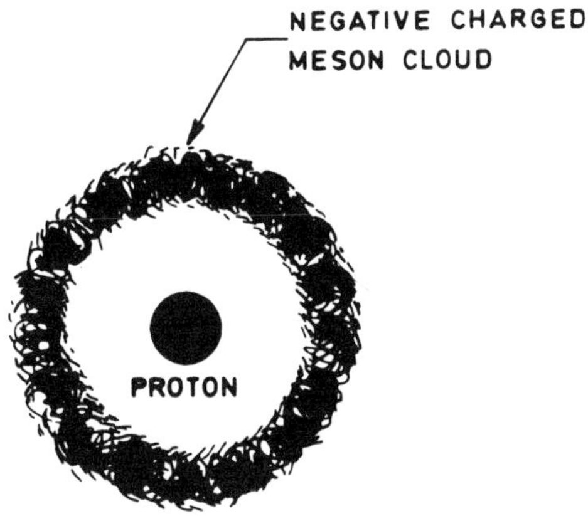

Fig. 6.2: A neutron concept.

For reactor calculations and the study of neutron interactions with matter, it is convenient, and necessary, to classify neutrons according to their kinetic energies. For example, a neutron with a kinetic energy of one electron volt is called a one electron volt neutron. Thus we refer to fast neutrons as neutrons with kinetic energies in the kiloelectron volt (keV) range and above. The resonance neutrons have an energy range from 0.1 to approximately a few keV. Thermal neutrons have their kinetic energies extending from thermal equilibrium with their surroundings up to 0.4 eV — the so-called cadmium cut-off point.

Table 6.1 gives a summary of neutron properties and characteristics.

Table 6.1: Neutron properties

Symbol	${}_0^1n$
Charge	0
Mass (amu)	1.00893
Half-life (min)	13
Magnetic moment	- 1.91 nuclear magneton
Resonance range (eV)	0.1 to 3000
Thermal range (eV)	0.025 to 0.4
Fast range (eV)	1000 to 10^8

Two physical terms which are used to indicate the number and the state of motion of neutrons are the neutron flux density ϕ and the neutron current J. The former is a scalar quantity and may be defined by

$$\phi \equiv nv, \tag{6.1}$$

where n is the neutron density and v is the neutron speed. Neutron current may be defined as

$$\vec{J} \equiv n\vec{v}, \tag{6.2}$$

where $\vec{v}$ the neutron velocity. The scalar property of the flux density and the vector property of the current must be carefully observed. For instance, if 10^6 neutrons are passing perpendicularly through a unit area and another 10^6 neutrons are simultaneously passing in the opposite direction with the same speed, as in Figure 6.3, the flux density is 2×10^6 while the current is zero.

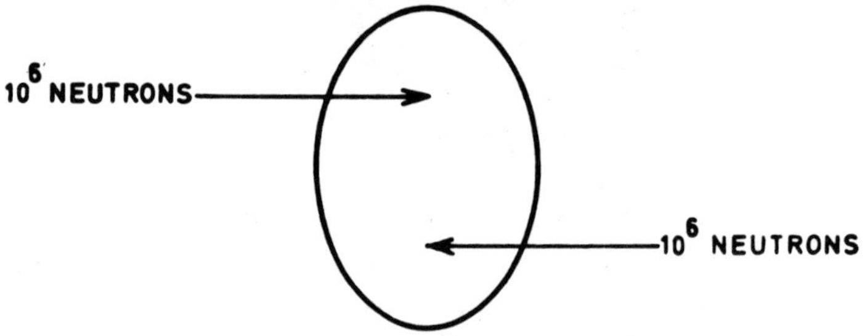

Fig. 6.3: Flux density and current.

In general the neutron density is a function of the spatial coordinate, the neutron energy and the neutron speed. For thermal neutrons, the neutron density follows a Maxwellian distribution.

$$n(v) = 4\pi n\left(\frac{m}{2\pi kT}\right)^{3/2} v^2 e^{-mv^2/2kT} \tag{6.3}$$

or

$$n(E) = \frac{2\pi n}{(\pi kT)^{3/2}} \sqrt{E}\, e^{-E/kT}, \tag{6.4}$$

where m is the rest mass of a neutron and k is the Boltzmann constant. Then n, n(v), and n(E) are the neutron density (neutron/unit volume), the neutron density per unit speed (neutron/unit volume/unit speed), and the neutron density per unit energy (neutron/unit volume/unit energy), respectively. The so-called neutron temperature, in °K, needs further explanation. This is the temperature of the neutrons having a Maxwellian energy distribution. It is a random and averaging process involving the bulk behavior of the neutrons which appear to have a temperature T. On the basis of individual neutron speeds, the following relationships can be derived:

The average neutron speed $\bar{v} = \sqrt{\frac{8kT}{\pi m}}$.

The most probable neutron speed $v_p = \sqrt{\frac{2kT}{m}}$.

The root-mean-square neutron speed $= \sqrt{\overline{v^2}} = \sqrt{\frac{3kT}{m}}$.

Similarly, one obtains the relationships between the kinetic energy and temperature of the neutron:

The average neutron kinetic energy, $\bar{E} = \frac{3}{2} kT$.

The most probable neutron kinetic energy, $E_p = ½kT$.
The energy associated with v_p, $E_p = kT$.

These expressions are calculated from the following. In order to obtain the average speed, the speed distribution and the following definition of average speed are used

$$\bar{v} \equiv \frac{\int_0^\infty v n(v) dv}{\int_0^\infty n(v) dv} = \frac{4\pi n \left(\frac{m}{2\pi kT}\right)^{3/2} \int_0^\infty v^3 e^{-mv^2/2kT}\, dv}{4\pi n \left(\frac{m}{2\pi kT}\right)^{3/2} \int_0^\infty v^2 e^{-mv^2/2kT}\, dv}$$

$$= \sqrt{\frac{8kT}{\pi m}}. \quad (6.5)$$

Similarly, the average kinetic energy of this group of neutrons is

$$\bar{E} \equiv \frac{\int_0^\infty E n(E) dE}{\int_0^\infty n(E) dE} = \frac{\frac{2\pi n}{(\pi kT)^{3/2}} \int_0^\infty E^{3/2} e^{-E/kT} dE}{\frac{2\pi n}{(\pi kT)^{3/2}} \int_0^\infty E^{1/2} e^{-E/kT}\, dE} = \frac{3}{2} kT. \quad (6.6)$$

When the thermal neutrons are referred to as a single group, as is often done in reactor analysis, the average speed and the most probable speed are related by

$$\bar{v} = \frac{2}{\sqrt{\pi}} v_p \cong 1.13\ v_p. \tag{6.7}$$

This is an important factor for the cross section calculations because most of the data on thermal neutron cross section are tabulated for neutrons at the most probable speed

$$v_p = 2200\ \text{m/s}.$$

6.2 Neutron Sources

As noted in the previous section, there are few free neutrons (~ 10 neutrons/cm^3) existing in air due to cosmic radiation and the spontaneous fission of some radioisotopes. Because of this low density, these neutrons have only a background effect. For practical applications, a neutron source must have a half-life of at least several years. It must also be intense ($>10^5$ neutrons/s) and free from other radiations such as beta particles and gamma rays. A commonly used terminology is to identify a neutron source with the activities of the projectile material. For example, a one curie neutron source means that the activity of the alpha, beta or the gamma emitters contained within the source is one curie. Some special nuclear reactions which can be used to produce neutrons are as follows:

a) (α, n) neutron sources: Alpha particles from Ra-226, Rn-222, Po-210, or Pu-239, may be used to interact with light elements such as F, Be, Li, or B. The neutron energies from these reactions range from 1 MeV to 14 MeV. A neutron source strength in the range of 10^5-10^8 neutrons/s may be achieved. Some of the useful (α, n) reactions are

$$^7\text{Li} + {}^4\text{He} \rightarrow {}^{10}\text{B} + \text{n} - 2.9\ \text{MeV} \tag{6.8}$$

$$\text{and}\quad {}^9\text{Be} + {}^4\text{He} \rightarrow {}^{12}\text{C} + \text{n} + 5.5\ \text{MeV}. \tag{6.9}$$

The first reaction requires alpha particles with kinetic energies larger than the threshold energy. The second reaction is exoergic and is most frequently used. A typical example is a one curie plutonium-beryllium neutron source. The alpha activity of about 15 g of plutonium is approximately one curie and it has a half-life of 24360 yr, which is a very desirable feature. The neutron yield, also called the emission rate or simply the source strength, is approximately 2×10^6 neutrons/s for this source. The average neutron kinetic energy is approximately 5 MeV. The plutonium-beryllium mixture is housed in a stainless steel cylinder with a threaded rod on the top for handling the source. This is illustrated in Figure 6.4.

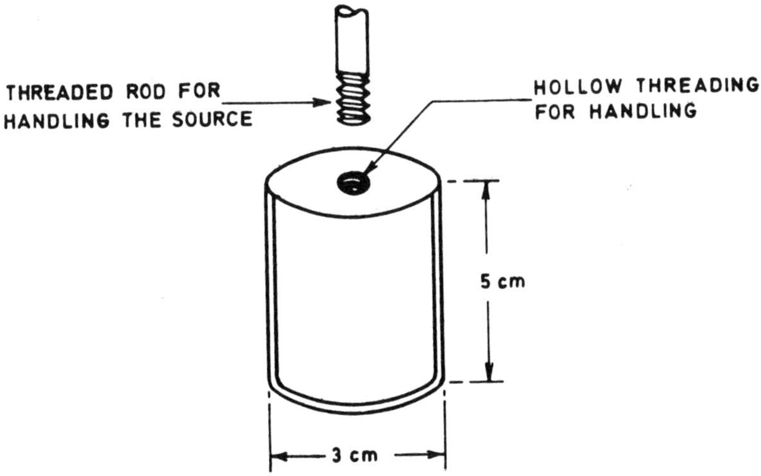

Fig. 6.4: Neutron source.

b) (γ, n) reaction: Most of the (γ, n) reactions require a gamma threshold energy which is higher than the energy range of gammas from common radioisotopes (~ 5 MeV). The only exceptions are reactions with beryllium and deuterium on which the threshold energies are 1.7 MeV and 2.2 MeV respectively. Some of the gamma sources are Na-24, Mn-56, Ga-72, and La-140. The neutron yield is approximately in the same range as the (α, n) neutron sources. One common disadvantage of (γ, n) neutron sources is the comparatively shorter half-lives of the gamma emitters. The longest half-lives are approximately 100 days. But they do have the advantage of producing monoenergetic neutrons and since high energy gamma rays are available from reactors, intense monoenergetic photoneutrons are frequently obtained from reactors or high energy particle accelerators. A commonly used (γ, n) source for power plant reactor startup is the antimony beryllium carbide (ABC) source.

c) (p, n), (d, n) and (T, n) reactions: Protons, deuterons, and tritium atoms can be accelerated to high energies in an accelerator. These are actually produced from nuclear fusion reactions such as

$$^{2}H + {}^{2}H \rightarrow n + {}^{3}He + 3.3 \text{ MeV}, \tag{6.10}$$

$$^{3}H + {}^{2}H \rightarrow n + {}^{4}He + 17 \text{ MeV}. \tag{6.11}$$

As long as power is available from the outside to run the accelerators, these fusion neutrons can be obtained in high source strengths.

There are many types of commercially available neutron generators using $^{7}Li\ (p, n)\ ^{7}Be$ or $^{9}Be\ (d, n)\ ^{10}B$. The protons or deuterons are accelerated by an accelerator such as a Van de Graaf or a Cockroft-Walton and strike Li-7 or Be-9 targets in the form of metal discs attached to the target tube. The neutron yield, depending on the energy of the protons or deuterons, can be adjusted to a range of 10^7 to 10^{10} neutrons/s/microamp. The neutron energy distribution is usually not widespread. It depends, however, on the energy of the bombarding particles. In most of the applications the disc containing Li-7 or Be-9 may be considered to be a fast neutron point source. Figure 6.5 illustrates the various components of such a neutron generator.

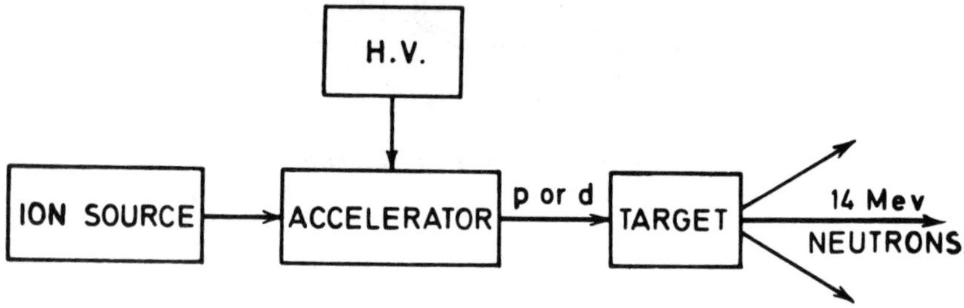

Fig. 6.5: Schematic of a neutron generator.

d) Spontaneous fission source: There are several radioisotopes which emit spontaneous fission neutrons in the order of 10^6 per second or more. Most of these radioisotopes, while having long half-lives of spontaneous fission, emit alpha particles at the same time and the alpha half-life is usually short, a few days at the most. Two of the few that have a comparatively longer alpha decay half-life are Cf-252 (2.6 yr) and Cm-244 (18 yr). We shall describe the Cf-252 neutron source in detail because its spontaneous fission neutron spectrum is very similar to the reactor fission neutron spectrum. Cf-252 is produced in a high flux ($\sim 10^{15}$ neutrons/cm^2s) isotope production reactor from the U-238 chain of neutron capture and radiation decay transmutation as described in Figure 6.6. The latest technique is to start this chain from Cm-244. This has the advantage of neutron economy, and also part of the errors in production estimation due to uncertainties of the activation cross sections of some of the intermediate radioisotopes are avoided. Cf-252 has a spontaneous fission half-life of 86 yr. The average fission neutron energy is 2.35 MeV. The alpha particles have average energies of 6.12 MeV. The most important feature of the Cf-252 source is its high specific fission neutron emission, 2.3×10^{12} neutrons/s/g. The Savannah River Operations Office of USDOE has been producing and test marketing the Cf-252 source for several years. These sources were fabricated by uniformly electroplating CfOH on a 90% platinum-10% iridium cathode which was oxidized to form Cf_2O_3. The

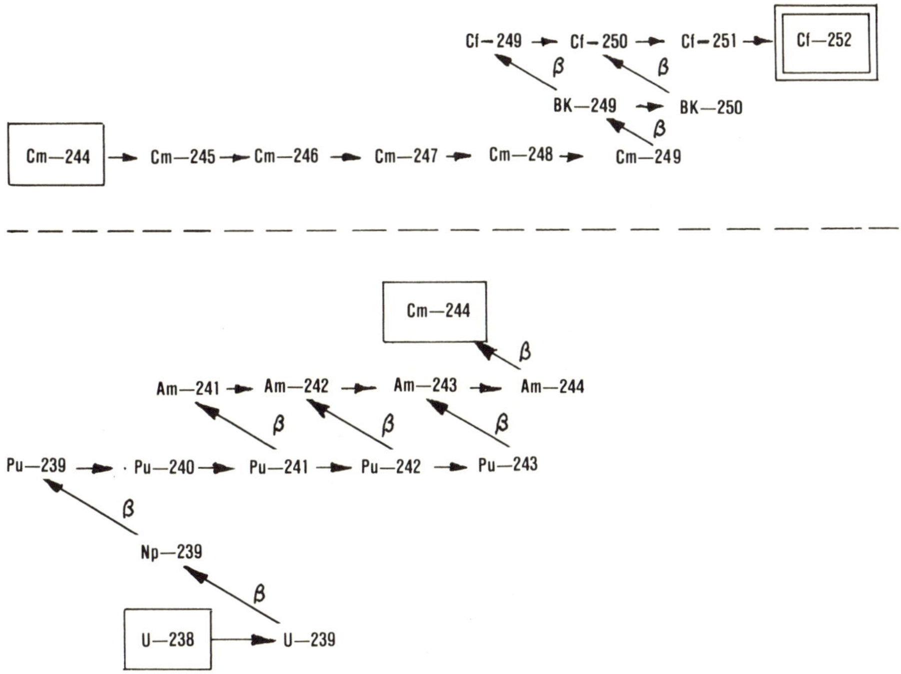

Fig. 6.6: Cf-252 production chain.

Cf_2O_3 was then sealed in a 90% platinum-10% iridium cell. Small quantities of Cf-252 (~ 1 milligram) have been produced. Possible uses of this particular source are neutron radiography, biomedical applications, dosimetry, geological applications such as well logging or chlorine log, pollution control, and neutron activation analysis.

e) Reactor neutrons: The fission neutrons produced in a nuclear reactor have an energy distribution ranging from a few keV to approximately 20 MeV, averaging about 2 MeV. This neutron energy distribution may be roughly represented by the following relation

$$n(E) = 2\sqrt{E}\, e^{-E/1.3}, \qquad (6.12)$$

where n(E) is the relative number of neutrons with energy E in MeV. The ln n(E) versus E plot is illustrated by Figure 6.7.

f) Thermal neutron sources: In general, neutrons are born with

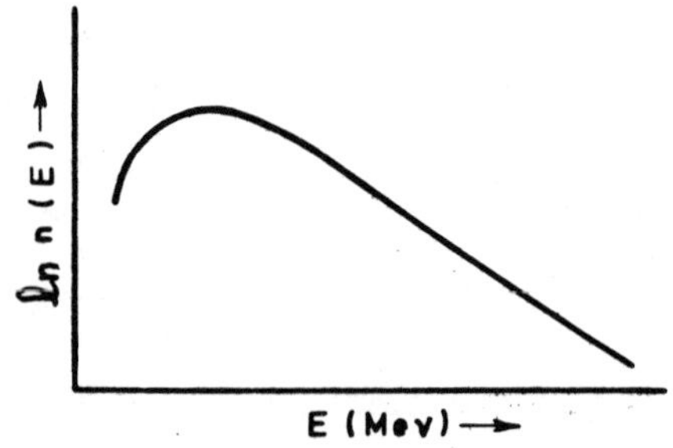

Fig. 6.7: Fission neutron spectrum.

energies in the keV to MeV range. To obtain thermal neutrons one must slow down these fast neutrons by passing them through a material that acts as an energy moderator. Many research reactors have a thermal column of carbon or water just outside the reactor core which acts as the thermal neutron source. Thermal neutrons can also be obtained by surrounding any fast neutron source with a moderating material such as water or paraffin.

6.3 Neutron Interaction with Matter

The interaction between neutrons and nuclei of a material depends on the number of neutrons involved, the energies and the states of motion of these neutrons, and the properties of the material. The macroscopic cross section, $\Sigma = N\sigma$, depends on the material nuclei density N and the microscopic cross section σ, which is a measure of the probability for the interaction to occur. The reaction rate R (per unit volume) is given by

$$R = \phi\Sigma = nvN\sigma, \tag{6.13}$$

where ϕ is the neutron flux density. It is obvious that the reaction rate is proportional to the number of neutrons and the number of nuclei of the interacting material.

From equation (6.13) it can be shown that the microscopic cross section σ has the dimension of length squared (cm^2). Thus it has the meaning of area. One way of interpreting this fact is to consider that σ represents an "effective" area of the target nucleus which is exposed to the incident neutron. Therefore σ is an indication of the probability of the occurrence of an interaction between the incident particle and the target nucleus. Due to nuclear dimensions, $r^2 \sim 10^{-26} cm^2$. Since cm^2 is too large a unit for σ, a common practice is to use a new unit called the barn. This unit for the microscopic cross section is defined as 1 barn $\equiv 10^{-24} cm^2$. Since the target nuclei density N has a dimension of length^{-3}, it follows that the unit for the macroscopic cross section Σ is length^{-1} (cm^{-1}).

Most of the interactions of neutrons with matter are absorption and scattering reactions. Absorption of a neutron may lead to some particle or gamma ray emission or fission reaction. The total microscopic cross section $\sigma = \sigma_a + \sigma_s$ of matter for these interactions is neutron energy dependent. It is convenient to study the variations of these microscopic cross sections in the thermal, resonance and fast neutron energy ranges.

Microscopic Absorption Cross Section

The absorption cross section is the sum of the fission cross section and the capture cross section, i.e. $\sigma_a = \sigma_f + \sigma_c$. Its variation with neutron energy for most materials is illustrated in Figure 6.8.

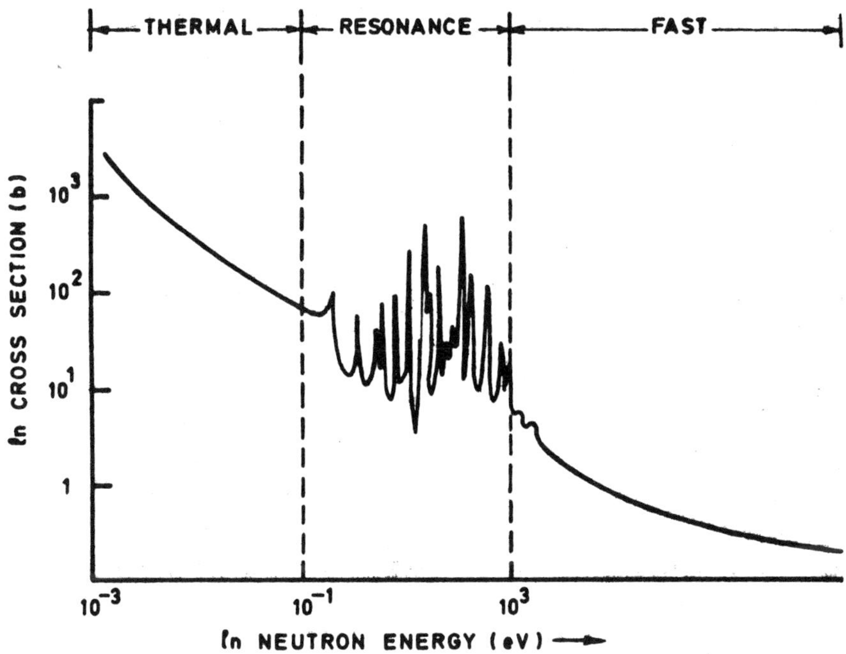

Fig. 6.8: Absorption cross section.

In general, in each of the three neutron energy ranges a certain pattern exists that may be used to estimate the value of these cross sections. In the low energy range, the microscopic absorption cross section of matter is inversely proportional to the square root of the neutron energy, i.e.

$$\sigma_a \sim \frac{1}{\sqrt{E}}. \tag{6.14}$$

Therefore the slope of the straight line as shown in the thermal region is -½. This is a very useful rule of thumb for calculating cross sections at different neutron energies in this region. In the resonance neutron energy range, sharp rising peaks occur. The most striking examples of the resonance peaks are those of Cd-113, 2×10^4b at 0.17 eV, and Xe-135, 3.4×10^6b at 0.7 eV. Resonance absorption cross sections can be calculated by the Breit-Wigner formula

$$\sigma(E) = \frac{\sigma(E_0)}{1 + \left(\frac{E - E_0}{\Gamma/2}\right)^2},$$

where E_0 and E are the resonance energy and the neutron energy respectively, and Γ is the full width at half maximum (FWHM) of the resonance peak. In the fast neutron region, the absorption cross section decreases slowly with increasing neutron energy. For neutrons with energies above a few keV, the absorption cross sections are usually less than 10b.

Microscopic Scattering Cross Section

Except for hydrogen, the microscopic scattering cross sections of materials are below 20b for practically all neutron energies. For reasons of special interest, we shall record the hydrogen microscopic scattering cross section for thermal neutrons at 0.025 eV as $\sigma_s = 38$b. This is an indication of the neutron slowing down ability of hydrogen or any hydrogen containing material such as paraffin and water.

Neutron Reactions

A neutron reaction can be written as

$$n + X \rightarrow Y + W + Q, \tag{6.15}$$

where X is the interacting nucleus while Y and W are the product nuclei. For example, when boron is bombarded by neutrons, the end products are Li-7 and alpha particles. The complete representation of this reaction is

$$n + {}^{10}B \rightarrow {}^{7}Li + {}^{4}He + 2.31\ MeV. \tag{6.16}$$

This is the basis for one method of neutron detection. In certain cases isotopes of the same element can be obtained. The product is usually radioactive. One specific example is the indium foil activation for neutron detection

$$n + {}^{115}In \rightarrow {}^{116}In \rightarrow {}^{116}Cd + \beta^-. \tag{6.17}$$

The beta particles can then be detected to yield a measure of the neutron flux density.

Perhaps, the most famous nuclear reaction is the thermal neutron fission of U-235

$$n + {}^{235}U \rightarrow {}^{A_1}Y + {}^{A_2}W + \nu n + Q, \quad (6.18)$$

where $\nu = 2$ or 3. This reaction will be discussed in detail in the next section.

6.4 Nuclear Fission Reaction

In a nuclear fission process the target nucleus, upon absorbing a neutron, forms an excited compound nucleus which splits into two fission fragments accompanied by 2 or 3 neutrons. The fission fragments due to thermal neutrons are usually asymmetric in mass numbers. Figure 6.9 shows the fission fragment groups, the light group with mass numbers ranging from 70 to 120, and the heavy group with A from 120 to 160.

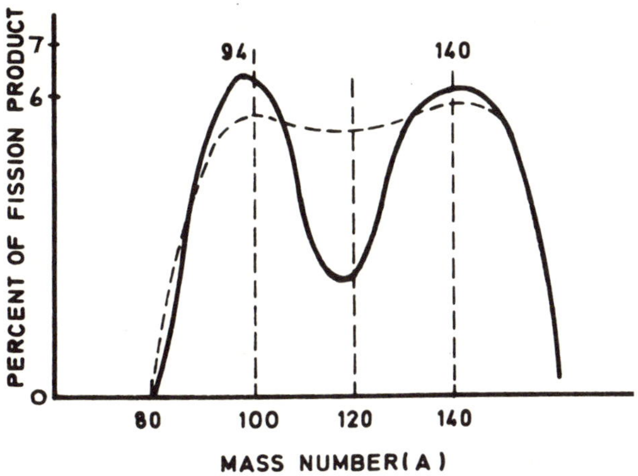

Fig. 6.9: Fission fragments.

A typical neutron fission reaction may be represented by

$$n + {}^{235}U \rightarrow {}^{236}U \rightarrow {}^{94}Sr + {}^{140}Xe + 2n + Q. \quad (6.19)$$

For fast neutron fission such as

$$n + {}^{238}U \rightarrow {}^{239}U \rightarrow {}^{118}Sn + {}^{118}Mo + 3n + Q, \quad (6.20)$$

the fission fragments are more symmetrical (dotted lines of Figure 6.9). There is little chance that both fission fragments will be stable. This fact can be easily deduced from the conservation of mass number and atomic number of a fission reaction and the stable nuclide chart.

Most of the neutrons (~ 90%) emitted from either fast or thermal fission reactions are the so-called prompt neutrons. They are produced within

$\sim 10^{-12}$s after the fission process. The remaining portion of the fission neutrons are produced from the decay of fission fragments. These fission fragments are called delayed neutron precursors, the half-lives of which range in seconds. Although the delayed neutrons have very small yields, they play a very important role in nuclear reactor control. They prolong the lifetime of fission neutrons from the prompt neutron lifetime to a more manageable effective neutron lifetime. This point is discussed in more detail in chapter 8.

A very important feature of a fission reaction is its Q-value. We can estimate this fission energy release by considering a hypothetical symmetric fission model. In this example a compound nucleus (A, Z) is split into two symmetric fission fragments $(\frac{A-2}{2}, \frac{Z}{2})$ and two neutrons. From the semi-empirical mass formula, the energy release Q can be calculated, assuming Z = 92 and A = 236.

$$Q = M(A, Z) - 2M\left(\frac{A-2}{2}, \frac{Z}{2}\right) - 2m_n$$

$$= Zm_H + (A - Z)m_n - a_vA + a_sA^{2/3} + a_c\frac{Z^2}{A^{1/3}} + a_a\frac{(A-2Z)^2}{A} - \frac{33.5}{A^{3/4}}$$

$$- 2\left\{\frac{Z}{2}m_H + \left(\frac{A-2}{2} - \frac{Z}{2}\right)m_n - a_v\frac{A-2}{2} + a_s\left(\frac{A-2}{2}\right)^{2/3} + a_c\frac{(Z/2)^2}{\left(\frac{A-2}{2}\right)^{1/3}}\right.$$

$$\left. + a_a\frac{\left(\frac{A-2}{2} - Z\right)^2}{\left(\frac{A-2}{2}\right)}\right\} - 2m_n$$

$$\cong 140 \text{ MeV}.$$

This energy appears in the form of the kinetic energies of the fission fragments (82%), the gamma ray energy (3.5%), the kinetic energy of the fission neutrons (2.5%), and the energies associated with other radiations (12%).

As stated before, the fission neutron energies range from a few keV up to 20 MeV. This energy distribution (fission neutron energy spectrum) can be represented by the semi-empirical formula

$$n(E) = 2\sqrt{E}\, e^{-E/1.3}. \qquad (6.21)$$

From the above equation we can calculate the average energy of the fission neutrons

$$\bar{E} = \int_0^\infty E\, n(E)dE / \int_0^\infty n(E)dE$$

$$= \int_0^\infty E^{3/2} e^{-E/1.3}\, dE / \int_0^\infty E^{1/2}\, e^{-E/1.3}\, dE$$

$$\cong 1.95 \text{ MeV}.$$

Note that the upper integration limit is artificially extended to infinity in order to facilitate the evaluation of the integrals. Although two or three fission neutrons are produced in most fission reactions, the number may actually range from 0 to 5. The average neutron yield (number of neutrons produced per fission) is $\nu = 2.43$.

The most attractive feature of a (n, 2n) or (n, 3n) fission reaction is the fact that it produces more neutrons. In this way, as long as the fissionable material is plentiful, no extraneous neutron sources are required to initiate other fission reactions.

lost due to leakage

$$n + {}^{235}U \rightarrow [{}^{236}U] \rightarrow {}^{94}Sr + {}^{140}Xe \quad + Q + n + n$$

$$n + {}^{235}U \rightarrow [{}^{236}U] \rightarrow \ldots \qquad (6.22)$$

To see how the fission chain reaction can be initiated and maintained, examine the microscopic fission cross section of the fissionable materials given in table 6.2. The microscopic fission cross section (σ_f) for most of the

Table 6.2: Microscopic fission cross section of some fissionable materials for thermal neutrons.

Isotope	σ_f **(barns)**
U-235	577
Pu-239	742
Pu-241	1020
U-233	525

fissionable materials such as U-235 or Pu-239 is very large for thermal neutrons as compared with fast neutrons, where it is only a few barns. For comparison, the fission cross section (σ_f) of U-235 for neutrons in the MeV energy range is always less than 2b. For neutrons with kinetic energies from 10 to 20 MeV, the U-235 fission cross section is still only slightly above 2 barns. Because of this the high energy fission neutrons are slowed down to thermal energies in order to increase the chance of another fission reaction. This is achieved by using a hydrogeneous material as a moderator. When a fast neutron collides with a moderator nucleus it is theoretically possible for it to lose all of its kinetic energy. In any case, the high scattering cross section (σ_s) of hydrogen for neutrons at thermal energies makes it one of the most frequently used moderators. Neutron slowing down is discussed in more detail in the next chapter.

Example 6.1

For accurate reactor physics calculations, the effective neutron cross section σ should be used. The expression for the effective cross section is generally written in the following form

$$\bar{\sigma} = \sigma_{2200} \frac{\sqrt{\pi}}{2} \sqrt{\frac{293}{T}} \times \left(\frac{1}{v} \text{ correction factor}\right).$$

Explain all the terms involved in the expression and calculate the value of the effective cross section for natural uranium and plutonium for neutrons thermalized at 40°C.

Solution

Most of the tabulated values of the thermal neutron cross sections correspond to the most probable neutron speed for thermal neutrons, i.e. 2200 meters per second. A speed correction factor is therefore necessary. The expression for the cross section averaged over all possible neutron energies is given by

$$\bar{\sigma} = \frac{v_{2200}}{\bar{v}} \sigma_{2200},$$

where $\bar{\sigma}$ is the neutron cross section corresponding to the average neutron speed, v. From the Maxwell-Boltzmann distribution it can be shown that

$$\frac{v_{2200}}{\bar{v}} = \frac{\sqrt{\pi}}{2}$$

$$\therefore \bar{\sigma} = \frac{\sqrt{\pi}}{2} \sigma_{2200}.$$

The neutron temperature is generally assumed to be equal to the moderator temperature when the neutrons are in thermal equilibrium with the moderating medium. In all practical cases, however, the moderator temperature is higher than the reference temperature for which all the cross section values are tabulated. Since the neutron cross sections decrease linearly with increasing speed, i.e. follow the 1/v law, and the speed is directly proportional to the square of the temperature, a temperature correction equal to $\sqrt{293/T}$ has to be applied, where T is the moderator or neutron temperature. The value of the effective cross section with speed and temperature corrections is therefore

$$\bar{\sigma} = \sigma_{2200} \frac{\sqrt{\pi}}{2}\sqrt{\frac{293}{T}}$$

The thermal absorption cross sections of some of the elements doe not follow the 1/v law very strictly. A correction factor is accordingly applied for such elements in order to obtain the value of neutron cross sections for use in reactor calculations. The final expression for the effective neutron cross section is given in the example.

For natural uranium, the 1/v correction factor is 0.99 and σ_{2200} is 7.7 b.

$$\therefore \bar{\sigma} = 7.7 \times 0.866 \times \sqrt{\frac{293}{313}} \times 0.99\text{b} = 6.5\text{b}.$$

For plutonium-239, the 1/v correction factor is 1.073 and σ_{2200} is 1025 b.

$$\therefore \bar{\sigma} = 1025 \times 0.886 \times \sqrt{\frac{293}{313}} \times 1.073 \text{ b} = 943 \text{ b}.$$

Example 6.2

The average distance that a neutron travels in a target material before undergoing a specified reaction is known as the neutron mean free path, λ. Derive an expression for λ and calculate the values of the absorption, scattering and total mean free paths for graphite.

Solution

$$\lambda = \frac{\text{Total path covered by all neutrons}}{\text{Total number of neutrons}}$$

$$= \frac{\int_0^\infty x \, dI}{I_0},$$

where x dI is the total path covered by neutrons that travel a distance x and then collide while travelling an additional distance x. Now

$$I = I_0 e^{-\Sigma x}.$$

$$\therefore dI = -I_0 \Sigma e^{-\Sigma x} dx$$

$$\lambda = -\Sigma \int_0^\infty x e^{-\Sigma x} dx$$

or

$$\lambda = \frac{1}{\Sigma} \cdot$$

Since neutrons undergo absorption or scattering reactions, the absorption and scattering mean free paths must be examined.

These are given as:

$$\lambda_s = \frac{1}{\Sigma_s}$$

and

$$\lambda_a = \frac{1}{\Sigma_a} \cdot$$

The total macroscopic cross section and total mean free path are given here:

$$\Sigma_T = \Sigma_a + \Sigma_s$$

and

$$\frac{1}{\lambda_T} = \frac{1}{\lambda_a} + \frac{1}{\lambda_s} \cdot$$

For graphite

$$N = \frac{\rho N_0}{A} = \frac{(2.2 \text{ g/cc})(6.03 \times 10^{23} \text{ atoms/mole})}{12 \text{ (g/mole)}}$$

$$= 1.1 \times 10^{23} \text{ atom/cc.}$$

$$\therefore \Sigma_a = N\sigma_a = (1.1 \times 10^{23} \text{ atoms/cc})(3.2 \times 10^{-27} \text{ cm}^{-2})$$

$$\therefore \lambda_a = \frac{1}{\Sigma_a} = \frac{1}{3.52 \times 10^{-4} \text{ cm}^{-1}} = 2.8 \times 10^3 \text{cm.}$$

Also

$$\Sigma_s = N\sigma_s = (1.1 \times 10^{23} \text{ atoms/cc})(4.8 \times 10^{-24} \text{ cm}^{-2})$$

$$\therefore \lambda_s = \frac{1}{\Sigma_s} = 1.9 \text{ cm.}$$

The value of the total mean free path can be calculated using the expression for $1/\lambda_T$.

Example 6.3

Calculate roughly how many grams of uranium-235 would yield one megawatt-day of thermal energy if only 185 MeV per fission is utilized for thermal energy production in power reactors.

Solution

1 megawatt day = 8.64×10^4 megawatt-s = 8.64×10^{10} watt-s.

Useful energy per fission = 185 MeV = 3×10^{-11} J.

Hence the number of fissions required to yield one megawatt day of energy is

$$\frac{8.64 \times 10^{10}}{3 \times 10^{-11}} = 2.88 \times 10^{21}.$$

Therefore the amount of uranium required is

$$\frac{235 \times 2.88 \times 10^{21}}{6.03 \times 10^{23}} = 1.1 \text{ g}.$$

Problems

6.1 Sometimes it is convenient to specify neutrons by some property other than their energy. The properties generally encountered are speed, wavelength and temperature. Derive expressions for the neutron energy in electron volts in terms of neutron speed in meters per second; wavelength in Angstroms; and temperature in absolute degrees.

Ans. $0.523 \times 10^8 v^2$; $\dfrac{8.18 \times 10^{-2}}{\lambda^2}$; $\dfrac{T}{1.16 \times 10^4}$

6.2 If the average neutron energy released in the fission process is 1.95 MeV, calculate the neutron speed, wavelength and temperature.

6.3 Determine the thermal neutron microscopic and macroscopic cross sections (absorption and scattering) of light and heavy water. On the basis of your results discuss the importance of heavy water in reactor technology as well as the effect of light water impurity in heavy water.

6.4 Derive the expressions for the average and most probable neutron speeds. Calculate these speeds for neutrons in thermal equilibrium with a moderating medium at 20° C.

6.5 For the expression $I = I_0 e^{-\Sigma x}$ and the corresponding $dI = -I\Sigma dx$ explain the physical significance of the following terms:

$e^{-\Sigma x}$; Σdx; $1 - \Sigma dx$

$\Sigma dx e^{-\Sigma x}$; $(1 - \Sigma dx)e^{-\Sigma x}$

6.6 Enriched uranium dioxide fuel is being used presently in most of the nuclear power reactors. Calculate the values of the absorption and scattering mean free paths in the oxide fuel with 3 percent U-235 enrichment.

6.7 A sheet of natural uranium 4 cm wide, 10 cm long and 1 cm in thickness was exposed to a thermal neutron flux of 10^{10} neutrons per cm^2 per second. Calculate roughly the thermal energy released in the sheet as a result of fissions of U-235.

6.8 A thin foil of Pu-239, 0.5 cm in diameter and 0.015 cm thick, was irradiated in the graphite thermal column of a nuclear reactor. The neutrons have a Maxwellian speed distribution corresponding to a temperature of 20° C. If the thermal neutron flux density at the location of the foil was 10^9 neutrons per cm^2 per second, calculate the absorption and fission rates in the foil. Will these rates be significantly influenced if the moderator temperature is raised to 200° C?

6.9 A polonium beryllium neutron source gives off neutrons at a rate of 5×10^7 neutrons per second. Polonium has a half-life of 138 days. Calculate the time in days when the source strength becomes 10^6 neutrons per second.

6.10 Show that the condition for stability with respect to spontaneous symmetric fission is

$$Z^2/A < 45.$$

6.11 Calculate the energy released per fission for the following reaction from the rest mass consideration of the interacting and the resulting nuclei and particles, as well as the binding energy per nucleon of fissionable nuclides and the fission products.

$$^{235}U + n \rightarrow n + {}^{235}U \rightarrow {}^{98}Mo + {}^{136}Xe + 2n + 4\beta^{-}.$$

6.12 Calculate the mass of uranium-235 which would yield about the same amount of energy as 20 kilotons of TNT. Assume 1 g of TNT yields 1000 calories.

6.13 A large nuclear powered tanker is capable of expending 40,000 horsepower. Calculate roughly the daily consumption of uranium-235 if the conversion efficiency of fission energy into useful work is 50 percent. What would be the daily coal consumption of the same tanker at 50 percent efficiency, assuming that one kilogram of coal releases 7000 calories?

6.14 Two beams of neutrons ($\phi_1 = 10^6 n/cm^2s$ and $\phi_2 = 5 \times 10^5 n/cm^2s$) are at 90° from each other, see Figure 6.10.

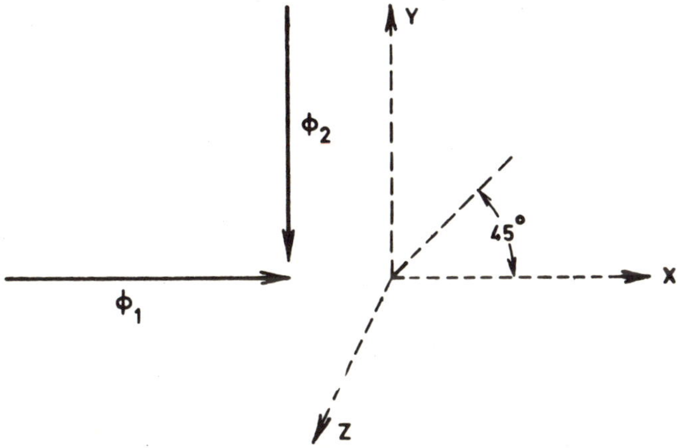

Fig. 6.10: Neutron flux density and current.

a) Find the neutron flux density at the point where they cross each other.

b) What is the neutron current along the x-axis? y-axis? in the plane at 45° from both axes?

Ans: a) 1.5×10^6 n/cm²s.
b) $10^6 n/cm^2s$; -5×10^5 n/cm²s;
1.06×10^5 n/cm²s to the right.

CHAPTER 7

NEUTRON SLOWING DOWN AND DIFFUSION

7.1 Introduction

The methods used to reduce the kinetic energy of a fission neutron to the thermal energy range are based on the collision processes between the neutron and the nucleus of a moderating material. For neutrons with energies between 30 eV and a threshold energy of a few MeV, elastic scattering is the dominant process. In order to excite the target nucleus for neutrons with energies above the threshold energy ($\sim$ 6 MeV for oxygen and $\sim$ 4 MeV for carbon), inelastic scattering is utilized. A part or all of the kinetic energy of a neutron in an inelastic collision is used to excite a nucleus and, subsequently, a gamma ray is emitted to carry off the excitation energy. Water is most frequently used for neutron slowing down because of its high scattering (Σ_s) and low absorption (Σ_a) cross sections. In fact, the criterion for choosing a moderator is based precisely on these two factors. This is reflected in the moderating ratio (MR), defined as

$$MR \equiv \xi \frac{\Sigma_s}{\Sigma_a}, \tag{7.1}$$

where ξ is the average logarithmic energy decrement per collision. It will be shown later that ξ is a function of the atomic mass number only. For a mixture or compound of two or more types of nuclei, the effective ξ is calculated from the scattering cross section (Σ_s) weighted average, i.e.

$$\xi = \frac{\Sigma_{s1}\,\xi_1 + \Sigma_{s2}\,\xi_2 + \cdots}{\Sigma_{s1} + \Sigma_{s2} + \cdots}. \tag{7.2}$$

After a neutron has reached thermal energy in a medium, i.e. thermal equilibrium with its surroundings, it keeps drifting about in a random manner until it is finally absorbed or leaked out of the medium. To examine thermal neutrons and their distribution in a medium, the diffusion theory is used.

The neutron lifetime (l) and the migration length (M) are two parameters frequently used to indicate the state of motion of a fission neutron in a medium. Fission neutrons live through the slowing down period for a time l_s by travelling, on the average, a slowing down distance. One measure of this distance is the so-called Fermi age τ. This is based on the Fermi age theory or continuous slowing down model discussed later in this chapter. In diffusion, the corresponding quantities are the diffusion time (l_d) and the diffusion length (L). Therefore

$$l = l_s + l_d \tag{7.3}$$

and

$$M = \sqrt{\tau + L^2}. \tag{7.4}$$

The migration length, M, is obtained from the migration area

$$M^2 = \tau + L^2, \tag{7.5}$$

where the diffusion length $L \cong \frac{1}{\sqrt{3\Sigma_t\,\Sigma_a}}$ (the term is formally defined later by equation 7.22). The slowing down time of a fission neutron is usually very small compared with the diffusion time which can be calculated from

$$l_d = \frac{1}{\Sigma_a v}, \tag{7.6}$$

where v is the thermal neutron speed. Table 7.1 lists the neutron slowing down parameters of some of the frequently used moderators.

Table 7.1: Neutron slowing down parameters

Moderator	D (cm)	τ (cm^2)	L (cm)	σ_s (b)*	σ_a (b)*	ξ
H_2O	0.18	30	2.7	50	0.66	0.93
D_2O	0.85	125	100	11	0.003	0.51
C	0.92	360	50	4.8	0.005	0.16
Be	0.61	110	24	6.0	0.009	0.21
Be oxide	0.56	110	30	9.8	0.009	0.17

***thermal neutron data.**

7.2 Neutron Kinetic Energy Loss

For fission neutrons the major process through which slowing down occurs is elastic scattering. Consider a neutron with mass m_n colliding with a nucleus with mass M at rest as shown in Figure 7.1.

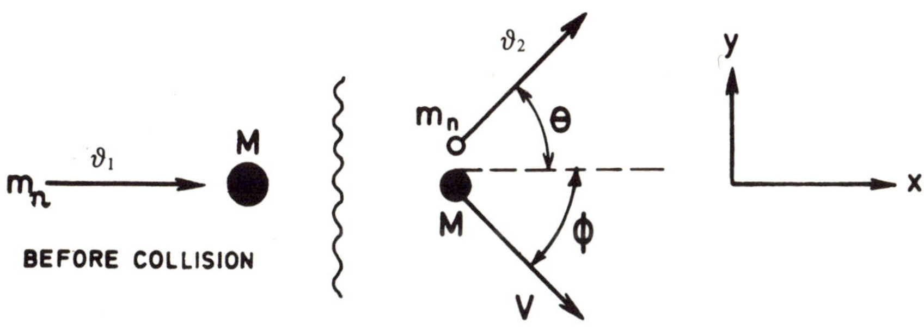

Fig. 7.1: Neutron elastic scattering.

Using the conservation principles of kinetic energy and momentum,

$$\tfrac{1}{2}m_n v_1^2 = \tfrac{1}{2}m_n v_2^2 + \tfrac{1}{2}MV^2.$$

$$m_n v_1 = m_n v_2 \cos\theta + MV\cos\phi.$$

$$0 = m_n v_2 \sin\theta - MV\sin\phi.$$

Eliminating V and ϕ from the above equations and introducing $A \equiv M/m_n$,

$$\sqrt{\frac{E_2}{E_1}} = \frac{\cos\theta \pm \sqrt{\cos^2\theta + A^2 - 1}}{A + 1} . \tag{7.7}$$

Note that the negative sign preceding the square root in the numerator must be discarded.

$$\therefore \frac{E_2}{E_1} = \frac{1}{(A + 1)^2} \quad (\cos\theta + \sqrt{\cos^2\theta + A^2 - 1})^2. \tag{7.8}$$

For a fixed value of E_1, the largest and smallest values of E_2 are

$$(E_2)_{max} = E_1 \qquad \text{when } \theta = 0° \tag{7.9}$$

and

$$(E_2)_{min} = \alpha E_1 \qquad \text{when } \theta = 180°, \tag{7.10}$$

where $\alpha = (A - 1)^2/(A + 1)^2$. This is the energy range, $\alpha E_1 \rightarrow E_1$, of the elastically scattered neutrons. It is obvious that the energy decrement per collision is a fraction, from α up to 1, of the original neutron energy. Because of this reason, in dealing with detailed neutron slowing down problems, it is more convenient to use a term 'lethargy' defined by

$$U \equiv \ln \frac{E_0}{E} , \tag{7.11}$$

where E_0 is a reference energy. For fission neutrons E_0 is usually chosen as 2MeV, the average fission neutron kinetic energy. Although lethargy will not be used in our discussions, note that when a neutron loses energy from E_0 to zero, the lethargy varies from zero to infinity like a conventional variable. This is illustrated in Figure 7.2.

The average logarithmic energy decrement can be obtained from neutron energetics. A neutron with energy E_1 would have a probability of $dE/(1 - \alpha) E_1$ to have an energy E - dE after an elastic collision.

$$\xi \equiv \overline{\ln \frac{E_1}{E}} = -\int_{E_1}^{\alpha E_1} \ln \frac{E_1}{E} \frac{dE}{(1 - \alpha)E_1} = 1 + \frac{\alpha}{1 - \alpha} \ln\alpha.$$

$$\therefore \xi = 1 + \frac{(A - 1)^2}{2A} \ln \frac{A - 1}{A + 1} . \tag{7.12}$$

The calculated values of ξ for some elements that can be used as moderators

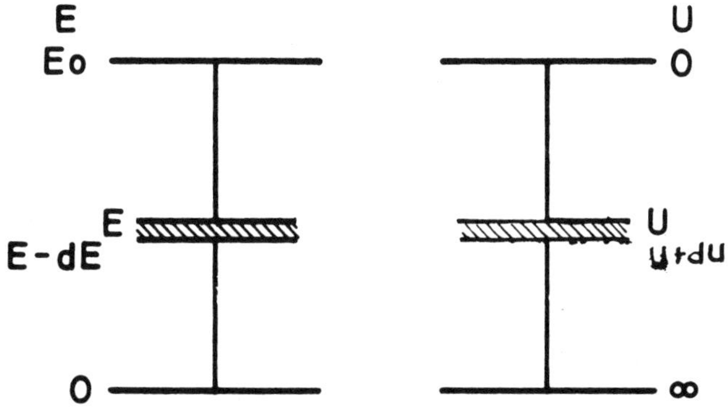

Fig. 7.2: Lethargy and energy scales.

are listed in table 7.2. Since the value of ξ is an indication of the neutron energy loss per collision, this quantity can be used to estimate the number of elastic collisions required to thermalize ($\sim$ 0.025 eV) a fission neutron. For carbon, it is

$$\frac{\ln \dfrac{2 \times 10^6}{0.025}}{\xi} = \frac{18.2}{\xi} = 114.$$

Table 7.2: Values of ξ for various moderators.

Element	A	α	ξ
H	1	0	1.00
D	2	0.11	0.73
He	4	0.36	0.43
C	12	0.72	0.16

But for hydrogen it is only 18. Since leakage occurs in a finite medium, the neutron slowing down gets more complicated. Here, the slowing down neutron density is a spatial as well as energy dependent quantity.

7.3 Fermi Age Slowing Down Theory

For a non-absorbing medium, the Fermi age equation may be written as

$$\nabla^2 q = \frac{\partial q}{\partial \tau}, \tag{7.13}$$

where the slowing down density q is the number of neutrons per unit volume per second that has been slowed down past the energy E. It is a function of space and the Fermi age τ defined by

$$\tau(E) \equiv \int_{E_0}^{E} \frac{D}{\xi \Sigma_s E}\, dE = \int_0^{\infty} \frac{D}{\xi \Sigma_s}\, dU, \tag{7.14}$$

where D is the diffusion coefficient, which is one-third of the transport mean free path. Its value is approximately given by $D = 1/3\Sigma_s$. Fermi age has a special meaning in that it is related to the average neutron slowing down distance.

The Fermi age equation resembles the usual heat conduction equation. For a point source of fast neutrons, energy E_0, the age is $\tau = 0$. In this case, using spherical geometry equation (7.13) reads

$$\frac{1}{r^2}\frac{\partial}{\partial r}\left(r^2 \frac{\partial q}{\partial r}\right) = \frac{\partial q}{\partial \tau}. \tag{7.15}$$

It can be shown that the solution of this equation is

$$q(r, \tau) = \frac{s e^{-r^2/4\tau}}{(4\pi\tau)^{3/2}}, \tag{7.16}$$

where s is the source strength in neutrons/s. From this solution, we can calculate the mean square slowing down distance

$$r^2 = \int_0^{\infty} r^2 q(r, \tau)\, 4\pi r^2 dr \Big/ \int_0^{\infty} q(r, \tau)\, 4\pi r^2 dr$$

$$= \int_0^{\infty} r^4 e^{-r^2/4\tau}\, dr \Big/ \int_0^{\infty} r^2 e^{-r^2/4\tau}\, dr$$

$$= 6\,\tau. \tag{7.17}$$

The age dependence of the slowing down density is a very important reactor analysis parameter. If we set $q(r, \tau) = q(r)\, q(\tau)$, then on rearranging the age equation to

$$\frac{\Delta^2 q(r)}{q(r)} = \frac{dq(r)}{d\tau} = -B^2, \tag{7.18}$$

we get $q(\tau) = q(0)e^{-B^2\tau}$, where B^2 is constant. This factor, $e^{-B^2\tau}$ is the fast leakage factor frequently used in reactor criticality calculations.

7.3 Neutron Diffusion

For thermal neutrons, in a narrow energy range in which the material coefficients such as the absorption and scattering cross sections can be considered as constants, the mathematical model based on neutron balance, neutron rate of change = - leakage - absorption + source, can be written as

$$\frac{\partial n}{\partial t} = - (- D\nabla^2\phi) - \Sigma_a\phi + S, \tag{7.19}$$

where S is the extraneous neutron source term, D is the diffusion coefficient, and $\partial n/\partial t$ is the time rate of change of the neutron density n. At steady state and in the absence of an extraneous neutron source, S, in a non-multiplying medium the diffusion equation is

$$D\nabla^2\phi - \Sigma_a\phi = 0 \tag{7.20}$$

or

$$\nabla^2\phi - \frac{1}{L^2}\phi = 0, \tag{7.21}$$

where

$$L \equiv \sqrt{\frac{\phi}{\nabla^2\phi}} = \sqrt{\frac{D}{\Sigma_a}} \tag{7.22}$$

is the thermal neutron diffusion length. For an infinite medium with a plane source at the origin, the diffusion equation is

$$\frac{d^2\phi}{dx^2} - \frac{1}{L^2}\phi = 0$$

with the boundary condition $\phi(\pm\infty) = 0$ and the source condition $\lim_{x \to 0}(D\frac{d\phi}{dx}) = S$. The solution to this diffusion equation is

$$\phi(x) = A_1 e^{x/L} + A_2 e^{-x/L}.$$

Here the constants A_1 and A_2 must be determined by the boundary and source conditions. Since $\phi(+\infty) \to 0$, set A_1 to zero and use the source condition to evaluate A_2. The final solutions are

$$\phi(x) = \frac{SL}{2D} e^{-x/L}, \text{ for } x < 0$$

and

$$\phi(x) = \frac{SL}{2D} e^{x/L}, \text{ for } x > 0.$$

If the medium is finite both terms, $e^{x/L}$ and $e^{-x/L}$, will appear in the solution with appropriate coefficients.

In a multiplying medium with thermal neutrons due to the slowing down of the fission neutrons, one can use the slowing down density to rewrite the neutron source term in the diffusion equation.

$$D \nabla^2\phi - \Sigma_a\phi + pq = 0, \tag{7.23}$$

where q is the slowing down density and p is the resonance escape probability. Solutions of this equation are carried out in the next chapter.

Example 7.1

Extrapolation Distance

When neutrons are diffusing in a finite medium, the neutron flux density at the boundary is not really zero but has a finite value. Some distance away from the boundary in the vacuum surrounding the medium, the neutron flux density drops to zero. This distance is the extrapolation distance. In plane geometry this is illustrated by Figure 7.3.

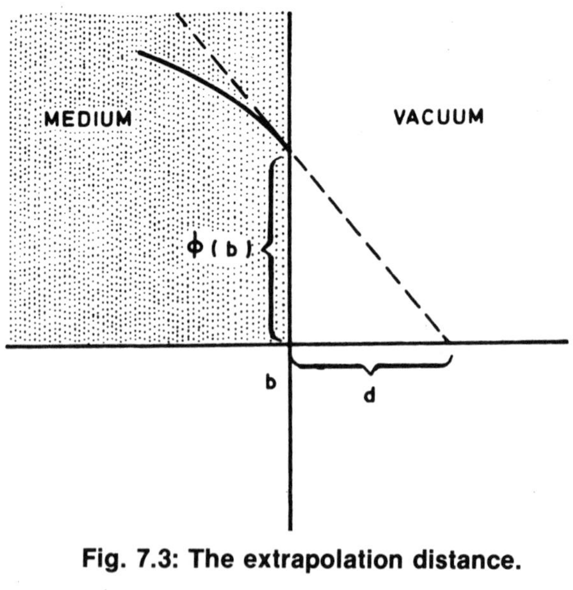

Fig. 7.3: The extrapolation distance.

Find a working formula for calculating the extrapolation distance. Calculate this distance in the case of thermal neutrons if the medium is a) graphite or b) water.

Solution

According to Fick's law, the neutron current at the boundary, x = b, going into the medium is

$$J = \frac{\phi(b)}{4} + \frac{D}{2}\left(\frac{d\phi}{dx}\right)_{x=b}.$$

Since the right hand side of the boundary is a vacuum, J = 0 at the boundary.

$$\therefore \phi(b) = -2D\left(\frac{d\phi}{dx}\right)_{x=b}.$$

From Figure 7.3, by geometry, we have

$$d = -\frac{\phi(b)}{\left(\frac{d\phi}{dx}\right)_{x=b}}.$$

Therefore $d = 2D = (2/3)\lambda_{tr}$, where λ_{tr} is the transport mean free path. This is an approximation compared with the correct value of $d = 0.7104\ \lambda_{tr}$. For isotropic scattering in a homogeneous medium such as graphite and water $\lambda_{tr} \cong \lambda_s = 1/\Sigma_s$

$$\therefore d = \frac{2}{3\Sigma_s}.$$

a) For graphite, $\Sigma_s = 0.385\ cm^{-1}$,

$$\therefore d = \frac{2}{3 \times 0.385 cm^{-1}} = 1.732\ cm.$$

b) For water, $\Sigma_s = 3.45\ cm^{-1}$,

$$\therefore d = \frac{2}{3 \times 3.45\ cm^{-1}} = 0.193\ cm.$$

Example 7.2

In many textbooks, the average logarithmic energy decrement is approximated by

$$\xi = \frac{2}{A + 2/3} \cdot$$

From the expression of ξ given by equation (7.12), prove this approximation by assuming $A \gg 1$.

Solution

$$\xi = 1 + \frac{(A - 1)^2}{2A} \ln \frac{A - 1}{A + 1}$$

$$= 1 - \frac{(A - 1)^2}{A} \text{Tanh}^{-1} (\frac{1}{A})$$

$$= 1 - \frac{(A - 1)^2}{A} (\frac{1}{A} + \frac{1}{3A^3} + \frac{1}{5A^5} + \ldots)$$

$$\cong 1 - \frac{(A - 1)^2}{A} (\frac{1}{A} + \frac{1}{3A^3}) \text{ for } A \gg 1.$$

$$\therefore \xi = 1 - \frac{1}{3A^4} (A - 1)^2(3A^2 + 1)$$

$$= \frac{6A^3 - 4A^2 - 2A - 1}{3A^4}$$

$$= \frac{2(1 - \frac{2}{3}\frac{1}{A} + \frac{1}{3A^2})}{A}$$

$$= \frac{2}{A + \frac{2}{3}} \cdot$$

Example 7.3

Indium foils absorb both thermal and resonance ($E \sim 1.4$ eV) neutrons. When they are placed between two cadmium covers so that all thermal neutrons are absorbed by the shielding, the indium foils would then detect resonance neutrons only. These resonance neutrons in a medium are proportional to the slowing down density. Therefore, based on the principle of neutron activation, the neutron activated indium foil activity is proportional to the neutron slowing down density. When such an experiment is performed for fission neutrons in beryllium oxide, the saturated indium activity at various distances from the neutron source is as follows:

Distance, r(cm)	**Saturated activity, A (counts/m)**
4.0	950
8.0	820
12.0	650
16.0	469
20.0	300
24.0	174
28.0	100
32.0	36
36.0	25
40.0	10

What is the Fermi age of the medium?

Solution

Equation (7.17) can be used to calculate the Fermi age. The basic steps to carry out these calculations are:
a) evaluate Ar^2 and Ar^4,
b) plot Ar^2 vs r and Ar^4 vs r curves,
c) integrate the curves to obtain the areas under the respective curves, and
d) caluculate τ from equation (7.17).
When these calculations are performed, $\tau = 100$ cm.

Problems

7.1 Calculate the number of collisions required for hydrogen to slow down a fission neutron (2 MeV) to a thermal energy (0.025 eV).

7.2 In an infinite medium the diffusion solution is

$$\phi(x) = A_1 e^{-x/L} + A_2 e^{-x/L},$$

the constants A_1 and A_2 are to be determined by boundary and source conditions. Prove that

$$\phi(x) = \frac{L\, e^{-x/L}}{2D} \quad \text{for } x < 0$$

and $\phi(x) = \dfrac{L\, e^{x/L}}{2D}$ for $x > 0$.

7.3 Solve the diffusion equation for a point neutron source S in an infinite medium. Hint: In spherical geometry the diffusion equation is

$$\frac{d^2\Phi}{dr^2} + \frac{2}{r}\frac{d\phi}{dr} - \frac{1}{L^2}\phi = 0.$$

This equation can be transformed to

$$\frac{d^2U}{dr^2} - \frac{1}{L^2}U = 0.$$

if $\phi(r) = U(r)/r$.

Ans: $\phi(r) = \dfrac{S}{4\pi D} \cdot \dfrac{e^{-r/L}}{r}$.

7.4 If the source strength in the preceding problem is 10^{10} neutrons per second, calculate the neutron flux density in graphite and water at the following distances: r = 5 cm, 10 cm, 20 cm, and 100 cm.

7.5 Calculate the average logarithmic energy decrement per collision of a) hydrogen, b) deuterium, c) carbon, and d) uranium.

Ans: a) 1, b) 0.73, c) 0.16, d) 0.00084.

7.6 Calculate the diffusion coefficient of thermal neutrons in graphite. Also calculate the same in graphite for 14 MeV neutrons.

7.7 Derive a similar relationship as equation (7.17) between L^2 and $\overline{r^2}$.

7.8 Discuss the meaning of each term of the diffusion equation $D\nabla^2\phi$ -

$\Sigma_a \phi$ + pq = 0. Is this strictly correct when applied to a fission reaction occuring in a medium? Explain your answer.

7.9 Neutrons are slowed down in graphite by elastic collision with carbon nuclei. Calculate the maximum and minimum energy of a 2 MeV neutron after one collision.

7.10 Calculate the migration area of fission neutrons in a) water, b) heavy water, and c) graphite.

CHAPTER 8

THERMAL FISSION REACTOR ANALYSIS

8.1 Introduction

A self-sustaining chain reactor may be classified as a power, reasearch, or isotope production reactor depending on its usage. According to the energies of the neutrons that cause the fission process, reactors are categorized as fast, intermediate, and thermal reactors. The manner in which fuel and moderator (for thermal reactors) are assembled to form a reactor core determines whether a reactor is homogeneous or heterogeneous.

Historically, the basic treatment of reactor theory starts with the Age-diffusion model applied to a homogeneous thermal research reactor. This will be the basis of the discussion through the entire treatment on reactor analysis, in which the following will be examined:

a. Multiplication factor and criticality calculations
b. Neutron flux distribution and heat transfer
c. Reactor kinetics and control.

8.2 Neutron Cycle — The Multiplication Factor

In a neutron chain reaction, it is necessary to know the neutron multiplication factor (K) which is the ratio of the numbers of neutrons of two successive generations. For simplicity, one may consider an infinite homogeneous mixture of fuel and moderator within which thermal neutron fission chain reactions occur. In this system, the infinite multiplication factor, K_∞, is given by the following four-factor formula

$$K_\infty = \eta \epsilon p f. \tag{8.1}$$

The reproduction factor η is the number of fission neutrons produced per thermal neutron absorbed by the fuel.

$$\therefore \eta = \nu(\Sigma_f / \Sigma_a)_{fuel}. \tag{8.2}$$

The symbol ϵ represents the fast fission factor which is the ratio of the number of fission neutrons due to neutrons of all energies to the number of fission neutrons due to thermal neutrons alone. The resonance escape probability, p, is the fraction of all fission neutrons that are slowed down to thermal energies. The thermal utilization factor, f, is the ratio of the thermal neutrons absorbed by the fuel to the thermal neutrons absorbed by all reactor core materials (in the simplified approach, these are fuel and moderator)

$$\therefore f = \frac{(\Sigma_a)_{fuel}}{(\Sigma_a)_{fuel\ +\ moderator}}. \tag{8.3}$$

Considering an idealized example of a homogeneous mixture of U-235 and water for thermal fission (infinite dimensions),

$$\eta = \nu\left(\frac{\Sigma_f}{\Sigma_a}\right)_{fuel} = 2.43\frac{577}{683} = 2.05,$$

$$\epsilon = 1.00,$$
$$p = 1.00,$$

and

$$f = \frac{(\Sigma_a)_{fuel}}{(\Sigma_a)_{fuel} + (\Sigma_a)_{water}}$$

$$= \frac{1}{1 + 9.66 \times 10^{-4}\,(N_{water}/N_{fuel})}.$$

If the moderator-fuel atom number ratio is

$$N_{water}/N_{fuel} = 400,$$

$$f = 0.721.$$

$$\therefore K_\infty = \eta\epsilon pf = (2.05)(1.00)(1.00)(0.721) = 1.48.$$

For a finite medium, the neutron leakages must be considered. Assuming that the homogeneous mixture is contained in a spherical volume and using age-diffusion theory, the fast neutron and thermal neutron nonleakage probabilities, L_f and L_{th} respectively, can be calculated as

$$L_f = e^{-B^2\tau} \quad \text{and} \quad L_{th} = \frac{1}{1 + B^2L^2}. \tag{8.4}$$

The first term was derived in the last chapter and the thermal leakage factor will be discussed in the next section. The parameter used for examining the neutron number is the effective multiplication factor K_e which differs from the infinite multiplication factor simply in the non-leakage factors, i.e.

$$K_e = L_f\, L_{th}\, K_\infty. \tag{8.5}$$

For a homogeneous water mixture contained in a sphere of radius R = 39.3 cm, the non-leakage factors (see the criticality calculations in section 8.4) are

$$L_f = 0.704 \quad \text{and} \quad L_{th} = 0.960.$$

The neutron cycle is illustrated in the diagram in Figure 8.1 by taking 100 thermal neutrons as the starting point. The chain reaction would be self-sustaining in this case because there are 100 thermal neutrons occuring at the end of the first generation, or at the beginning of the second generation. The effective multiplication factor is unity.

8.3 Age-Diffusion Model

In a thermal reactor, a fast neutron slows down by successively colliding with the moderator nuclei. After it reaches thermal energy it diffuses until it is absorbed in the fuel or moderator. In a finite volume, the number of thermal neutrons present and their average speeds are the most important quantities for reactor calculations. This is usually referred to as the thermal neutron flux density distribution of the reactor. In one-group steady state diffusion theory,

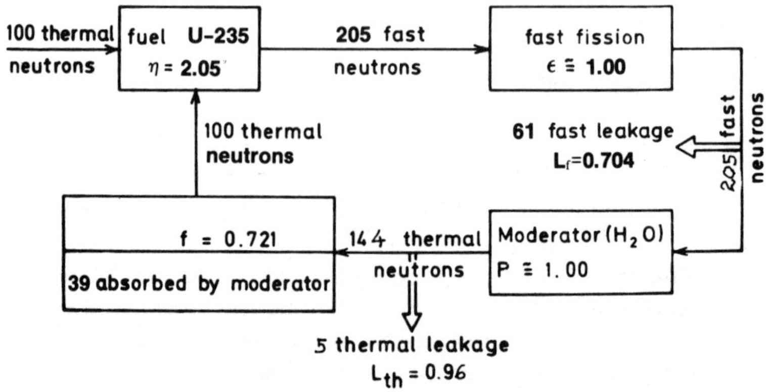

Fig. 8.1: Neutron cycle.

$$D \nabla^2 \phi - \Sigma_a \phi + S = 0. \qquad (8.6)$$

The source term must be derived from an appropriate neutron slowing down model. When age theory is used, it can be written as

$$S = K_\infty \Sigma_a \phi e^{-B^2 \tau}.$$

$$\therefore D\Delta^2 \phi + \Sigma_a (K_\infty e^{-B^2 \tau} - 1)\phi = 0. \qquad (8.7)$$

In spherical geometry, the equation is

$$\frac{d^2\phi}{dr^2} + \frac{2}{r}\frac{d\phi}{dr} + \frac{1}{L^2}(K_\infty e^{-B^2\tau} - 1)\phi = 0. \qquad (8.8)$$

The boundary conditions are

1) $\phi(r) = \phi_0$ is finite at $r = 0$ (8.9)

and

2) $\phi(R) = 0$ at $r = R$. (8.10)

The solution to the diffusion equation is

$$\phi(r) = \frac{A \sin Br}{r} + \frac{C \cos Br}{r}. \qquad (8.11)$$

Boundary condition 1) requires that we set C = 0. From boundary condition 2),

$$\frac{\sin B_n}{R} = 0 \text{ or } B = n\pi/R. \tag{8.12}$$

$$\therefore B_n^2 = (\frac{n\pi}{R})^2. \tag{8.13}$$

This quantity, when n = 1, is called the geometrical buckling. In both the slowing down theory and the diffusion theory, the spatial variations of the slowing down density $\Delta^2 q/q$ and the thermal neutron flux density $\Delta^2\phi/\phi$ are set to be equal to a constant $-B^2$. The equations thus obtained are called Helmholtz equations and are in the forms of

$$\nabla^2\phi + B^2\phi = 0 \text{ or } \nabla^2 q + B^2 q = 0. \tag{8.14}$$

The constant $B^2 = -\nabla^2\phi/\phi$ is called the buckling because it is a measure of the neutron flux density bending. Replacing $\nabla^2\phi$ by $-B^2\phi$ in the diffusion equation,

$$\frac{K_\infty e^{-B^2\tau}}{1 + B^2L^2} = 1. \tag{8.15}$$

This is the critical condition. Taking the limit of $r \rightarrow 0$ for the remaining first term of the right hand side of equation (8.11),

$$\lim_{r\to o} \phi(r) = A \lim_{r\to o} \frac{\sin Br}{r} = AB \lim_{r\to o} \cos Br = \phi_0.$$

$$\therefore A = \phi_0/B. \tag{8.16}$$

Thus the flux density is

$$\phi(r) = \frac{\phi_0 \sin Br}{Br}. \tag{8.17}$$

This is illustrated in Figure 8.2.

8.4 Reactor Power

The power available from a nuclear reactor depends on the neutron flux density. Since one watt of power can be obtained from 3.3×10^{10}

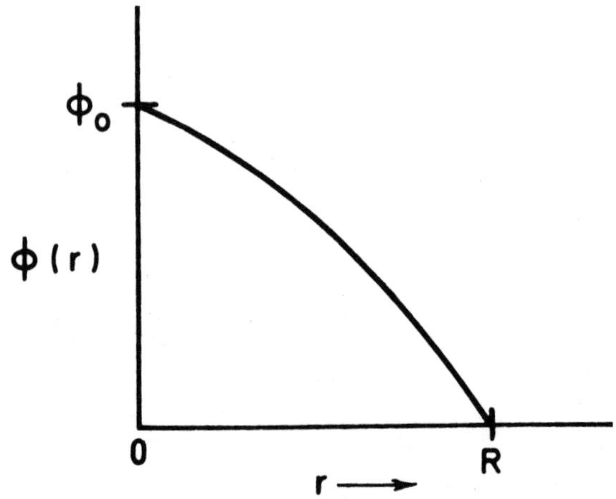

Fig. 8.2: Neutron flux density distribution.

fissions/s, the reactor power density, p, can be calculated in terms of the flux density, ϕ, from

$$p = \frac{\phi\Sigma_f}{C} \quad W/cm^3, \tag{8.18}$$

where $C = 3.3 \times 10^{10}$ fissions/W·s, and Σ_f is the macroscopic fission cross section of the fuel. Since the flux density varies with space in a reactor, the power density varies also. Therefore, one usually refers to the power density distribution in a reactor. The total reactor power is then given by

$$P = \int_V p dV = \frac{1}{C}\int \Sigma_f \phi dV, \tag{8.19}$$

where V is the reactor volume. For example, in a spherical reactor with a flux density $\phi = \phi_0 \frac{(\sin Br)}{Br}$, for constant Σ_f,

$$P = \frac{1}{C}\int_0^R \Sigma_f \phi dV = \frac{\Sigma_f}{C}\phi_0 \int_0^B \frac{\sin Br}{Br} 4\pi r^2 dr = \frac{4\pi\phi_0\Sigma_f}{CB^2} R.$$

$$\therefore P = \left(\frac{3\Sigma_f}{\pi^2 C}\right)\phi_0 V. \tag{8.20}$$

For U-235, $\Sigma_f = 24.5\ cm^{-1}$

$$\therefore P = 2.23 \times 10^{-10} \phi_0 V, \qquad (8.21)$$

where ϕ_0 is the neutron flux density at the center of the reactor.

8.5 The Criticality Calculations

From the Age-diffusion model, using the Helmholtz wave equation $\nabla^2\phi + B^2\phi = 0$, the critical condition

$$\frac{K_\infty e^{-B^2\tau}}{1 + L^2B^2} = 1 \qquad (8.22)$$

has been derived. This expression is simply the definition of the effective multiplication factor, K_e, equal to unity.

The buckling B^2 can be found from this equation in terms of the material properties K_∞, L^2, and τ. Therefore, it is sometimes designated as the 'material buckling', B_m^2. When buckling is obtained by solving the diffusion equation it is always expressed in terms of the physical dimensions of the reactor. This buckling is usually referred to as the geometrical buckling, B_g^2. For example, for a spherical reactor of radius R,

$$B^2_g = -\frac{\nabla^2\phi}{\phi} = \left(\frac{\pi}{R}\right)^2. \qquad (8.23)$$

In a critical reactor the number of neutrons is constant from one generation to the other. This fact is indicated by the critical condition, $K_e = 1$, given by equation (8.22). For a critical system, the material buckling must be equal to the geometrical buckling, $B_m^2 = B_g^2$,

$$\frac{K_\infty e^{-B_m^2\tau}}{1 + B_m^2L^2} = \frac{K_\infty e^{-B_g^2\tau}}{1 + B_g^2L^2} = 1. \qquad (8.24)$$

For a supercritical system, $K_e > 1$. This fact can be restated mathematically, on the basis of the diffusion equation, by

$$\nabla^2\phi + B_m^2\phi > 0$$

i.e.

$$B_g^2 = \frac{\nabla^2\phi}{\phi} < B_m^2.$$

Therefore, $B_g^2 < B_m^2$ is a supercriticality condition. Similarly, for a subcritical system, $K_e < 1$ leads to the condition of

$$B_g^2 > B_m^2.$$

In reactor calculations, in order to obtain a critical system for a certain type of fuel-moderator mixture, the following parameters would be varied:

a. fuel to moderator atom ratio, $N_{fuel}/N_{moderator}$ (this is a parameter indirectly indicating the fuel mass) and

b. reactor dimensions.

In both cases, the basic criterion for criticality is the equality of the material and the geometrical bucklings. For a fixed reactor material composition, the material buckling can be obtained from the critical condition. Since this involves exponential and algebraic terms, iterative or approximate procedures, such as the graphical method illustrated by Figure 8.3, must be used. From these

$$y_1 = K_\infty e^{-B_m^2 \tau} = 1 + L^2 B_m^2 = y_2.$$

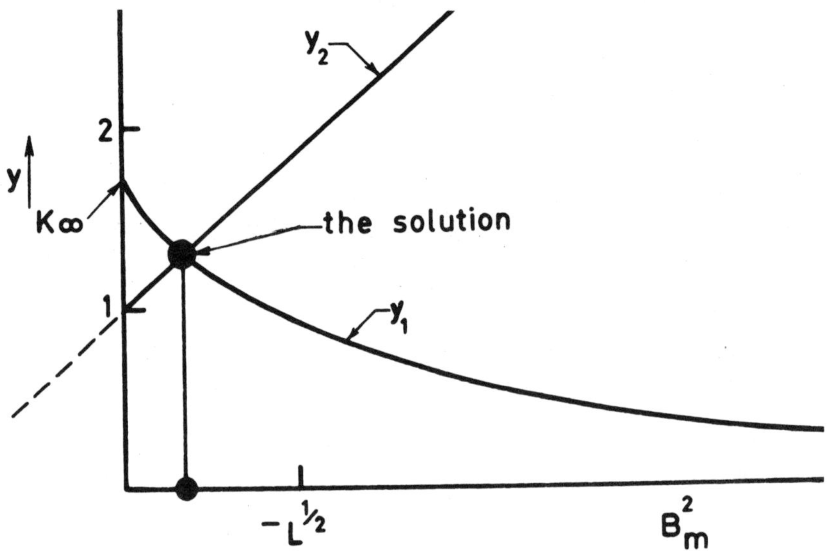

Fig. 8.3: Graphical solution of the critical equation.

Equating this value of B_m^2 to the critical geometrical buckling, an equation with the reactor dimension as the only unknown is obtained. This yields the critical dimension of the reactor. For example, if $N_{mod.}/N_{fuel} = 400$, the Fermi age $\tau = 2.54$ cm,

$$\frac{\nu(\Sigma_f)_{fuel}}{(\Sigma_a)_{fuel} + (\Sigma_a)_{mod.}} \quad e^{-B_m^2\tau} = 1 + B_m^2L^2.$$

$$\frac{\nu\,(\sigma_f)_{fuel}}{(\sigma_a)_{fuel} + (N_{mod.}/N_{fuel})(\sigma_a)_{mod.}} \quad e^{-B_m^2\tau} = 1 + B_m^2L^2\,.$$

$$\frac{2.43 \times 577}{683 + (400)0.66} \; e^{-B_m^2(55\ cm^2)} = 1 + (2.54\ cm)^2B_m^2.$$

$$\therefore\ 1.48\ e^{-(55.0)\,B_m^2} = 1 + 6.45\ B_m^2 \qquad (8.25)$$

with B_m^2 in cm^{-2}.

$$y_1 = 1.48\ e^{-55\ B_m^2},$$

$$y_2 = 1 + 6.45\ B_m^2.$$

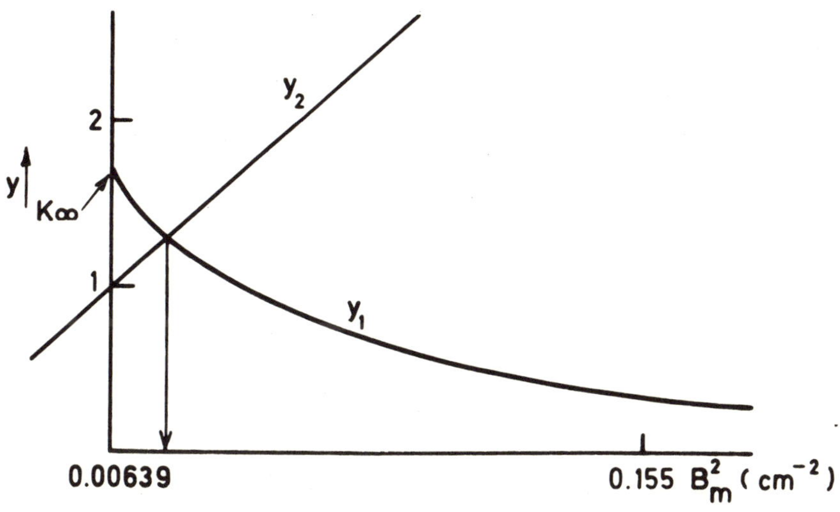

Fig. 8.4: Graphical solution of equation (8.25)

The graphical solution yields $B_m^2 = 0.00639\ cm^{-2}$. This is illustrated in Figure 8.4. Therefore, the critical radius of the spherical reactor is

$$R = \frac{\pi}{B} = \frac{\pi}{0.00639}\ cm = 39.3\ cm.$$

If, on the other hand, the critical dimension is given so that the geometrical buckling is known, one can obtain the critical mass as follows. Replacing the material buckling in the critical condition by the known value of the geometrical buckling, $B^2 = 0.00639\ cm^{-2}$,

$$\frac{2.43(577)}{683 + (\frac{N_{mod.}}{N_{fuel}})\,0.66}\, e^{-(55.0)(0.00639)} = 1 + (2.54)^2(0.00639)$$

$$\therefore \frac{N_{mod.}}{N_{fuel}} \cong 400.$$

Note that $N_{fuel}/N_{mod.} = 1/400$ which indicates that the mixture contains so little uranium-235 that values of τ and L can be considered to be equal to those of pure water.

8.6 Reactor Kinetics and Control

Reactor kinetics deals with the transient behavior of a reactor. The major concern is the time variation of the neutron flux density caused by any of the following factors:

a. Control rod motion for changing reactor operating conditions or control,
b. Fuel burnup and conversion,
c. Isotope production such as fission products and nuclei,
d. Accidents.

The effect of these factors on the neutron flux density can be described in terms of the reactivity ρ defined by

$$\rho = \frac{K_e - 1}{K_e} \cong K_e - 1, \tag{8.26}$$

where K_e is the effective multiplication factor. The reactivity value is a direct indication of the reactor behavior. For example, for $\rho > 0$, $K_e > 1$. This means that the reactor is supercritical. For a subcritical reactor, $\rho < 0$. The reactivity variation depends on and can be initiated by many factors, both natural and artificial as shown in Table 8.1.

Table 8.1: Factors responsible for reactivity variation.

Type of effect	Description	Net effect
Natural	Temperature	negative
	Fission products (Xe & Sm)	negative
	Pressure	positive
	Void	positive or negative

Type of effect	Description	Net effect
Artificial	Control rod insertion	negative
	Poisoning	negative
	Fuel depletion	negative
	Reflector addition	positive
	Moderator	positive or negative

From the definition of K_e and the neutron generation time l_p, which is the average lifetime of successive generations of neutrons, the time variation of the neutron flux density ϕ can be calculated from the following neutron balance equation

$$\frac{d\phi}{dt} = \frac{K_e - 1}{l_p}\phi \approx \frac{\rho}{l_p}\ \phi. \tag{8.27}$$

The solution of this equation is

$$\phi = \phi_0 e^{\rho/l_p\ t} = \phi_0\ e^{t/T}, \tag{8.28}$$

where the reactor period $T = l_p/\rho$ is the time required for the neutron flux density to change by a factor of e.

Examine the neutron flux density increase of a typical reactor with $l_p = 10^{-4}$s operating at 1000 MWe ($\phi_0 \cong 10^{14}/cm^2s$) due to a positive reactivity addition $\rho = 0.001$. In one second, the power would reach

$$P = (1000)e^{(0.001)(1s)/(10^{-4}s)}\ \text{MWe} \cong 2 \times 10^7\ \text{MWe!}$$

This is obviously not true! The answer to this dilemma is that nature has provided us with an inherent safety factor for reactor operation — the delayed neutrons.

In a fission reaction many of the fission products such as Br-87, I-137, and Kr-93 decay to isomeric states. The excitation energies of these are above the binding energies of the 'last' neutron in their nuclei. As a consequence they become neutron emitters. In order to distinguish these neutrons from prompt fission neutrons, they are called delayed neutrons. The delayed neutron emitters are frequently referred to as delayed neutron precursors. In general there are six distinctive groups of delayed neutron precursors in a typical fission reactor. They are characterized by their half-lives T_i and delayed neutron fractions β_i. For example, for U-235 fission the delayed neutron data is given in Table 8.2.

Table 8.2: U-235 delayed neutron data

Group	Half-life, T_i (s)	Fraction, β_i
1	56	0.000215
2	23	0.00142
3	6.2	0.00127
4	2.3	0.00257
5	0.61	0.000748
6	0.23	0.000273

The total delayed neutron fraction for the U-235 fission reaction $\beta = \Sigma_i \beta_i = 0.0065$ represents the fraction of neutrons which appear as delayed neutrons. Detailed delayed neutron data can be found in *Physics of Nuclear Kinetics* by G.R. Keepin and other references.

The important feature of delayed neutrons is their effect on reactor operations; for a reactor to operate without delayed neutrons, power excursion is a very real problem. With delayed neutrons, the mean lifetime of all the fission neutrons, prompt and delayed, is given by

$$l = (1 - \beta)\, l_p + \sum_{i=1}^{6} \beta_i \tau_i. \qquad (8.29)$$

Note that l_p is the prompt neutron mean lifetime. The mean lifetime of the ith-group of delayed neutrons is given by

$$\tau_i = \frac{1}{\lambda_i}, \qquad (8.30)$$

where λ_i is the corresponding decay constant. It can be easily calculated that the mean lifetime of all fission neutrons for U-235 is $l = 0.085$ s as compared with the prompt neutron lifetime of $l_p = 10^{-4}$s. Also, it can be seen that for a reactivity addition of $\rho = 0.001$, the power of a 1000 MWe reactor would increase in one second to, with delayed neutrons,

$$P = (1000 \text{ MWe})\, e^{(0.001)(1s)/(0.085)} \cong 1012 \text{ MWe}.$$

In fact, based on this model, it takes 85 seconds for the reactor power to have an e-fold increase. This is of course manageable and hence a very fortunate safety factor in reactor operations.

The above method of estimating the reactor power is only qualitative in nature. In order to analyze the transient behavior of a reactor, more detailed analytical methods must be used. An approximation method will be used here. This is based on considering the reactor to be infinite in dimension such that the neutron flux density is independent of the space coordinates. Due to these assumptions, the equations representing this model are called the point reactor kinetics equations (PRKE).

The time behavior of the neutron density, n, in an idealized case of a 'point' reactor in which the neutron density is independent of spatial coordinates, can be obtained by considering the neutron balance at a given time. Firstly, the prompt neutron multiplication rate is given by

$$\frac{K_e(1 - \beta) - 1}{l_p} n \cong \frac{\rho - \beta}{l_p} n , \tag{8.31}$$

where l_p is the mean prompt neutron lifetime and β is the total delayed neutron fraction. Secondly, the contribution to the neutron density variation due to delayed neutrons is given by

$$\sum_{i=1}^{6} \lambda_i C_i,$$

where λ_i and C_i represent the decay constant and the density (usually referred to as the precursor concentration) of the ith group of the delayed neutron precursors, respectively. The summation from i = 1 to 6 is due to the experimental evidence that the delayed neutron precursors can be classified into six distinctive groups. The net neutron density change with respect to time is therefore

$$\frac{dn}{dt} = \frac{\rho - \beta}{l_p} n + \sum_i \lambda_i C_i. \tag{8.32}$$

The ith group of the delayed neutron precursor density obeys

$$\frac{dC_i}{dt} = -\lambda_i C_i + \frac{\beta_i}{l_p} n. \tag{8.33}$$

These coupled point reactor kinetics equations can be solved by assuming

$$n = n_0 e^{t/T} \quad \text{and} \quad C_i = C_{i0}\, e^{t/T} ,$$

where T is the reactor period. Substituting these into equations (8.32) and (8.33) and rearranging, the inhour equation is obtained,

$$\rho = \frac{l_p}{T} + \sum_i \frac{\beta_i}{1 + \lambda_i T} . \tag{8.34}$$

This equation provides the direct relationship between the reactivity of a reactor ρ and the reactor period T. For large reactivity, the reactor period is small, $\lambda_i T \ll 1$, and the inhour equation can be approximated by

$$\rho = \frac{l_p}{T} + \sum_i \beta_i = \frac{l_p}{T} + \beta$$

i.e.

$$T = \frac{l_p}{\rho - \beta} \cdot \tag{8.35}$$

In this case, the rate of change of the neutron density is

$$\frac{dn}{dt} = \frac{1}{T} n + \sum_i \lambda_i C_i. \tag{8.36}$$

This evidently is contradictory to the assumption of $n = n_0 e^{t/T}$ and $C_i = C_{i_0} e^{t/T}$ unless $C_i = 0$. Such a contradiction can be explained on the basis that, when ρ is large, the delayed neutron effect on the neutron density is insignificant. However, for small values of reactivity, $\lambda_i T \gg 1$,

$$\rho = \frac{l_p}{T} + \sum_i \frac{\beta_i}{\lambda_i T} = (l_p + \beta\tau)\frac{1}{T}, \tag{8.37}$$

where τ is the average mean lifetime of all the delayed neutron precursors. This means that the lifetime is modified by the term $\beta\tau$. The numerical value of the reactivity that distinguishes the two extreme cases mentioned above is the delayed neutron fraction. When the reactivity is equal to this value, i.e. $\rho = \beta = 0.0065$, the critical reactor is said to be prompt critical. In other words, no delayed neutron contribution is needed to maintain the criticality of the reactor. This is sometimes described as a dollar of reactivity. Reactivity in dollars is hence defined as ρ/β. Due to the extremely short neutron lifetime in the prompt critical state, this is an operating condition to be avoided.

The reactivity changes due to the changes of pressure, temperature and void, are indicated by the reactivity coefficients, defined by

$$\alpha_p \equiv \frac{d\rho}{dx}, \tag{8.38}$$

where x is the cause. For temperature,

$$\alpha_T = \frac{d\rho}{dT} \cdot \tag{8.39}$$

Typically, α_T lies between $-10^{-5}/°C$ to $10^{-3}/°C$.

A reactor control problem is frequently termed as 'poisoning', defined by

$$\psi \equiv \frac{\Sigma_p}{\Sigma_u}, \tag{8.40}$$

where Σ_p and Σ_u are the macroscopic absorption cross sections of the "poisoning" material and the fuel respectively. It is a measure of the degree of neutron absorption by the "poisoning" material compared to that due to the fuel. Some fission products have high thermal neutron absorption cross sections and behave like "poisons". The most important are Xe-135 and Sm-149. They are produced in a reactor either directly from fission as fission products or from the disintegrations of other fission products.

$$^{135}\text{Te} \xrightarrow[29\text{s}]{\beta^-} {}^{135}\text{I} \xrightarrow[6.7\text{hr}]{\beta^-} {}^{135}\text{Xe} \xrightarrow[9.2\text{hr}]{\beta^-} {}^{135}\text{Cs} \xrightarrow[2\times 10^6\text{yr}]{\beta^-} {}^{135}\text{Ba (stable)}. \tag{8.41}$$

$$^{149}\text{Nd} \xrightarrow[2\text{hr}]{\beta^-} {}^{149}\text{Pm} \xrightarrow[53\text{hr}]{\beta^-} {}^{149}\text{Sm (stable)}. \tag{8.42}$$

Detailed calculation and experiment reveal that the concentrations of fission poisons (number of atoms/unit volume) reach equilibrium levels after a long time of steady state reactor operation. These are given by

$$(N_{Xe})_{eq.} \cong \frac{(r_I + r_{Xe})\Sigma_f \phi}{\lambda_{Xe} + \sigma_{Xe}\phi} \tag{8.43}$$

and

$$(N_{Sm})_{eq.} \cong \frac{r_{Nd}\Sigma_f}{\sigma_{Sm}}, \tag{8.44}$$

where $r_I = 0.061$, $r_{Xe} = 0.003$, and $r_{Nd} = 0.014$ are the fission yields of I-135, Xe-135 and Sm-149 respectively; Σ_f is the macroscopic fission cross section of the fuel; λ_{Xe}, σ_{Xe}, σ_{Sm} are the decay constant and the microscropic absorption cross sections of Xe-135 and Sm-149, respectively; and ϕ is the thermal neutron flux density. It is evident that the Xe-135 equilibrium concentration increases with increasing neutron flux density. Therefore, xenon poisoning is more serious for power reactors. The equilibrium samarium poisoning is, however, independent of neutron flux density.

The Xe-135 and Sm-149 concentrations after the reactor shutdown also behave differently. After reactor shutdown, the xenon concentration increases and reaches a maximum value given by the following equation

$$(N_{Xe})_{max} \cong \frac{r_I \Sigma_f \phi}{\lambda_{Xe}} e^{-\lambda_I t_{max}} \quad \text{for } \phi \ll 10^{13} \text{ n/cm}^2\text{s}, \tag{8.45}$$

where

$$t_{max} \cong \ln(\lambda_{Xe}/\lambda_I)/(\lambda_{Xe} - \lambda_I). \tag{8.46}$$

The time after reactor shutdown at which the Xe-135 concentration reaches its maximum value is t_{max}. The value of t_{max} is ~ 11 hours when decay constants of xenon and iodine, $\lambda_{Xe} = 2.1 \times 10^{-5}\ s^{-1}$ and $\lambda_I = 2.9 \times 10^{-5} s^{-1}$, are used. This means that 11 hours after reactor shutdown, the Xe-135 concentration reaches its maximum value, which is approximately proportional to the neutron flux density as shown in Figure 8.5. To overcome this poisoning, it is a usual practice in reactor operation to reduce the power before reactor shutdown if it is desirable to restart the reactor soon. The Sm-149 concentration after reactor shutdown increases from its equilibrium value to an asymptotic value

$$(N_{Sm})_{asym} = \frac{r_{Nd}\ \Sigma_f}{\sigma_{Sm}} \ \frac{(\sigma_{Sm}\ \phi + 1)}{\lambda_p}, \tag{8.51}$$

where λ_p is the decay constant of Pm-149. Notice that the added Sm-149 concentration after reaction shutdown is simply due to the decay of promethium.

Reactor control deals mainly with reactivity variation. There are several ways to manipulate reactivity such as adding or removing fuel, moderator, reflector, or absorbers. The most frequently used method is the insertion or withdrawal of rods made of neutron absorbing materials such as cadmium, boron, steel, gadolinium, etc. Depending on the function of the absorbing rods they are called control or regulating rods, shim rods, or safety rods. Regulating and shim rods are used for fine and coarse adjustments of reactivity, respectively. Safety rods are mainly used for quick shutdown of a reactor in an emergency or accident. This is called a reactor scram.

Example 8.1

Find the critical mass of a fast reactor using pure Pu-239 as fuel. The reactor is a circular cylinder, the height is equal to the diameter. (See problem 8.5 for selecting optimum cylinder radius and height).

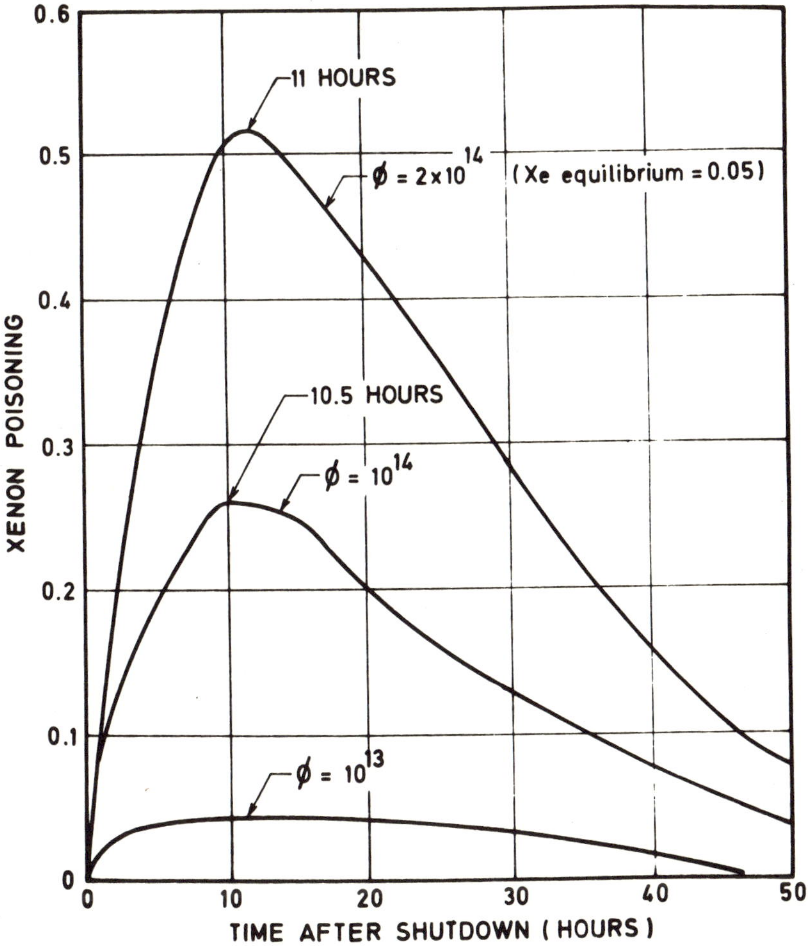

Fig. 8.5: Xenon poisoning after reactor shutdown.

Solution

In cylindrical geometry, it can be shown that the neutron flux density is

$$\phi(r, z) = \phi_0 J_0 \left(\frac{j_0}{R} r\right) \cos \frac{\pi z}{H},$$

where R and H are the extrapolated radius and height of the cylinder, respectively, $J_0(x)$ is Bessel's function of the first kind of zero order and

$j_0 = 2.405$. ϕ_0 is the neutron flux density at the center of the cylinder. Based on this flux density distribution, the critical buckling of the reactor can be calculated,

$$B^2 = \left(\frac{\pi}{H}\right)^2 + \left(\frac{j_0}{R}\right)^2 = \left[\left(\frac{\pi}{2}\right)^2 + j_0^2\right]\frac{1}{R^2} \cong \frac{8.3}{R^2}.$$

The group constants are $\sigma_{tr} = 8.0$ b, $\nu = 3.11$, $\sigma_f = 1.95$ b, and $\sigma_a = 2.05$ b. The infinite multiplication factor is

$$K_\infty = \eta f = \nu \frac{\sigma_f}{\sigma_a} = \frac{(3.11)(1.95b)}{(2.05b)} = 2.96.$$

The one-group critical condition

$$\frac{K_\infty}{1 + B^2L^2} = 1$$

is now used to calculate the critical buckling.

$$B^2 = \frac{K_\infty - 1}{L^2} = (K^\infty - 1)\ 3N^2\sigma_{tr}\ \sigma_a$$

$$= (2.96 - 1)\ 3\ (0.05 \times 10^{24}cm^{-3})^2 \times 9.5 \times 10^{-24}cm^2 \times 2.05 \times 10^{-24}cm^2$$

$$= 0.29\ cm^{-2}.$$

$$\therefore\ R^2 = \frac{8.3}{B^2} = \frac{8.3}{0.29}\,cm^2 = 29\ cm^2\ .$$

Thus the critical radius of the cylinder is $R = (29\ cm^2)^{1/2} = 5.3$ cm. From this, the height of the cylinder can be calculated $H = 2R = 10.6$ cm. Therefore the critical volume is

$$\bar{V} = \pi R^2 H = \pi(5.3\ cm)^2(10.6\ cm)$$

$$= 953\ cm^3.$$

Since the density of Pu-239 is $\rho = 19.7\ g/cm^3$, we can calculate the critical mass as

$$m = \rho\bar{V} = (19.7\ g/cm^3)(953\ cm^3)$$

$$= 18.8\ kg.$$

The final result from this series of calculations is highly dependent on the values of the cross sections which in turn are sensitive to neutron energy. Although rough estimates can be made in this way, to obtain accurate results multigroup methods must be used for this type of calculation for a fast reactor.

Example 8.2

Age-diffusion theory is a method of examining the neutron motion in a reactor by combining the Fermi age slowing down model with thermal neutron diffusion. Another approach to the same problem is the two-group method. In this method two diffusion equations are used, one for each of the two chosen energy groups. For thermal reactors these are the fast and thermal neutron groups. For fast reactors specific energy ranges of these two groups are specified. Within each group the neutrons are a) moving about with a kinetic energy within the group energy range or b) scattered into the other energy group. When these two diffusion equations are solved, the two-group critical condition can be derived,

$$\frac{K_\infty}{(1 + B^2L_1^2)(1 + B^2L_2^2)} = 1 ,$$

where L_1 and L_2 are the diffusion lengths of the medium for neutrons in groups one and two, respectively. Show that under certain conditions this is equivalent to the Age-diffusion critical condition.

Solution

The two group critical condition can be written as

$$1 = \frac{K_\infty}{(1 + B^2L_1^2)(1 + B^2L_2^2)} = \frac{K_\infty}{1 + B^2M^2}$$

if $L_1^2L_2^2B^2 \ll L_1^2 + L_2^2$, i.e. $B^2 \ll \frac{1}{L_1^2} + \frac{1}{L_2^2}$.

This is true for a medium in which the dimensions are large as compared with the neutron transport mean free path.

The Fermi age slowing down model gives the fast non-leakage probability as $e^{-B^2\tau} \cong 1 - B^2\tau \cong \frac{1}{1 + B^2\tau}$ for small B^2. Thus the Age-diffusion critical condition reads

$$1 = \frac{K_\infty}{(1 + B^2L^2)(1 + B^2\tau)} = \frac{K_\infty}{1 + (L^2 + \tau)B^2 + L^2\tau B^4}$$

$$\cong \frac{K_\infty}{1 + B^2M^2},$$

where M^2 in both expressions is the migration area.

Example 8.3

Calculate the thermal neutron diffusion and the migration lengths of water. Given: D = 0.17 cm and τ = 31 cm^2.

Solution

The macroscopic thermal neutron absorption of water = 0.022 cm^{-1}.

$$\therefore L^2 = \frac{D}{\Sigma_a} = \frac{0.17 \text{ cm}}{0.022 \text{ cm}^{-1}} = 7.73 \text{ cm}^2.$$

The diffusion length is $L = \sqrt{7.73 \text{ cm}^2} = 2.78$ cm. The migration length

$$M = (L^2 + \tau)^{1/2} = (7.73 \text{ cm}^2 + 31 \text{ cm}^2)^{1/2} \cong 6.2 \text{ cm}.$$

This is an estimate of the distance from a fast neutron source where thermalized neutrons are obtained. It is considerably less than the total distance a diffusing neutron travels before it is absorbed. In water, this absorption mean free path is

$$\lambda_a = \frac{1}{\Sigma_a} = \frac{1}{0.022 \text{ cm}^{-1}} \cong 45 \text{ cm}.$$

The migration area $M^2 = 38.7 \text{ cm}^2$ is a measure of the extent of the neutron motion. It has a direct bearing on reactor criticality. For instance, if the migration area is large compared with the cross sectional area of the reactor, the thermal non-leakage probability is small. If this is the case, the reactor dimension must be made larger. On the other hand, if the reactor dimension is large compared with the neutron migration area, B^2M^2 is small. Thus $\frac{1}{1 + B^2M^2}$ is large so that the value of K_e can be reduced.

Problems

8.1 For a large critical system, prove that the fast non-leakage probability $e^{-B^2\tau}$ can be approximated by $\frac{1}{1 + B^2L^2}$ and therefore it has essentially the same form as the thermal non-leakage probability.

8.2 If the fuel-to-moderator ratio is kept constant for a homogeneous bare reactor with mixed fuel and moderator, prove that there is a minimum critical mass of fuel. In other words, as the critical volume decreases below an optimum value, the critical mass increases. This means that, if the same fuel moderator ratio is maintained, its density must be increased.

8.3 Calculate the thermal utilization factor of a homogeneous mixture of a) U-235 and water and b) Pu-239 and graphite.

8.4 What is the critical volume of a homogeneous spherical system the buckling of which is 0.0005 cm^{-2}?

8.5 It is intended to vary the radius and the height of a circular cylinder so that its volume is constant but the total surface area is a minimum. Prove that an optimum condition is achieved if the height is equal to the diameter.

8.6 Calculate the neutron reproduction factor η of natural uranium, U-235, and Pu-239 for a) thermal neutron fission and b) fast neutron fission.

8.7 In a critical or a supercritical reactor the neutron flux density distribution is concave downward. Why? Is the reverse true for a subcritical system?

8.8 In a critical spherical reactor, the center flux density is $\phi_0 = 10^{14} n/cm^2 s$. Calculate the reactor power.

8.9 Calculate the fission mean lifetime for an infinitely large thermal fission reactor fueled by a) U-235, b) Pu-239, and c) U-233.

8.10 Use one-group delayed neutron approximation of the point reactor model to show that the steady state neutron density n of a subcritical reactor is equal to $\frac{Sl_p}{1 - K_e}$, where S is the source strength of the extraneous neutron source and l_p is the prompt neutron lifetime.

8.11 Show that the temperature coefficient of reactivity α_p is equal to the sum of the temperature coefficients of the four factors (η, ϵ, p and f) and that of the fast and thermal non-leakage probabilities of a thermal reactor. Note that $\alpha_p = \frac{d\rho}{dT}$, not $\frac{1}{\rho}\frac{d\rho}{dT}$. Why?

8.12 Calculate the thermal non-leakage probability of a homogeneous U-235-water system, the buckling of which is $B^2 = 0.0002\ cm^{-2}$.

8.13 From the inhour equation, assuming one-group of delayed neutrons, show that

a) if $\rho \ll \beta$, $T \cong \frac{l_p + \beta\tau}{\rho} \cong \frac{\beta}{\lambda\rho}$.

b) if $\rho \gg \beta$, $T \cong \frac{l_p}{\rho - \beta} \cong \frac{l_p}{\rho}$.

c) if $\rho = \beta$, $T \cong (l_p/\lambda\beta)^{1/2}$.

Discuss each of these cases.

8.14 A reactivity of $\rho = 0.0065$ is inserted into a thermal critical reactor fueled by a) U-235 and b) Pu-239. Find the reactivity dollar value in each case. Draw a conclusion from these results.

8.15 The reactivity value $\rho = 0.0025$ for a U-235 reactor has a special meaning. It is a value which satisfies the condition $\beta - \rho + \lambda l_p \gg \sqrt{2\lambda\rho l_p}$. Solve for n and C of the point reactor kinetics equations by using the one-group delayed neutron approximation and assuming an initial steady state, i.e.
$n = n_0$ and $C = C_0$ when $t = 0$.
From the neutron density, n, and delayed neutron precursor concentration, C, show that the stable period is approximately $T = 32$ s.

8.16 Calculate the equilibrium xenon concentration of a reactor with neutron flux density a) $\phi = 10^8 n/cm^2s$, b) $\phi = 10^{13} n/cm^2s$, and c) $\phi = 10^{16} n/cm^2s$.

8.17 What is the relationship between the reactor flux density, hence the reactor power, and the maximum value of the xenon concentration after reactor shutdown? For example, is the xenon concentration a major factor on reactor restart if the original steady power is two watts?

8.18 The linear temperature coefficient of expansion of a reactor is 1.5×10^{-5} per C°. What is the temperature coefficient of the buckling, B^2, if the reactor is homogeneous?

8.19 The thermal fission neutron reproduction factor η for Pu-239 is a function of temperature. Its values are 2.035, 1.998, 1.947, 1.860, 1.811, 1.785, and 1.770 when the temperatures are 20°C, 100°C, 200°C, 400°C, 600°C, 800°C, and 1000°C, respectively. Calculate the temperature coefficient of η from this set of data.

8.20 What does reactor deadtime mean? What are the factors that affect the starting and the duration of a reactor deadtime?

8.21 Use group constants to show that for fast reactors the xenon poisoning is not a factor in a reactor startup.

8.22 Calculate the smallest value of the fuel-to-moderator atom ratio for the following homogeneous critical assemblies:
a) U-235 and graphite,
b) U-235 and water,
c) U-235 and heavy water,
d) Pu-239 and water, and
e) natural uranium and water.
Assume that the critical assemblies are infinitely large.

CHAPTER 9

REACTOR HEAT TRANSFER

9.1 Introduction

From the energy point of view, in a nuclear power plant the events leading to the generation of electricity may be described by the flow diagram given in Figure 9.1. A reactor replaces the conventional boiler as a heat source. Therefore, except for radiation precautions, the working stages between the heat exchanger and the ultimate electric energy output are similar to the fossil fuel power plants. It is the method of extracting heat from a reactor core that is a major departure from the conventional power plant design and operation.

In a nuclear reactor, fuel elements serve as the main heat source. The heat generation is proportional to the neutron flux density. The relative temperature distribution can be obtained from the heat source distribution. In general there are three ways in which heat can be transferred. These are (a) conduction, (b) convection, and (c) radiation. Basic equations for these modes of heat transfer are:

a) Conduction $Q = -kA\nabla T$, (9.1)

b) Convection $Q = hA\nabla T$, and (9.2)

c) Radiation $Q = \sigma A T^4$, (9.3)

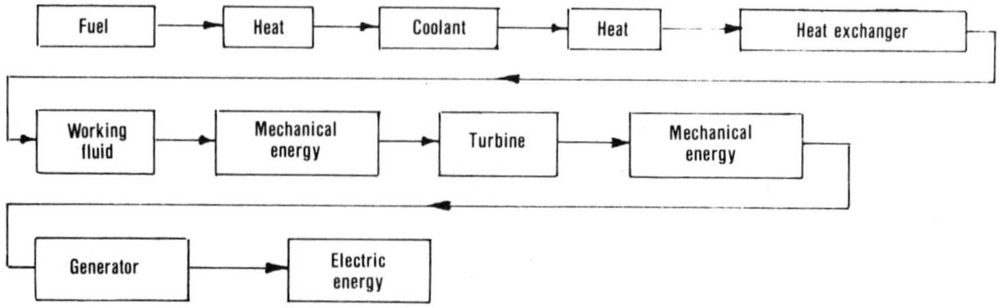

Fig. 9.1: Energy transformation flow chart.

where

ρ = density of the conductor,
C_p = specific heat of the conductor,
K = thermal conductivity of the conductor,
T = temperature,
A = surface area,
Q = energy or heat transfer rate,
h = heat transfer coefficient,

and

σ = Stefan-Boltzmann constant.

In a conventional heterogeneous reactor, heat conduction occurs inside a fuel element. Through convection, heat is transferred from the fuel element surface to the coolant in order to establish a coolant temperature distribution. For extracting heat, the coolant must be pumped through the coolant channels in the reactor core. The pumping power is usually a function of the pressure drop in the coolant system and the coolant flow rate.

9.2 Heat Generation

To simplify the discussion assume that the fuel elements form the only heat source in a reactor. Actually, the moderator and shield also produce heat from neutron slowing down and bombardment. However, the total amount of heat generated in moderator and shield is only a small percentage of the total heat generation. Furthermore, we shall assume a concentric unit cell model, Figure 9.2, for our calculations.

Inside the fuel element the radial neutron flux density may be assumed to be constant. The longitudinal neutron flux density is approximately given by

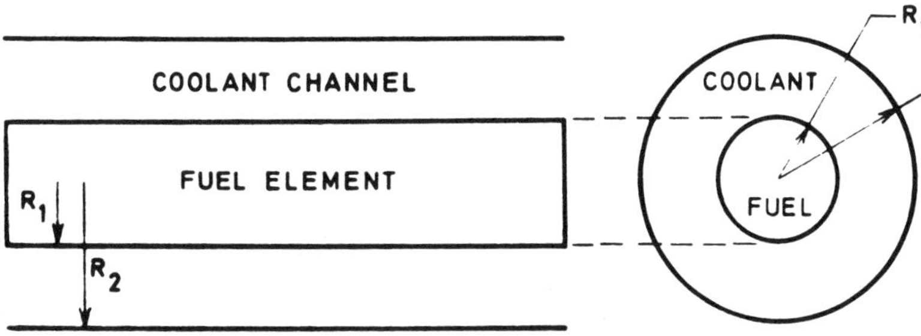

Fig. 9.2: A heterogeneous reactor unit cell.

$$\phi(z) = \phi_{max} \sin \frac{\pi z}{L}, \tag{9.4}$$

where ϕ_{max} is the flux density at mid-point of the element. The heat generation rate distribution is therefore given by

$$q(z) = \pi R_1^2 \, E\Sigma_f \, \phi(z), \tag{9.5}$$

where E is the energy per fission and q(z) is the heat generation rate per unit length. Therefore the power density is given by

$$q = \frac{1}{\pi R_1^2 L} \int_0^L q(z)dz = \frac{E\Sigma_f \, \phi_{max}}{L} \int_0^L \sin \frac{\pi z}{L} \, dz .$$

$$\therefore \; q = \frac{2}{\pi} E\Sigma_f \, \phi_{max}. \tag{9.6}$$

Note that q has a unit of watt per unit volume.

Coolant channel for a plate type of fuel element is shown in Figure 9.3. The longitudinal neutron flux density for such an element is also given by equation (9.4).

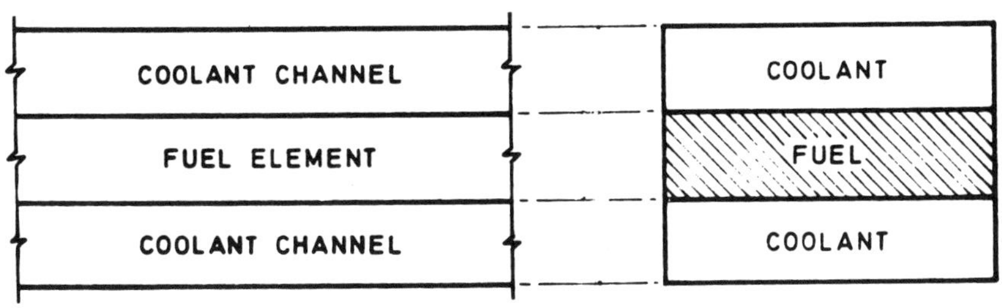

Fig. 9.3: A fuel plate-coolant channel.

In reactor design, the limiting factor is the maximum permissible fuel temperature. This is directly related to the maximum neutron flux density, ϕ_{max}, or power density. To obtain higher power output from a reactor, the common practice is to minimize the maximum to average power density ratio. This process is called power flattening, and can be achieved by nonuniform fuel loading and burnable poison distribution. Varying coolant speed is another way to increase the reactor power output.

9.3 Heat Transfer and Temperature Distribution

In a fuel element the center temperature is the highest and the fuel surface temperature is the lowest. Heat is transferred in the element by conduction. In cylindrical coordinates, the heat generation rate can be considered as a constant with respect to r, at height z, because the fuel element dimension is very small compared with the radial dimensions of a reactor, Figure 9.4.

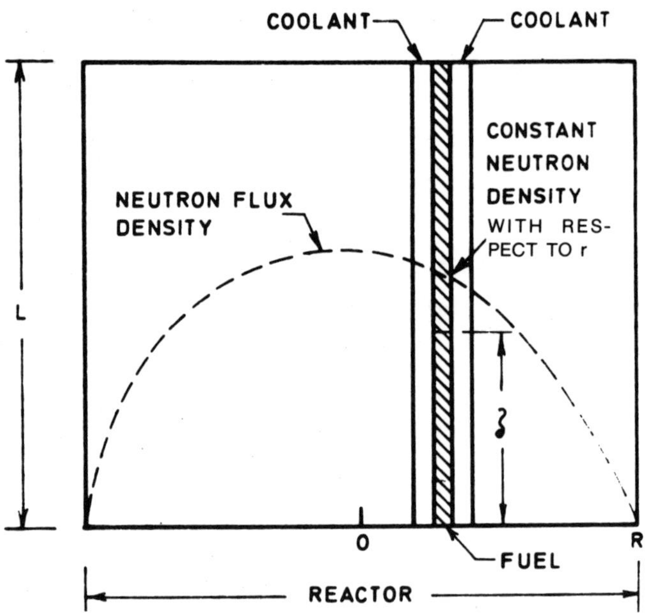

Fig. 9.4: Fuel element and the reactor.

Along the z direction, the neutron flux density and therefore the heat generation rate would vary. Applying equation (9.1) at a position z in the fuel element,

$$q(z)dz = -k\,(2\pi r dz)\,\frac{dT}{dr}. \tag{9.7}$$

Using q(z) for a fuel element of radius r, substitute equation (9.5) into equation (9.7) to obtain

$$\frac{E\Sigma_f \phi(z)}{2k} \, rdr = dT. \tag{9.8}$$

Integrating this equation with the integration limits $T = T_0$ at $r = 0$, and $T = T(r)$ ar $r = r$,

$$T(r) = T_0 - \frac{E\Sigma_f r^2}{4k} \phi(z). \tag{9.9}$$

Since the total channel power

$$Q = L\pi R_1^2 \cdot q = 2LR_1^2 \, E\Sigma_f \, \phi_{max},$$

by setting $r = R_1$ in equation (9.9), the fuel surface temperature can be calculated as a function of z, i.e.

$$T_s(z) = T(R_1) = T_0(z) - \frac{Q}{8Lk} \sin \frac{\pi z}{L} \; . \tag{9.10}$$

The transfer of heat from the fuel surface to the coolant can be considered as convection through a thin film separating the coolant and the fuel element surface. If the coolant temperature at z is designated as $T_c(z)$, equation (9.2) yields

$$q(z)dz = h\,(2\pi R_1 dz)\,[T_s(z) - T_c(z)].$$

$$\therefore \; T_s(z) = T_c(z) + \frac{Q}{4LhR_1} \sin \frac{\pi z.}{L} \tag{9.11}$$

Finally, to calculate the heat extraction by the coolant, consider the fuel element heat generation to be equal to the heat absorption by the coolant in raising its temperature from the coolant inlet temperature $T_c(0)$ to the outlet temperature $T_c(L)$. At a point z, the conservation of energy yields

$$q(z)dz = CM \, dT_c, \tag{9.12}$$

where C and M are the specific heat and mass flow rate of the coolant. Integration of equation (9.12) from $z = 0$ to $z = z$ yields

$$T_c(z) - T_c(0) = \frac{Q}{2CM} \left(1 - \cos \frac{\pi z}{L}\right). \tag{9.13}$$

The temperature at the coolant outlet is

$$T_c(L) = T_c(0) + \frac{Q}{CM} \cdot \tag{9.14}$$

This is simply a confirmation of the fact that the total channel power is used for heating the coolant from $T_c(0)$ to $T_c(L)$. For plate fuel geometry the analysis is similar and equation (9.14) would hold in this case also.

In the calculation of heat transfer from the fuel element surface to the coolant, the heat transfer coefficient h must be treated carefully. Its value depends on many factors. The usual practice is to relate the heat transfer coefficient to the Reynolds number N_R which serves as a guide to distinguish a laminar flow from a turbulent flow. Generally, in high power reactors, forced convection causes the coolant flow to become turbulent flow. Only in a few low power reactors is the cooling process conduction or free convection. In these cases the coolant flow is essentially laminar.

For laminar flow the Reynolds number, N_R, is less than 2100. In this case all the particles flow parallel to the direction of flow. In turbulent flow, $N_R > 4000$. A thin layer of laminar flow exists near the pipe wall. A transition region, the Reynolds number of which is between 2100 and 4000, exists between this thin layer and the central region of completely turbulent flow, Figure 9.5.

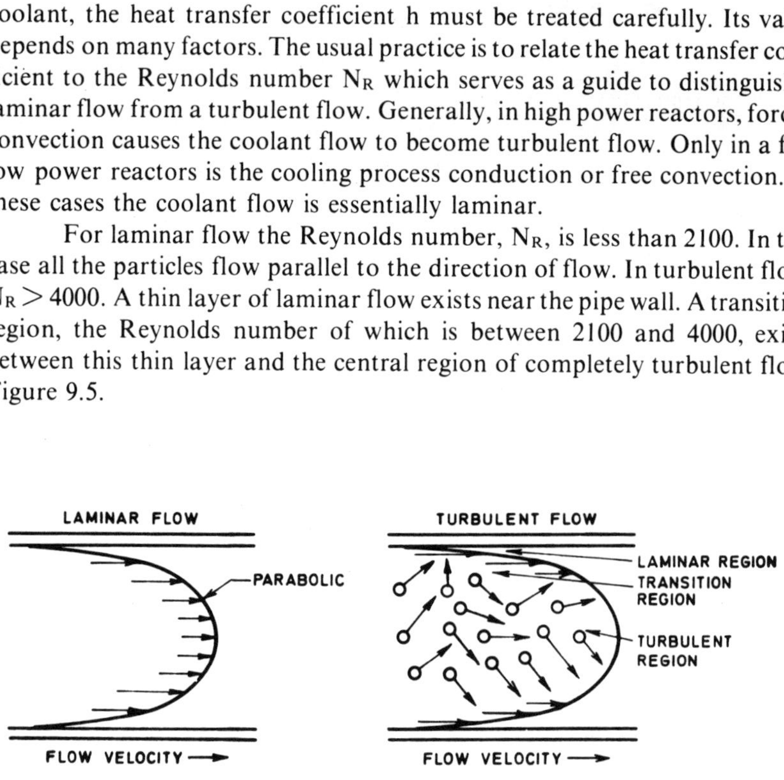

Fig. 9.5: Laminar and turbulent flows.

Equations (9.15) to (9.17) give the definitions of the Reynolds number and two other frequently used dimensionless numbers, the Nusselt number, Nu, and Prandtl number, Pr. The heat transfer coefficient can be calculated from these numbers. For a circular channel of diameter D, flow speed v, coolant density ρ, coolant viscosity μ, coolant heat capacity C_p, coolant thermal conductivity k, the dimensionless numbers are defined by

$$N_R \equiv \frac{Dv}{\mu}, \tag{9.15}$$

$$Nu \equiv \frac{hD}{k}, \tag{9.16}$$

and

$$Pr \equiv \frac{C_p \mu}{k}. \tag{9.17}$$

For geometries other than a circular pipe, an approximate effective diameter of the pipe D_e may be used in place of D in the above equations. This is calculated from

$$D_e = 4 \frac{\text{cross sectional area}}{\text{wetted perimeter of the pipe}} \tag{9.18}$$

Note that for a circular pipe of radius r,

$$D_e = \frac{4\pi r^2}{2\pi r} = 2r \tag{9.19}$$

as expected. For laminar coolant flow Newton's equation applies, i.e.

$$F = \mu A \frac{dv}{dn}, \tag{9.20}$$

where F is the shearing force, A is the area between two parallel layers of the fluid, dv/dn is the velocity gradient perpendicular to the direction of flow.

Since most cases of reactor coolant flow fall in the turbulent flow category, this shall be discussed in some detail. In this case, the heat transfer coefficient is calculated from the three dimensionless numbers i.e. N_R, Nu, and Pr. A frequently used approximation relating these numbers for a straight channel is

$$Nu = 0.023 (N_R)^{0.8}(Pr)^{0.4}. \tag{9.21}$$

If the fluid is a gas,®

$$Nu = 0.02 (N_R)^{0.8}. \tag{9.22}$$

For a liquid metal

$$Nu = 7 + 0.025\,(Pr)^{0.8}. \tag{9.23}$$

If liquid metal flows through an annular channel, then

$$Nu = 5 + 0.02\,(Pr)^{0.8}\left(\frac{r_o}{r_i}\right)^{0.3}, \tag{9.24}$$

where r_o and r_i are the outer and inner radii of the channel. Depending on the coolant flow characteristics, one of the above equations may be used to calculate the Nusselt number. The value of the heat transfer coefficient may then be estimated by using equation (9.16).

9.4 Coolant Pumping Power

The pumping power P_p needed to move the coolant through the coolant channel is an important reactor design parameter. It depends on the coolant mass flow rate and the pressure drop.

$$P_p = (\Delta p)(Av), \tag{9.25}$$

where A is the cross-sectional area, v is the flow speed, and Δp is the pressure drop. Then (Av) is the volume flow rate. The coolant pressure drop is effected by many factors such as the flow characteristics, compressibility of the coolant, flow friction, and the channel cross sectional area.

In general, the pressure drop needed to move a fluid of density ρ in a cricular pipe of diameter D and length L with a flow speed of v may be expressed as

$$\nabla p = \rho f \frac{L}{D}\frac{v^2}{2g}, \tag{9.26}$$

where g is the gravitational acceleration. The quantity f needs to be examined carefully. This is called the friction factor. Its value depends on the type of flow. For the laminar flow of an incompressible fluid

$$f_L = \frac{64}{N_R}. \tag{9.27}$$

For turbulent flow the following empirical formula may be used,

$$f_T = 0.05\,(N_R)^{-0.2}. \tag{9.28}$$

The coupling of equation (9.27) or (9.28) to (9.26) gives the working formulae for calculating the pumping power.

$$P_p = 32 \frac{\rho A v^2 L}{N_R \; Dg} \text{ for laminar flow.} \qquad (9.29)$$

$$P_p = 0.025 \, (N_R)^{-0.2} \frac{L v^3 A}{D g^2} \text{ for turbulent flow.} \qquad (9.30)$$

Example 9.1

The fuel element temperature distribution along the central axis has a maximum value near but not at the center. Find the fuel temperature distribution and the location at which maximum fuel temperature occurs by using the coolant inlet temperature as a reference point.

Solution

Substitute equations (9.13) and (9.11) into (9.10) to obtain

$$T_0(z) = T_c(0) + \frac{Q}{2}\left[\left(\frac{1}{4Lk} + \frac{1}{2LhR_1} \right) \sin\frac{\pi z}{L} + \frac{1}{CM}\left(1 - \cos\frac{\pi z}{L}\right) \right].$$

Note that when $z=0$, $T_0(z)=T_c(0)$. This means that at the coolant inlet the fuel and the coolant temperature are the same. When $z=L$, $T_0(z)= T_c(0) + Q/CM$ as expected. Differentiating $T_0(z)$ with respect to z,

$$\frac{dT_0(z)}{dz} = \frac{Q\pi}{2L}\left[\left(\frac{1}{4Lk} + \frac{1}{2LhR_1} \right) \cos\frac{\pi z}{L} + \frac{1}{CM} \sin\frac{\pi z}{L} \right].$$

This derivative is zero when

$$z = \frac{L}{\pi} \tan^{-1}\left[-CM\left(\frac{1}{2LhR_1} + \frac{1}{4Lk} \right) \right].$$

The second derivative of $T_0(z)$ at this value of z is

$$\frac{d^2T_0(z)}{dz^2} = - \frac{Q\pi^2}{2L^2} \sin\frac{\pi z}{L}\left[\left(\frac{1}{4Lk} + \frac{1}{2LhR_1} \right) + \frac{1}{C^2M^2\left(\frac{1}{4Lk} + \frac{1}{2LhR_1} \right)} \right].$$

Since the second derivative of $T_0(z)$ is negative, the maximum temperature occurs at the value of z for which $dT_0(z)/dz = 0$.

Example 9.2

The concept of hot channel in reactor thermal design is very important. The hot spot in the hot channel is an ultimate design limitation on reactor performance and power output. What are the major factors which contribute to the development of hot channel, hence the hot spot, in a reactor?

Solution

The most important factor to be considered in hot spot determination is the neutron flux density distribution. Other influencing factors are the coolant temperature and flow rate, the fuel temperature distribution, and the fuel cladding temperature. A basic parameter in hot channel thermal design is the ratio of the temperature variation of the coolant in the hot channel to that of an average channel, $\Delta T_{hot}/\Delta T_{average}$. This ratio is evaluated for all contributing factors or uncertainties. The overall hot channel factor is then the product of all the individual factors. For example, if the hot channel factor for the film temperature drop (T_s - T_c) is 1.29 due to fuel dimension and coolant flow variations and the same factor for the temperature drop due to power peaking is 1.3, the overall hot channel factor is $1.29 \times 1.3 = 1.7$. Other properties such as heat flux or coolant enthalpy can be considered in the same manner as temperature hot channel factors.

Example 9.3

Calculate the pumping power per channel through a water cooled reactor operating at 10 kW channel power. The pressure drop is assumed to be 1000 N/cm^2 and the volume flow rate is 250 m^3/hr.

Solution

The pumping power is

$$P_p = (\Delta p)(vA) = (1000\ N/m^2)(250\ m^3/hr) \cong 70\ W.$$

In this case, the power for pumping the coolant through the coolant channel is small compared with the total channel power of 10 kW. In this calculation only the power required for the passage of coolant through the reactor has been considered. When all parts of the coolant circuit are included, the total pumping power will be considerably higher. Usually, the pumping power is in the range of 1 to 5 percent of the total power.

Problems

9.1 Calculate the coolant outlet temperature of a liquid metal if molten sodium is the coolant. The coolant mass flow rate per channel is 5200 kg/hr. The coolant inlet temperature and the total channel power are $T_c(0) = 288°C$ and $Q = 80$ kW, respectively.

9.2 A circular reactor coolant channel has an inner diameter $D_i = 2$ cm and an outer diameter $D_o = 2.7$ cm. What should be the coolant flow rate to obtain a mass flow rate of 3000 kg/hr, if water is used as a coolant.

Ans: 326 cm/s.

9.3 Prove that for a cylindrical bare reactor the maximum to average power ratio, P_{max}/P_{av}, is equal to 3.64.

9.4 It is desired to establish a fuel center to surface temperature difference of 110°C in a cylindrical fuel element. The total channel power is 50 kW. The channel length $L = 2$ m and the coolant thermal conductivity $k = 0.22$ J/cm s°C. At what vertical distance from $z = 0$ in the fuel element is this realized?

9.5 Cooling water flows in a rectangular coolant channel of cross sectional area 2.5 cm^2 with a speed of 1.2 m/s. The heat removal rate from the flat fuel element surface is 158 kW/m^2. What type of coolant flow is this? What is the film to coolant temperature difference?

9.6 Consider a bare cylindrical reactor cooled by molten sodium. The reactor power is 137 MWth. The dimensions of the cylindrical fuel elements are $R = 1$ cm and $L = 2.5$ m. The coolant flows through annular channels with 1.5 cm outer radius and a coolant flow speed of 8m/s. The fuel temperature limit is set at 600° C. If the coolant inlet temperature is 350° C, calculate the number of coolant channels required and the coolant outlet temperature.

9.7 Water flows through a cylindrical coolant channel with an inner diameter of 2 cm and an outer diameter of 3 cm. The flow rate is 10^5 kg/hr. Calculate the pressure drop.

9.8 Calculate the heat transfer coefficient of a turbulent coolant (water) flow through a coolant channel similar to that described in problem 9.5 with a flow rate of 2 m/s.

9.9 To calculate the dimensionless numbers for the coolant flow through channels of different geometries, the effective diameter must be calculated. Use cquation (9.18) to prove the following results:

a) For annualr channel: $D_e = D_o - D_i$.
b) For rectangular channel of width w and gap d: $D_e = 4wd/2(w + d)$.
c) For square channel with side width a: $D_e = 2a$.

d) For a hexagonal channel of side d with inner circular diameter d: $D_e = (\frac{1}{2} - \pi/24)d$.

9.10 Plot the curves of $T_c(z) - T_c(0)$, $T_s - T_c$, and $T_o - T_s$ versus the z coordinate if the axial neutron flux density distribution is assumed to be $\sin \pi z/L$, where L is the length of the reactor.

CHAPTER 10

THERMAL REACTOR MATERIALS

10.1 Introduction

The materials for reactors must be chosen with great care; besides the physical and chemical properties, their nuclear properties must be considered. In general, reactor materials can be grouped into fuel, cladding, moderator, coolant, structural, control, and shielding materials. All of these must have high boiling points and low radiation damage rates due to induced radioactivity or structural variation. High machinability is also an important factor. The purification and enrichment of these materials must be easy to accomplish. Their corrosion resistance under all operating conditions must be high enough to insure reliable performance of the reactor.

The nuclear consideration deals with the neutron capture cross section of the structural materials which must be as low as possible in order to conserve neutrons. On the other hand, a high absorption cross section is required for control materials. The fission cross section of the fuel must be high, but must be practically zero for all other materials. Depending on the function of the material, it must have an appropriate scattering cross section. For shielding, moderating, and control purposes, a high scattering cross section is necessary, while fuel, cladding, coolant and structural materials must have comparatively low scattering cross sections.

Since ideal materials for reactor applications are rare, compromises must be made in selecting suitable ones. Conflicts in nuclear properties or cross sections must be reconciled. For example, in most of the light water reactors, water acts both as a moderator and coolant. Following is a list of materials frequently used:

Fuel:	Natural uranium, enriched uranium, plutonium-239, uranium-233.
Cladding:	Stainless steel, zircaloy, aluminum, magnesium, magnox.
Moderator:	Light water, heavy water, graphite, beryllium.
Coolant:	Light water, heavy water, organic coolants, liquid metals, carbon dioxide, helium.
Structural:	Stainless steel, zirconium, aluminum, magnesium, chromium, titanium, niobium, nickel.
Control:	Boron, cadmium, hafnium, europium, samarium, dysprosium, indium, silver.
Shielding:	Concrete, stainless steel, lead, iron, light water, boron, cadmium.

10.2 Nuclear Fuels

The three fissionable materials which may be used in nuclear power reactors are U-235, Pu-239, and U-233. However, most of the commercially available reactors today are based on natural or enriched uranium-235. The present discussion is therefore directed mainly towards this material.

Uranium is used for power production in one of the following forms:

a) metallic form,
b) alloyed form,
c) oxide and carbide fuel, and
d) dispersion type of fuel.

Metallic uranium has poor corrosion resistance in an oxidizing medium and poor resistance to radiation damage. When irradiated even at low temperatures the uranium metal fuel elements become disfigured due to phase changes and different properties of the individual phases (Table 10.1).

Table 10.1: Phases of uranium.

Phase	Temperatures range °C.	Crystal structure	Remarks
α	Below 663	Orthorhombic	Highly anisotropic
β	663-764	Tetragonal	Slightly anisotropic
γ	764-1133	B.C.C.	Isotropic

Anisotropy of alpha uranium leads to a coefficient of expansion ranging from slightly negative in one direction to positive in the other direction. This leads to severe distortions on temperature cycling or irradiations. These shape changes are cumulative with successive events and are undesirable.

Fuel alloys were developed mainly for improving the corrosion and radiation damage resistance of metallic uranium. The amount of alloying materials required to control corrosion and radiation damage seriously penalizes the neutron economy of the reactor. This conflict of requirements has made it necessary to seek a balance between the thermal and radiation resistance of alloyed uranium and the good neutron economy of the nearly pure uranium. Uranium dioxide and carbide have been found to meet most of the desired requirements of thermal nuclear power reactors. Uranium dioxide has outstanding properties as a nuclear fuel. The attractive features of UO_2 are:

a) high melting point (2750° C),
b) stability under irradiation at core temperatures,
c) compatibility with most of the cladding materials,
d) ease of fabrication to high densities,
e) better neutron economy than alloyed fuels,
f) chemical stability with many materials at high temperatures,
g) high resistance to radiation damage, and
h) high burnup.

However, uranium dioxide has a relatively low thermal conductivity which further decreases at higher temperatures, and, because of its brittleness, it is susceptible to thermal fracture. Uranium carbide (UC) has a much better thermal conductivity as compared to uranium oxide fuel. It also has a significantly higher uranium density and a low parasitic absorption cross section and compares favourably with UO_2 in mechanical stability, radiation damage resistance and thermal cycling behavior. Although its melting point (2380° C) is lower than that of UO_2, it is not low enough to seriously counteract the effect of its higher thermal conductivity. A major obstacle to the early widespread use of UC is its poor compativility with water. The fuel is being used in some non-aqueous reactors using liquid metals or gases as coolants.

Dispersion type of fuel consists of fuel material dispersed in a matrix material. The fissionable material is generally in the form of carbide or oxide fuels. Such fuel allows extremely high burnup rates.

Plutonium and thorium, although not used presently in most of the commercially available power reactors, are expected to be used in fast and thermal reactors respectively on a large scale. Since plutonium in the metallic form undergoes various phase changes, it is used only as an alloy or as PuO_2

or PuC. Thorium may be used either in the metallic form or again as ThO_2 or ThC fuel.

10.3 Control Materials

Control elements are very important components of a nuclear power reactor for reasons of safety and effective control of the dynamic behavior of the reactor. Materials for control are generally considered on the basis of

a) nuclear acceptability, i.e. high neutron absorption cross section,
b) metallurgical fabricability,
c) corrosion resistance,
d) thermal conductivity,
e) radiation damage resistance,
f) probable contribution to the overall radioactivity problem, and
g) economy.

The materials compatible with thermal reactors are boron, cadmium, hafnium, europium, gadolinium, indium, samarium, and silver. Control element materials in actual use include cadmium-indium-silver alloy, boron-steel, Eu_2O_3-steel, hafnium and canned B_4C pellets. The selection of a particular type of element depends on the reactor type and the requirements of control element life in the core.

10.4 Coolant and Moderator

An ideal power reactor coolant and moderator should combine the heat removal function with the neutron slowing down ability. Another necessary property for a reactor coolant is the ease with which it can be pumped through the reactor. For these reasons, water is a frequently used coolant and moderator for boiling water reactors (BWR) or pressurized water reactors (PWR). Water is a good coolant in thermal reactors because of its low atomic mass for slowing down fast neutrons. Although it is only a fair thermal conductor, the high specific heat and latent heat (for BWR) more than make up for this deficiency.

Lately, gas-cooled reactor experiments have been carried out in many parts of the world. In the U.S., helium is the most frequently used gas coolant. Although hydrogen has good performance characteristics as a reactor coolant, safety considerations simply rule out its use. Other materials such as oxygen, nitrogen, or air are not used either because of their tendency of corroding the reactor structural materials (oxygen and nitrogen) or due to their poor heat transfer characteristics. Helium has the following favorable properties for reactor coolant applications:

a) extremely low neutron capture cross section,
b) stability to heat and radiation,

c) one phase operation, without causing void and boiling as compared with BWR, and

d) chemical inertness, will not corrode the structural and fuel cladding materials.

Carbon dioxide has been used in some European countries. It is considerably cheaper than helium although its heat transfer characteristics are inferior. Graphite is a good moderator and reflector due to its low mass number and its high moderating ratio, and is frequently used, especially in research reactors. Heavy water, sometimes enriched to 99.8% of D_2O, has a very high moderating ratio and is therefore an excellent moderator. Due to their high scattering cross section, both graphite and heavy water have been used as moderators in reactors with natural uranium as the fuel such as the X-10 graphite reactor at ORNL and the KANUPP Candu power reactor. The moderating ratios of some moderators are listed in Table 10.2.

Table 10.2: Moderating ratios of some materials.

Moderator	Moderating ratio
Beryllium	150
Beryllia (BeO)	170
Graphite	200
Heavy water	21,000
Helium	45
Light water	69
Polysterene	62

10.5 Shielding and Structural Materials

The most important problem in selecting reactor shielding and structural materials is the possible damage to these materials by the intense radiation. All forms of radiation and fission fragments must be considered as they may lead to: (a) material atomic displacements, (b) thermal stresses, and (c) induced radioactivities.

For structural material and fuel cladding, fission fragments and neutron absorptions are the main concern. Using the microscopic thermal neutron cross section of 1 barn and 10 barns as the dividing lines, reactor structural materials can be classified into three categories, namely low absorption, intermediate absorption, and high absorption materials. The last class of materials is almost always ruled out. The intermediate absorption materials are sometimes used if they have excellent mechanical and chemical properties. In this case, more fuel would have to be used in order to compensate for the extra neutron absorption. Some examples of these materials are stainless steel, titanium, and vanadium. Due to its mechanical and chemical properties stainless steel is widely used in reactors. Aluminum and zirconium are very frequently used as structural or cladding materials. Zirconium alloys are excellent for fuel cladding due to their corrosion resistance and low neutron absorption.

Reactor shielding may be classified into biological shielding and thermal shielding. A biological shield prevents radiation of all types from escaping the reactor. The first part of the shield, nearest the reactor, is often called the thermal shield because it serves the additional purpose of providing a heat boundary to the reactor. About 10% of the reactor energy is carried off by gamma rays and neutrons out of the reactor; the bulk of this energy is deposited in the thermal shield, which therefore normally requires special cooling.

A typical thermal shield may utilize two concentric cylindrical shells of one to several inches in thickness separated by a water coolant channel of about one inch in spacing, Figure 10.1.

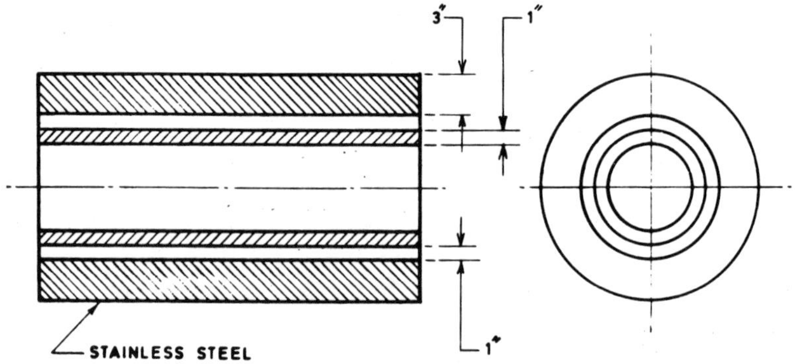

Fig. 10.1: A thermal shield.

A low atomic number material, such as water or concrete, is good for slowing down fast neutrons. For shielding against gamma radiation a high atomic number material, such as iron or lead, is necessary. An excellent biological shield is the combination of lead, cadmium, and water or some rich hydrogeneous material such as paraffin. Here fast neutrons are slowed down, thermal neutrons are absorbed and gamma rays are shielded in one material. Many of the reactors for navigation use such a combination because of its weight and volume advantages over concrete.

In most of the commercial nuclear power plants, a concrete biological shield, 3 to 10 ft is usually used. Near the reactor there is also a fuel storage pool of water for reactor loading or for spent fuel storage and cooling.

Example 10.1

Selecting Nuclear Materials

What are the major steps in selecting nuclear materials?

Solution

There are three general steps in nuclear material selection. The first does not pertain to the material itself. It is the service condition evaluation or the 'environmental' survey to define the situation in which the material is to be used. Then the material characteristics are evaluated, and, on this basis, the best material is chosen. The third step involves the actual service test. This is very important in that the performance of the selected material must not only satisfy selection criteria but also must meet the actual operating conditions. Some of the properties evaluated are

a) mechanical strength,
b) ductility,
c) stability,
d) availability,
e) fabricability,
f) corrosion resistance,
g) neutron and radiation absorption,
h) heat transfer properties,
i) radiation stability and activation, and
j) cost.

Example 10.2

What are the major radiation effects on materials?

Solution

Radiation damage occurs when large amounts of energy are transferred from the incident radiation to the materials. A direct consequence of such an energy transfer is the creation of material defects which can be classified into:

a) mechanical effects — strain and spikes formation, crystal structural variation, ductility change, etc.,
b) electronic effects — electric conductivity change, dielectric loss, paramagnetism, etc., and
c) rate process effects — diffusion property change, phase change, chemical reaction rate change, etc.

Example 10.3

Age and diffusion length are two important material properties to be considered when selecting a moderator. If the moderator is a mixture of species 1, 2, 3, ... etc., with volume fractions $\alpha_1, \alpha_2, \alpha_3, \ldots$, respectively, the age and the cross sections of the mixture can be calculated from the following formulas:

$$\frac{1}{\sqrt{\tau}} = \frac{\alpha_1}{\sqrt{\tau_1}} + \frac{\alpha_2}{\sqrt{\tau_2}} + \frac{\alpha_3}{\sqrt{\tau_3}} + \cdots$$

and

$$\Sigma = \alpha_1\Sigma_1 + \alpha_2\Sigma_2 + \alpha_3\Sigma_3 + \cdots$$

Develop formulas for calculating τ and L^2 for a H_2O and D_2O mixture with volume fractions of $1 - \alpha$ and α, respectively.

Solution

For H_2O: $\tau = 31\ \text{cm}^2$, $\Sigma_a = 0.02\ \text{cm}^{-1}$ and $\Sigma_{tr} = 2.3\ \text{cm}^{-1}$. For D_2O: $\tau = 125\ \text{cm}^2$, $\Sigma_a = 3.4 \times 10^{-5}\ \text{cm}^{-1}$, and $\Sigma_{tr} = 0.4\ \text{cm}^{-1}$. The macroscopic absorption cross section of the mixture is therefore given by

$$\Sigma_a = 0.02\ (1 - \alpha) + 3.4 \times 10^{-5}\alpha\ \text{cm}^{-1} = 0.02\ (1 - \alpha)\ \text{cm}^{-1}.$$

The macroscopic transport cross section

$$\Sigma_{tr} = 2.3\ (1 - \alpha) + 0.4\ \alpha = 2.3 - 1.9\ \alpha\ \text{cm}^{-1}.$$

The diffusion length can now be calculated

$$L = \frac{1}{\sqrt{3\Sigma_{tr}\Sigma_a}} = \frac{1}{\sqrt{3(2.3 - 1.9\alpha)(0.02(1 - \alpha))}}.$$

$$\therefore L \cong \frac{2.9}{1 - \alpha}\text{cm}.$$

Note that this expression yields $L = 2.9$ cm for pure H_2O when $\alpha = 0$, as expected.

The diffusion length of D_2O alone is

$$L = \frac{1}{\sqrt{3\Sigma_{tr}\Sigma_a}} = \frac{1}{\sqrt{3(0.4\ \text{cm}^{-1})(3.4 \times 10^{-5}\ \text{cm}^{-1})}} \cong 160\ \text{cm}.$$

The reason the approximate formula does not yield this answer when $\alpha = 1$ for pure D_2O is due to the approximation made in the calculation of the macroscopic absorption cross section of Σ_a.

The age of the mixture can be calculated from

$$\frac{1}{\sqrt{\tau}} = \frac{1 - \alpha}{\sqrt{31\ cm^2}} + \frac{\alpha}{\sqrt{125\ cm^2}}$$

$$\therefore \tau \cong \frac{123}{(2 - \alpha)^2}\ cm^2.$$

This expression is approximately correct in both extreme cases of $\alpha = 0$ and $\alpha = 1$.

Problems

10.1 What are the basic requirements of nuclear control materials? Discuss, using specific examples.

10.2 Sometimes the term pumping power index (PPI) is use to evaluate a coolant. This is defined as

$$PPI \equiv \frac{\mu^{0.6}}{\rho^2 C_p^{2.4} k^{0.6}} .$$

Here μ = viscosity, ρ = density, C_p = specific heat, and k = thermal conductivity. If this index is used prove, from the meanings of these four factors and the way they are combined to form the PPI, that the best coolant material has the least pumping power index. For example, $(PPI)_{water} \cong 2.4$, $(PPI)_{air} \cong 6.3 \times 10^7$, $(PPI)_{Nak} \cong 10$, and $(PPI)_{helium} = 2.1 \times 10^7$.

10.3 What are the requirements of reactor structural materials? Point out the consequences of structural material failure in reactor operations.

10.4 What are the thermal and nuclear properties of uranium and plutonium? Indicate the significance of each of these properties in nuclear fuel applications.

10.5 What are the radiation hazards in processing plutonium? thorium? uranium?

10.6 Discuss the advantages and disadvantages of using a) aluminum and b) stainless steel as reactor structural materials according to the list of example 10.1.

10.7 What are ceramics and cermets? Discuss the advantages and disadvantages of these materials for nuclear reactor applications.

10.8 Fuel densification is a recently discovered fuel material problem in reactor operation. What does this mean? Propose a remedy to this problem.

10.9 Calculate the moderating ratio of beryllium, beryllia (BeO), graphite, heavy water, helium, and light water.

10.10 Plot the curves of τ versus α and L versus α based on the results derived in example 10.3 for an H_2O and D_2O mixture.

CHAPTER 11

REACTOR SHIELDING

11.1 Introduction

One of the earliest applications of radiation was x-rays; due to ignorance and accidental circumstances quite a few persons received radiation injuries ranging from minor skin lesions to bone sarcoma; some resulted in death. In due course, radiation safety procedures were adopted, and by 1930 maximum permissible levels of exposure to radiation were proposed. Realization of the danger led to a substantial decrease in radiation injuries from the use of x-rays and radioisotopes.

Radiation hazards and the danger of radiation injuries increased manyfold as nuclear reactors were built. A nuclear reactor designed and built without proper shielding would not only cause radiation injuries to the workers and surrounding population but could also affect the satisfactory functioning of reactor operation and control instruments. Proper shielding has therefore become an important adjunct of reactor design, greatly influencing the cost of nuclear power.

The shielding requirements for a reactor are dependent on the type, size and design of the reactor core. Different types of shielding materials would be used for various reactors such as: (a) central power reactors, (b) package reactors, (c) zero power reactors, and (d) propulsion reactors, etc.

Reactor shielding design involves the study of the radiation distribution in different sections of a shield. This gives an idea about the radiation intensities at various points in the shield as well as the radiation levels on the outside surface. The shield thickness is so arranged that the radiation level at the outer shield surface is below the maximum permissible level. High energy neutrons and gamma rays produced in a reactor core penetrate the shield, causing internal heating and radiation damage. Therefore, materials used for shielding purposes should be stable under the temperature and radiation fields existing in the various sections of the shield.

Although a large quantity of radiation is emitted from an operating reactor, it actually does not create hazardous conditions for the personnel because great care is taken to erect multiple barriers between the fuel, which is the source of the radiation, and the personnel. These include the fission fragment containment by the fuel capsule, coolant radioactivity control, reactor containment vessel, radioactive waste management, and finally the biological shielding of the reactor system, Figure 11.1.

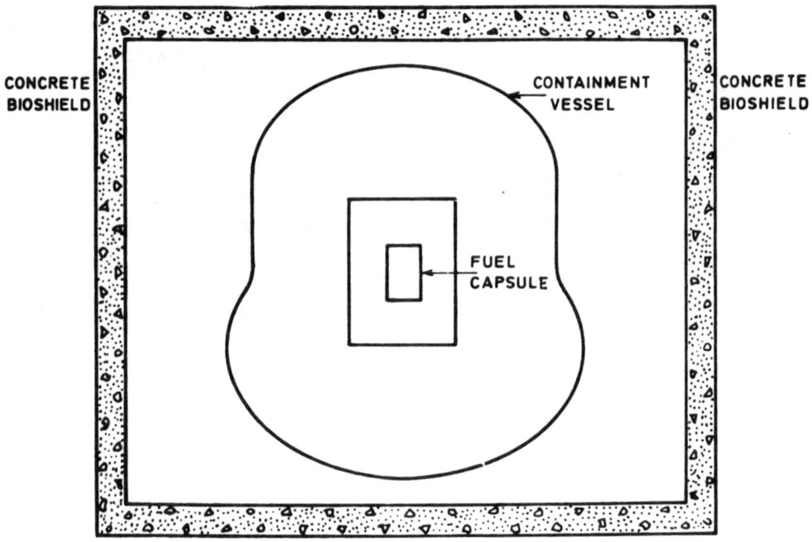

Fig. 11.1: Reactor shielding.

11.2 Reactor Radiation Sources

Almost all types of nuclear radiations are emitted when a reactor is operating. Table 11.1 outlines the nuclear radiations involved in shielding.

Table 11.1: Reactor radiation sources.

Nuclear reactor radiation sources	Comments
Fission fragments	These are very heavy ions and are generally radioactive. They emit other nuclear radiations outlined below. The range of the fission fragments is very small. They are contained inside the fuel element itself by a thin cladding material. However, other nuclear radiations emitted by them escape the fuel elements and require proper shielding.
Alpha particles (^4He)	These are helium nuclei and are produced as a result of nuclear reactions in the reactor core or shield. The range of alpha particles is also very small and they are stopped within a very small thickness of any shielding material. For example, the range of a 10 MeV alpha particle in a very light material like plastic is about 10^{-2} g/cm^2. Alpha particles therefore pose no shielding problems.
Tritons (T)	Tritons are tritium nuclei and are produced as a result of nuclear reactions in the reactor core or shield. Their range is also very small and hence they cause no shielding problems.
Deuterons (D)	These are deuterium nuclei produced in nuclear reactions and, like alpha particles and tritons, cause no shielding problems due to their short range.
Protons (H)	These are also produced in nuclear reactions and although their range is almost 10 times as large as that of alpha particles under the same conditions, it is still too small to cause any shielding problems.
Beta rays (β)	Beta rays may be positive or negative and are produced mainly by the radioactive decay of fission products and by the interaction of gamma rays with matter. Due to their very small mass as compared to the other types of radiation discussed above, the range of an electron of a given energy is much larger than that for the heavier particles. For example a 5 MeV energy electron has a range of 2.6 cm in water and 0.33

Nuclear reactor radiation sources	Comments
	cm in lead. Hence electron shielding does not pose any problems in itself. However, electrons of very high energies produce electromagnetic radiations when passing through materials of high atomic numbers. This process is known as bremsstrahlung. These gamma rays are more difficult to shield and their presence must be carefully studied in any reactor shield design.
Neutrinos (ν)	The neutrino is a neutral particle of great penetrating power. Initially postulated by Pauli, Fermi's theory of β-decay in 1934 predicted the existence of such a particle. Since a neutrino has no charge, practically no mass, and its interaction with nuclear matter is very weak, it is extremely difficult to detect. The presence of neutrinos from a nuclear reaction was detected with the aid of very large liquid scintillation counters containing several thousand gallons of liquid. Theoretical calculations indicate that the mean free path of a neutrino in lead would be about 10^{18}m. This is the thickness corresponding to 100 light years. It is therefore obvious that we cannot attenuate the neutrino fluxes by any kind of shielding. However, since neutrinos do not interact with matter they cannot cause biological damage, although they carry away almost 5 percent of the reactor power.
Gamma rays and neutrons	In addition to the above mentioned radiations, nuclear reactors also emit gamma rays and neutrons. Both of these interact strongly with matter and hence large quantities of shielding materials are required for attaining the desired attenuation. Gamma rays and neutrons therefore constitute the main burden of the shielding problem. There is a fundamental difference in the manner in which the neutrons and gamma rays are attenuated in comparison to charged particles. All the charged particles lose energy by a multitude of collisions without changing their direction significantly. This process continues

Nuclear reactor radiation sources	Comments
	until the charged particles lose all their energy and are neutralized. All these particles have definite ranges which depend on the particle type, initial energy and the medium. Hence if we put an absorber which is slightly greater in thickness than the average range of these particles, all of these particles are stopped. Neutrons and gamma rays, on the other hand, have entirely different interaction mechanisms. They might be completely absorbed in the first collision or may lose a fraction of their energy. A detailed study of the interaction of these radiations with matter indicates that there is no sharply defined distance beyond which they are completely eliminated. Hence the problems arising in shielding against neutrons and gamma rays are more difficult as compared to the charged particles. The gamma energies encountered in nuclear reactors are generally between a few keV to about 10 MeV. Neutrons of energy as high as 20 MeV have been observed in the fission spectrum.

11.3 Shielding Calculations

The attenuation of narrow beams of neutrons and gamma rays can be written in the form

$$I(x) = I_0 e^{-\mu x}, \quad (11.1)$$

where $I(x)$ = neutron or gamma ray intensity after passing through a shield of thickness x,
I_0 = intensity before entering the shield,
μ = linear attenuation coefficient of given gamma rays or neutrons, and
x = thickness of the shield.

The above expression is based on the assumption that the scattered radiation is completely removed from the beam. This assumption is correct only for very thin shields. As shown in Figure 11.2, the probability that any scattered radiation will reach the observation point after a single collision is very small

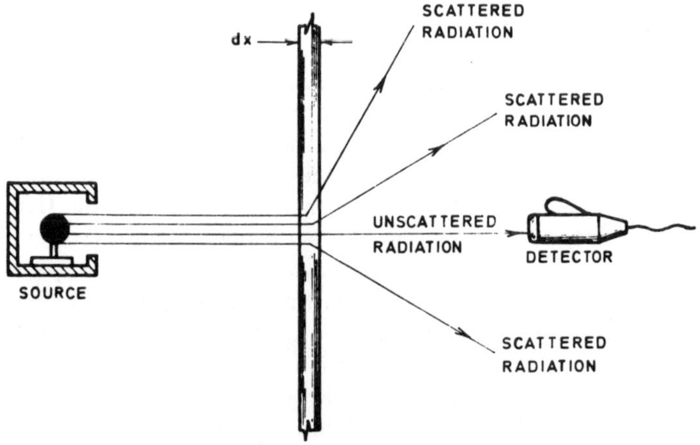

Fig. 11.2: Scattered radiation not reaching the detector.

and the total radiation intensity measured is the same as the total unscattered flux.

However, if the thickness of the shield is appreciable some of the radiation could reach the detector after suffering two or more scattering collisions in the shield. This is shown in Figure 11.3. When the scattered radiations reach the observation point, the expressions given before no longer hold. The influence of the scattered radiation in thick shields tends to increase the total flux at a given observation point. This is taken into account by means of a factor which is a function of shield thickness, shield material as well as the nature of the radiation. This factor is called the buildup factor (B). Since B depends on the three factors mentioned above, it is generally written as B (μx). It is called the buildup factor because the deeper a thick shield is penetrated, a larger amount of the energy in the shield will be in the form of scattered radiation, and therefore the value of B increases or builds up. The definition of B is as follows,

$$B = 1 + \frac{\text{scattered radiation reaching the observation point}}{\text{unscattered radiation}} .$$

The values of B are tabulated in various handbooks on shielding for gamma rays. For approximate calculations, the 'linear buildup factor' given below may be used.

$$B = 1 + b\mu x, \tag{11.2}$$

where x is the shield thickness, μ is the attenuation coefficient, the constant

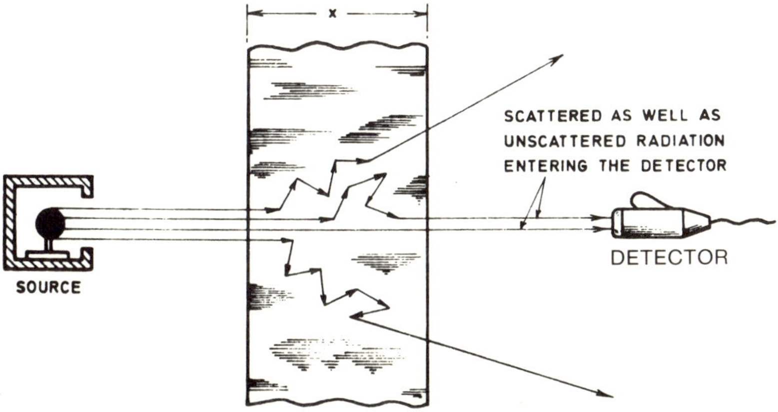

Fig. 11.3: Scattered as well as unscattered radiation reaching detector.

b varies with shield thickness, shield material and the energy of the gamma rays.

If the radiation source is a point source in a non-absorbing medium, the inverse square law can be used to calculate the intensity at a point P shown by Figure 11.4.

$$I = \frac{S_0}{4\pi r^2}, \tag{11.3}$$

where I is the radiation intensity at P, S_0 is the source strength, and r is the distance between the source and the point P.

If an absorbing medium of thickness t is symmetrically introduced between these two points as shown in Figure 11.4, an absorption factor must be added.

$$I = \frac{S_0}{4\pi r^2} e^{-\mu t}, \tag{11.4}$$

where μ is the linear absorption coefficient of the shielding ring of thickness t. Taking into account the buildup factor of the shield, the formula for calculating the radiation intensity at P may be written as

$$I = \frac{S_0 B(r)}{4\pi r^2} e^{-\mu t}, \tag{11.5}$$

where the buildup factor B(r) is a function of r.

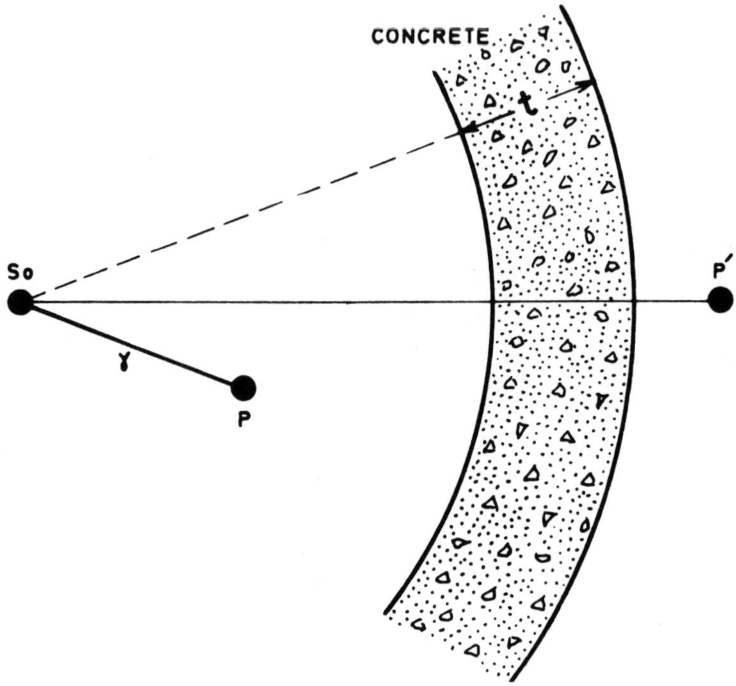

Fig. 11.4: Point radiation source.

In equation (11.5) the factor $\frac{e^{-\mu t}}{4\pi r^2}$ is called the point kernel. This serves as the building block for calculating the radiation distribution due to other types of radiation such as line sources and volume sources. This type of calculation may be summarized as follows (Figure 11.5):

First, a small volume dV of the source as a differential point radiation source is chosen. The 'point' source strength is then written as $ds = S_0 dV$, which is a function of the space coordinates. Due to the point source $ds = S_0 dV$, the radiation intensity at P is then given by

$$dI = \frac{ds}{4\pi r^2} B(t)\, e^{-\mu t}. \qquad (11.6)$$

Then the total radiation intensity at P due to the entire source in volume V_1 is calculated by integration

$$I = \int dI = \int_{V_1} \frac{B(t)\, e^{-\mu t}}{4\pi r^2}\, ds. \qquad (11.7)$$

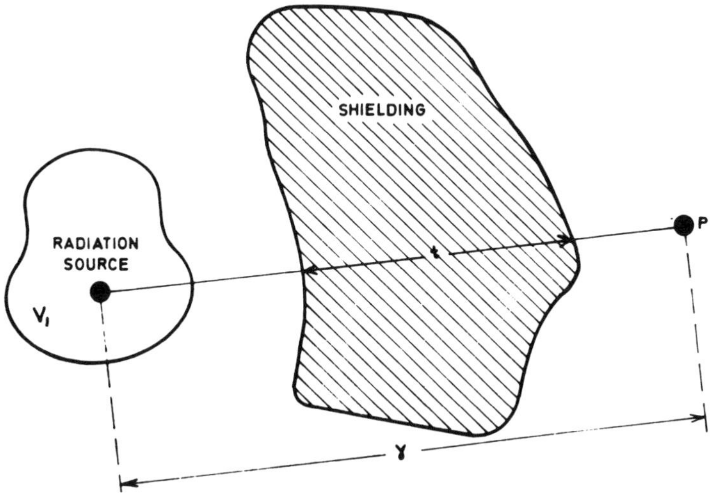

Fig. 11.5: An arbitrary geometry for shielding calculations.

Consider a simple example of a line source, Figure 11.6. The radiation intensity at P due to a small segment of the line source dl is given by

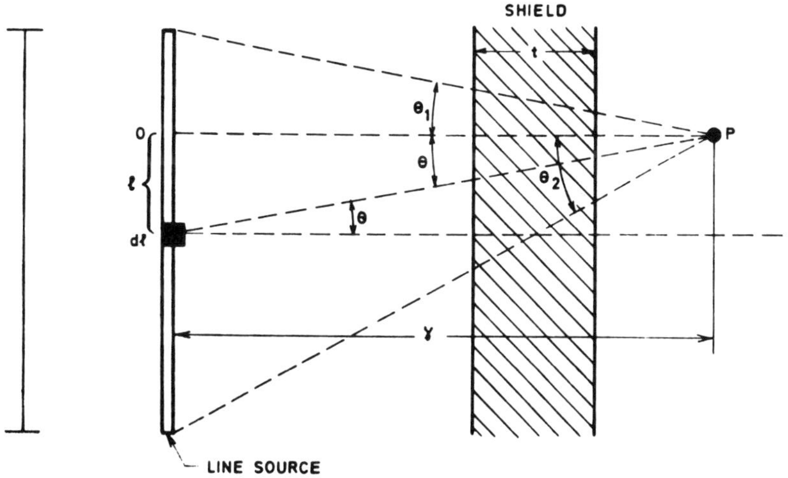

Fig. 11.6: A line source.

$$dI = \frac{S_l\, dl}{4\pi (r \sec\theta)^2}\, e^{-\mu(t \sec\theta)}, \tag{11.8}$$

where S_l is the line source strength per cm per second, and the buildup factor of the shield is taken as unity. To integrate dI, express all variables in terms of the angle θ as follows.

$$l = r\tan\theta.$$
$$dl = r\sec^2\theta\, d\theta.$$

$$\therefore dI = \frac{Sl}{4\pi r} e^{-\mu t \sec\theta} d\theta. \tag{11.9}$$

Integrating this equation from $-\theta_2$ to θ_1,

$$\begin{aligned} I &= \frac{S_l}{4\pi r}\left[\int_0^{\theta_1} e^{-\mu t\sec\theta} d\theta + \int_0^{\theta_2} e^{-\mu t\sec\theta} d\theta\right] \\ &= \frac{S_l}{4\pi r}\left[F(\theta_1, \mu t) + F(\theta_2, \mu t)\right]. \end{aligned} \tag{11.10}$$

where $F(\theta_i, \mu t) \equiv \int_0^{\theta_i} e^{-\mu t\sec\theta} d\theta$, $i = 1$ or 2, is the secant integral the values of which are tabulated.

For other geometries, the procedure of calculating the radiation intensity at a point outside a shield is similar. Detailed coverage of this subject is contained in *Reactor Shielding Design Manual,* edited by Theodore Rockwell, III, D. Van Nostrand Co. and other shielding textbooks.

11.4 Shield Design Considerations

The large weight and size of a reactor is mainly due to its shield. A reduction in shielding size can be made either by reducing the core size or by selecting the proper shielding design and materials. Since the core size is based on the heat transfer considerations, the only parameter which can be manipulated is the shield design. For economical and effective shields, the heavy gamma shielding materials like lead, iron, etc., are arranged closer to the core, and water or hydrocarbons are placed farther from the core. The capture gamma rays are reduced by placing some heavy materials in thin layers away from the core and by using neutron absorbers which emit relatively soft capture gamma radiation. Figure 11.7 illustrates a typical lightweight and compact shield design.

The primary gamma rays and neutrons from the reactor core decrease steadily as the distance away from the core increases. The production of secondary gamma rays from neutron capture and other nuclear reactions is, however, much more complex and requires a thorough study of all sources of secondary radiations in the reactor. Efforts are made to avoid all those materials in the outer layers of the shield which yield hard gamma rays from neutron capture.

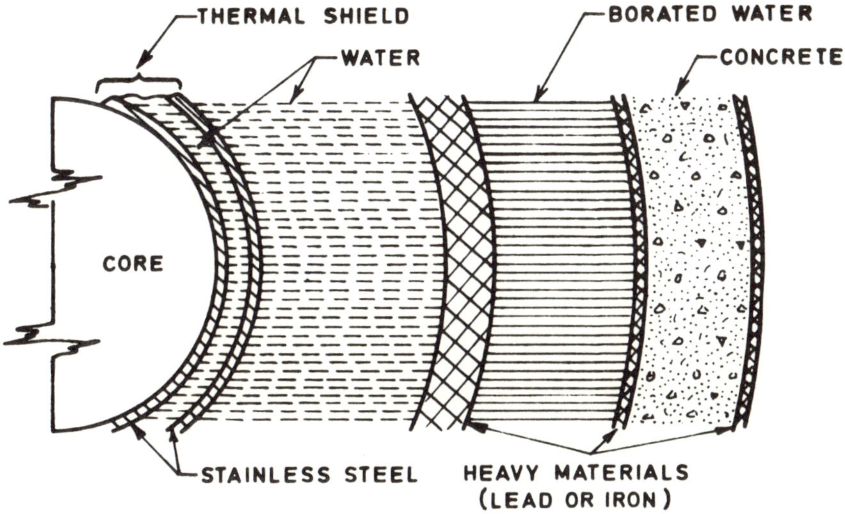

Fig. 11.7: Multiple layer shield.

Concrete is an excellent shielding and is favored particularly where fabrication difficulties arise due to curved surfaces interrupted by framing, piping and ducting because it can be poured in almost any desired shape and form. The heavier additives like barytes, iron punchings and boron frits in concrete make it a very effective shielding material. Such heavy concrete compares quite favourably with iron and lead for gamma ray attenuation. Another form of all purpose shielding is pelletized materials, consisting of compacted powders of high density materials. One example is a mixture of lead, titanium hydride and boron. The powder-mix is pressed at extremely high pressures into small diameter pellets. These pellets can be dumped into a shielding container of any shape and the void space is filled with an organic material or some hydrogeneous liquid. Figure 11.8 illustrates two of the configurations utilizing the concept of multilayers of various materials.

Example 11.1

The neutron removal cross section can be obtained from an absorption experiment using a parallel beam of neutrons. When a beam of neutrons ϕ (n/cm^{2}s) passes a thin slab of material, the transmitted neutron flux density ϕ is given by

$$\phi = \phi_0 \, e^{-\Sigma x} ,$$

where x is the slab thickness and Σ is the macroscopic removal cross section

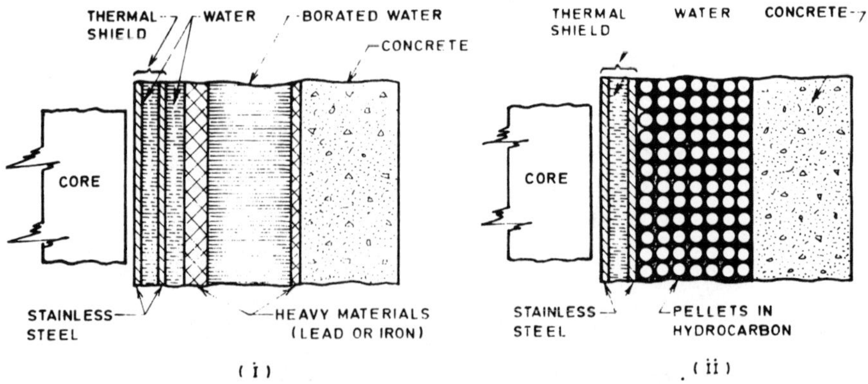

Fig. 11.8: ~~Some~~ shielding configurations.

which is a function of the neutron energy. A beam of 10 MeV neutrons, $\phi_0 = 10^7 n/cm^2s$, is incident normally upon a slab of water, thickness x = 10 cm. The emerging neutron flux density is $\phi = 3.7 \times 10^6 n/cm^2s$. Calculate the 10 MeV neutron removal cross section of water.

Solution

$$\text{Since } \phi = \phi_0 e^{-\Sigma x}, \text{ we have } \Sigma = \frac{1}{x} \ln \frac{\phi_0}{\phi}.$$

$$\therefore \Sigma = \frac{1}{10 \text{ cm}} \ln \frac{10^7}{3.7 \times 10^6} = 0.1 \text{ cm}^{-1}.$$

The inverse of Σ, $\lambda = 1/\Sigma$, is called relaxation length.

Example 11.2

Detailed reactor shielding calculations are usually quite complicated. To reduce this complexity, approximate by considering the reactor as a uniform volumetric fast neutron source. Derive a formula for source strength expressed in terms of reactor power P.

Solution

Since 3.1×10^{10} fissions/s is equivalent to 1 watt of power, if we assume 2.5 neutrons emitted per fission, the total number of fast neutrons is then

$$2.5 \times 3.1 \times 10^{10} P \text{ n/s}.$$

$$\therefore S \cong 7.8 \times 10^{10} P \text{ n/s},$$

where P is the reactor power expressed in watts. Note that this is the thermal power of the reactor. If the reactor power is in MWe, one can obtain the thermal power if the plant efficiency is known. Otherwise, for rough estimation, an efficiency of 30% may be assumed.

Example 11.3

Sometimes the buildup factor $B(\mu t)$ is expressed as

$$B(\mu t) = Ae^{-\alpha\mu t} + (1 - A)e^{-\beta\mu t},$$

where the constants A, α, and β are tabulated. Usually $A > 1$. α is negative, β is positive, and the absolute values of both are less than unity. Prove that the buildup factor increases as the thickness of the shield increases.

Solution

Calculate the first derivative of the buildup factor with the thickness as a variable.

$$\frac{dB(\mu t)}{dt} = -\mu[\alpha e^{-\alpha\mu t} + \beta e(1\text{-}A)]e^{-\beta\mu t}$$

This derivative is zero if

$$e^{-\beta\mu t} = -\frac{\alpha A}{\beta(1 - A)} e^{-\alpha\mu t}.$$

This relation cannot be valid in view of the values of A, α, and β. Therefore, the first derivative is not zero. It is always positive. This means that $B(\mu t)$ is always increasing with increasing thickness t. Furthermore, the rate of increase of $B(\mu t)$ decreases until a point of inflection is reached. This thickness is obtaind from

$$\frac{d^2B(\mu t)}{dt^2} = \mu^2[\alpha^2 Ae^{-\alpha\mu t} + \beta^2(1 - A)e^{-\beta\mu t}] = 0.$$

$$\text{or } t = \frac{1}{(\beta - \alpha)\mu} \ln \frac{\alpha^2 A}{\beta^2(A - 1)}.$$

After this, the rate of increase of $B(\mu t)$ increases as the thickness of the shield is increased. In shielding design this information is useful if high level radiation sources are involved.

Problems

11.1 A point source of Co-60 is fixed at the center of a large spherical water tank. The source strength is 5 Megacuries.

a) Calculate the energy flux density in W/cm^2 20 meters from the source if the tank is empty and if we assume that the gamma ray absorption by air is negligible.

b) At what distance from the source is the energy flux density the same as in a) if the tank is filled with water?

11.2 A plutonium-beryllium neutron source has a source strength of $S = 10^6 n/s$ at an average energy of 4.4 MeV. What should be the spherical shield thickness in order to reduce the neutron flux density at 100 n/cm^2s if water is used as the shielding material. The removal cross section of water is approximately $\Sigma = 0.2\ cm^{-1}$.

11.3 Define the mass absorption coefficient and discuss the advantages of using it in shielding calculations.

11.4 Lead is a frequently used gamma shield. Make up a list of all its physical, chemical, mechanical, and nuclear properties as a reference.

11.5 A swimming pool reactor is rated at 10 kW. The critical mass of uranium-235 is 3 kg. Calculate the thickness of water required on top of the reactor to reduce the gamma radiation to a safe level prescribed by DE = 5(A - 18).

11.6 Sometimes, the microscopic scattering cross section of hydrogen may be represented by

$$\sigma_s \cong \frac{33}{3E + 5},$$

where E is the neutron energy in MeV. This expression is approximately correct in the energy range from 1 MeV to 12 MeV. Plot the curve of σ_s versus E and compare it with a standard reference.

11.7 Calculate the fast neutron source strength of a reactor operating with a power of 4000 MWth.

11.8 Calculate the dose buildup factor by using the exponential formula provided by example 11.3 for 4 MeV gamma rays in water of a) 20 cm, b) 60 cm, c) 100 cm, d) 140 cm, e) 180 cm, and f) 220 cm. The constants are $A = 4.5$, $\alpha = -0.06$, and $\beta = 0.12$. Compare your results with the discussions presented in the example.

11.9 A beam of fast neutrons, $5 \times 10^4 / cm^2s$, might be received by a radiation worker. Discuss the precautions that must be taken by the worker.

11.10 A lead sphere has a point Co-60 source embedded in the center. If the radius of the sphere is $r = 10cm$ and the source strength is one Curie, what is the radiation dose rate on the surface of the sphere?

11.11 Sometimes the radiation source strength is expressed in terms of the mass of the radioisotope(s) contained in the source. What is the source strength in Curies of a source containing two milligrams of Cs-137?

CHAPTER 12

NUCLEAR POWER AND THE ENVIRONMENT

12.1 Radioctive Waste from a Nuclear Power Plant

Radioactive waste disposal is increasing the environmental radiation background and this may become a serious problem if careful control is not exercised. The disposal of radioactive waste must be done in such a way that it would neither cause any direct or indirect radiation hazard to the population nor upset the ecological conditions of the environment.

At each stage in the chain of operations leading to electric power generation from a nuclear reactor there are potential local and environmental radiation hazards. The largest source of radiation in a nuclear power plant is the reactor core; while a nuclear fuel reprocessing plant, by its very nature, is the largest source of radioactive wastes. A great majority of the radioactive fission products remain confined within the fuel capsule. However, a fraction of these diffuse through the capsule walls into the coolant. Neutron activation of the coolant is another source of radiation in the reactor core. In heavy water reactors the most serious radiation problem encountered is the production of tritium from deuterium. With a half-life of 12.3 years, tritium poses a serious hazard if it is not handled properly. Corrosion and erosion of structural material yields some long lived radioisotopes that are released from the reactor core into the coolant. Table 12.1 shows some of these radioisotopes.

Table 12.1 : Structural corrosion and erosion radioisotopes.

Isotope	Half-life (days)
Cr-51	27
Co-58	72
Co-60	5.3
Fe-59	45
Mn-54	300
Zr-95	65

In normal reactor operations effluent streams are processed and then released under the guidelines **prescribed** by Federal Regualtions. According to the data for various power plants operating during 1972 in the U.S.A., the maximum release of radioactivity from a single plant due to all types of wastes is summarized (20, 26, 52, 72) in Table 12.2. The limit of various types of radioactivities indicates the safety margin.

Table 12.2 : Maximum yearly radioactive waste release.

Item	Radioactivity (Curies and % of limit)
Liquid effluent-mixed fission and activation products	51.5 (7)
Liquid effluent-tritium	5890 (0.26)
Airborne effluent-noble gas	8.77×10^5 (5.3)
Airborne effluent-halogens and particulates	6.48 (5.1)
Solid	1.05×10^5 (shipped offsite).

It must be pointed out that the amount of radioactivity in the effluent released from, or the solid waste shipped off, a nuclear power plant varies from plant to plant. Different types of reactors also yield different amounts of radioactive wastes depending on

a) the operating power of the reactor;
b) the fuel in-core time, and
c) the elapsed time for radioactive decay.

The radionuclides released from an operating nuclear power plant are many. The important ones are those with long half-lives. They include strontium-90, iodine-131, cobalt-60, chromium-51, strontium-89, cobalt-58, arsenic-76, xenon-133, krypton-87, krypton-88, xenon-135, iodine-133, cesium-134, manganese-54, zinc-65, cesium-137, and others. Under the present technology, their discharge levels are well under the levels set up in the radiation protection guidelines stated in Chapter 5 of Part I. In fact, the radioactive release into the environment for most of the operating nuclear power plants is well within the background radiation fluctuations at the fence post of the plant.

The major radioactive waste problem in a nuclear power program rests with the spent fuel reprocessing. The spent fuel elements are usually stored in a swimming pool type of fuel storage tank where the short lived fission products decay for some months. They are then shipped to the reprocessing plant. The reprocessing methods depend on the type of reactor and the fuel used. After a spent fuel is sent to a reprocessing plant, it is put through a series of operations as shown in Figure 12.1.

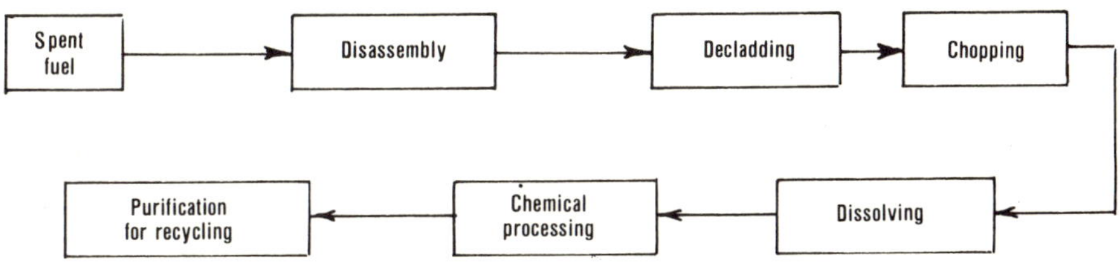

Fig. 12.1: Fuel reprocessing flow diagram.

The major chemical processes are precipitation, the solvent extraction, and fractional distillation. In these processes the amount of radioactivity is enormous. For example a 137 MWe reactor will produce 10^6 Curies of strontium-90 in a year. Most of these are retained in the fuel capsule and released as radioactive waste at the fuel reprocessing plant.

12.2 Treatment and Disposal of Radioactive Wastes

In a nuclear power plant radioactive waste management is an integral part of the design. Some typical designs of such waste management systems are given by Figures 12.2, 12.3, 12.4, and 12.5.

The effluent radioactive waste is usually treated by one or more of the following techniques: delay and decay, filtration, low temperature adsorption, evaporation or demineralization (ion exchange). The evaporated concentrates are solidified with cement and sealed in 55-gallon drums. These drums, together with other solid radioactive waste, are then shipped off for disposal.

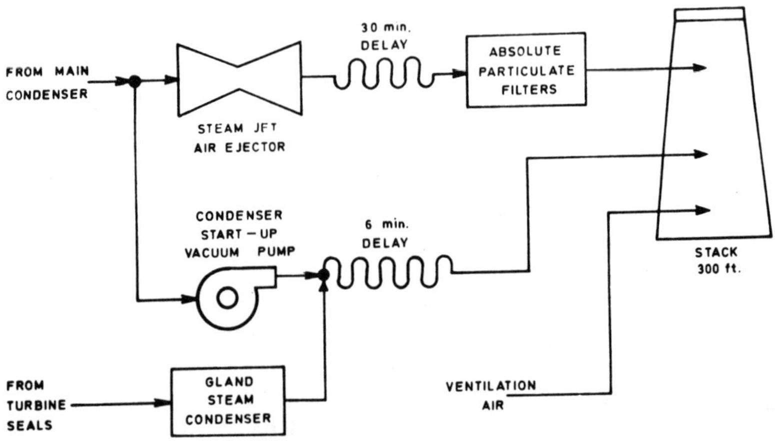

Fig. 12.2: Gaseous waste system of boiling water reactors

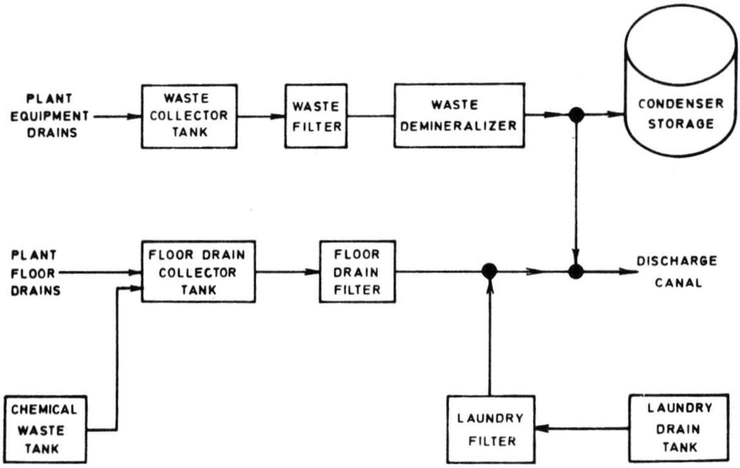

Fig. 12.3: Liquid radioactive waste system of boiling water reactors

Radioactive waste from a fuel reprocessing plant is a more difficult problem because of the large quantities of radioactivity involved. With the increasing number of nuclear power plants in operation, more spent fuel must be reprocessed. Before this problem of disposing large amounts of high intensity radioactive waste becomes a real threat to the environment, a viable solution must be found. A review of the radioactive waste control and management methods presently used in reprocessing plants is presented here. The methods are different for gaseous, solid, and liquid wastes. Liquid

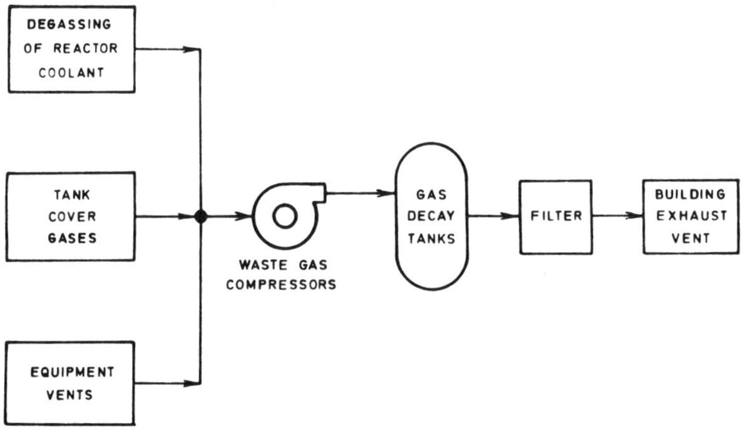

Fig. 12.4: Gaseous waste system of pressurized water reactors.

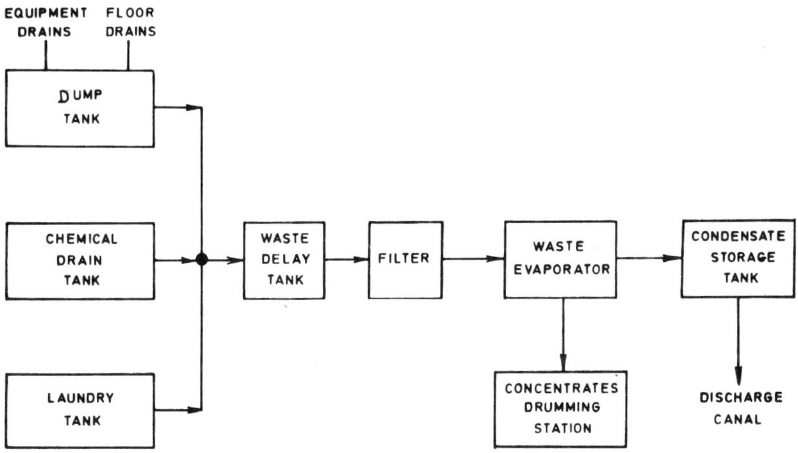

Fig. 12.5 : Liquid waste system of pressurized water reactors.

radioactive waste can in turn be divided into low, medium, or high activity wastes.

The main sources of solid radioactive waste are the metallic fragments and leached parts of fuel capsules. These are usually stored in concrete tanks. Sometimes low activity solid radioactive waste is buried in large trenches. Constant monitoring and control of these burial sites is carried out. Occasionally low level radioactive wastes have been encased in concrete poured into 55-gallon drums and buried in deep ocean areas under controlled conditions. If the waste is combustible, it may be incinerated. The ash would then be treated and stored. If the ash contains plutonium, it is sent through a plutonium separator. In any case the volume of the waste would be drastically reduced. The gaseous combustion products are treated by methods described in the next paragraph.

Gaseous radioactive waste may contain particulates, entrained radioactivities, iodine-129, iodine-131, and krypton-85. It is usually scrubbed and filtered, then sent through a series of absorption systems to trap radioiodine. Strict monitoring of the radioactivity level of the gaseous waste released into the environment is constantly done at the stacks. In almost all the fuel reprocessing plants, environmental monitoring in the surrounding area of the plant is carried out to check the airborne effluent radioactivity level. Here, the maximum permissible concentration of radionuclides (10 CFR part 20) is used as a guide.

The liquid radioactive wastes from a reprocessing plant are treated according to their radioactivity levels. Low level aqueous waste is first identified by its contents and level of radioactivity. Carefully controlled release into the environment is then administered. The pathway through which the low level aqueous radioactive waste propagates must be carefully studied. In particular the ecological disturbance caused by these discharges must be kept within the guidelines listed in 10 CFR part 20. Tritium presents a special problem associated with the aqueous waste. The tritiated water must be handled carefully. At the present time, dispersing the diluted tritiated water into the environment is the only viable method of disposal. If a method could be found to separate tritium from the nonradioactive hydrogen isotopes it would be advantageous not only for radioactive waste treatment but also for nuclear fusion reactors in the future.

Medium level liquid radioactive waste is treated by a combination of processes such as filtration, decay storage, precipitation, evaporation, or solidification. Usually the radioactivity level of the waste can be sufficiently reduced.

High radioactivity level radioactive waste presents the most difficult problem. Its composition is quite varied. It is first concentrated by evaporation. These concentrated wastes usually amount to about 500-1000 liters per ton of uranium fuel processed. They are then stored in steel tanks surrounded by concerte shields up to 2 meters. In most practical cases these tanks are built underground. Figure 12.6 shows such an underground storage tank. To prevent overheating, a set of cooling coils is immersed in each tank. Sometimes air jets are used to stir the mixture in order to prevent the precipitated solids from accumulating at the bottom of the tank. The underground storage must be monitored continuously for leakage and heating. In the last 20 years, intensive research has been carried out to perfect a solidification technique for converting high level liquid wastes into solids which have good thermal conductivity, low leachability, high chemical stability and high mechanical strength. The solidified waste can then be

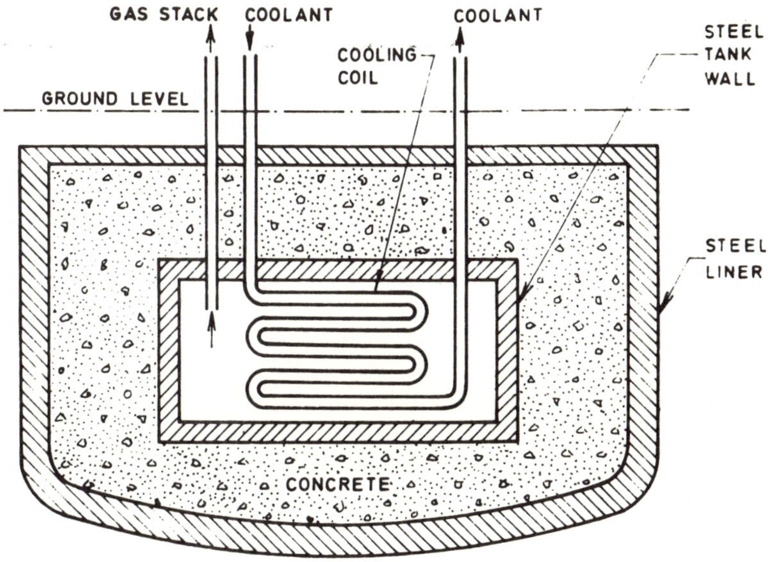

Fig. 12.6 : An underground storage tank.

stored. As compared with liquid waste storage, it has the advantage of requiring less monitoring and surveillance.

12.3 Thermal Discharge from a Nuclear Power Plant.

The unusable heat rejected from a nuclear plant is of the same order of magnitude as that of a fossil plant of the same size. For a 3,000 MWth plant, the thermal discharge is approximately 2,000 MWth. This is due to the fact that all plants are operating with a thermal to electric conversion efficiency of 30%.

Although the amount of the thermal discharge from a nuclear power plant is about the same as that from a fossil fuelled plant of the same capacity, the problem is more serious for nuclear power plants because they are larger in capacity due to economic reasons.

When the number of these plants increases, thermal discharges will increase accordingly. For example, if there are 1,000 nuclear plants (3000 MWth each) in operation, the thermal energy discharged into the water and the environment would be approximately equivalent to 2×10^6 MW. This amount of thermal discharge power is about 10^{-5} that of the total solar energy received by the earth. This could be a problem that needs to be thoroughly investigated.

Recently, efforts have been directed towards attaining higher operating temperatures in an attempt to increase the plant efficiency. While this is a step in the right derection, thermal discharge still remains a major ecological and environmental problem. To prevent any possible disturbance to the aquatic environment, a maximum limit of a 2.8 C° temperature rise at discharge has been imposed. Direct mixing of the heated cooling water and the passing current is almost impossible. Therefore cooling ponds (with or without spray), wet cooling towers, or dry cooling towers must be used.

A wet cooling tower is a device in which heated water is pumped to sprayers above a lattice work. Heat is dissipated to the air by drafts going up through the tower as water flows down the lattice. Sometimes the air flow may be increased by using draft fans. In a dry cooling tower, heated water is pumped through a set of cooling coils installed directly below a set of induced draft fans. This arrangement has the advantage of less cooling water loss so that a makeup water system may not be needed. However, the heat transfer is less effective in this case. For large nuclear power plants a dry cooling tower must be used in conjunction with a secondary water cooling system such as a spray condenser.

12.4 Nuclear Power Plant Safety

The safety designs and measures for a nuclear power plant are extremely elaborate. Same of the standard features are:

1) redundant power, coolant flow, and radiation level monitoring instrumentation systems,
2) multi-level fission product containment systems,
3) reactor shutdown capability and automatic scram features,
4) redundant coolant pumps and circuits, and
5) Elaborate testing procedures.

It was due to these features that no direct casualties were incurred in power reactor operations since the early 1960's. However, even with all of these precautions, absolute reactor safety is impossible! In order to estimate the possible hazards derived from potential nuclear accidents, the USAEC has conducted a detailed reactor safety study. This work began in 1972 involving some 60 experts headed by Norman C. Rasmussen of MIT. The summary results were published in the report *An Assessment of Accident Risks in U.S. Commercial Nuclear Power Plants, Summary Report,* WASH—1400, August 1974. Their conclusion was that the risks to the public from potential accidents in nuclear power plants are very small. Of special interest in this report is a comparison of the risks to individuals of being fatally injured by various types of accidents. These are taken from the 1973 U.S. Statistical Abstract as applied to the year 1969. Some of this data is reproduced here in table 12.3.

Table 12.3: Risk of fatality by various causes

Accident Type	Total Number	Individual Chance per Year
Motor Vehicle	55,791	1 in 4,000
Falls	17,827	1 in 10,000
Fires	7,451	1 in 25,000
Drowning	6,181	1 in 30,000
Firearms	2,309	1 in 100,000
Air Travel	1,778	1 in 100,000
Falling Objects	1,271	1 in 160,000
Electrocution	1,148	1 in 160,000
Lightning	160	1 in 2,000,000
Tornados	91	1 in 2,500,000
Hurricanes	93	1 in 2,500,000
Nuclear Reactor Accidents for 100 plants	0	1 in 300,000,000

Other health effects might be caused by nuclear accidents. These may include injuries, cancer, genetic effects, and thyroid gland illness. While definite quantitative results of these risks are difficult to obtain, it is believed that the number of cases of such nuclear accident induced injuries, genetic effects and long-term cancers would be much smaller than the normal rates caused by other factors.

One of the Kemeny commission recommendations is to improve the public communication effort. This is one area nuclear workers must strive to strengthen. Lately efforts in this direction have been exerted. As a starting point a **Q** and **A** list is presented here:

Q: Why are there many people against nuclear power plants?

A: There are many reasons for the seemingly large opposition to nuclear power. First of all, public reluctance to accept any new technology is quite understandable. Secondly, the first application of nuclear power came from a devastating destruction of properties and lives (the atomic bombs dropped in Japan at the near-end of the Second World War). Such psychological fear, no matter how unnecessary it is in relation to present nuclear technology and application, is real. This fear can be diminished only through education and communication. Thirdly, to some degree, the development of nuclear power plants is not very systematic. As a result, there are bona fide unresolved problems associated with this development. Some of these problems are the fuel supply, the waste disposal, and the risk-benefit analysis.

Q: Can a nuclear power plant behave like an atomic bomb?

A: No! It is impossible for this to happen because a nuclear power plant contains only a fraction, about 5%, of the necessary U-235 for an atomic bomb.

Q: What is the main cause of nuclear plant risks?

A: The major risk involved in a nuclear plant accident is the release of radioactivity from the plant. This radioactivity comes from the fission products which are normally confined in the fuel assemblies. If fuel rods rupture, there is a slight chance that some of these fission products might leak into the environment. This fuel rod rupture is almost certain in the case of a reactor core meltdown.

Q: What are the causes of a core meltdown?

A: The fuel rod cladding will melt at about 5000° F. This could occur if there is a loss of coolant accident. Or, if there is a power transient.

Q: What are the safety designs against losing coolant?

A: Redundant pumps and emergency core cooling systems (ECCS) could prevent core meltdown.

Q: What does power transient mean?

A: Normally the term transient means a temporary deviation from the normal pattern. In reactor operations, this term refers specifically to conditions which can occur in a reactor operation that require the reactor to be shutdown.

Q: How does one deal with transient conditions?

A: Reactor safety design has redundant reactor scram features. For example, any reactor achieving 120% of the demanded power level due to transient or any other factors is automatically shutdown.

Q: What is the chance of a core meltdown even with all the safety design features?

A: According to WASH-1400, the probability of a reactor core meltdown is about one in 17,000 per reactor per year.

Q: What are the likely consequences of a core meltdown?

A: Again, according to WASH-1400, the fatalities and other effects are less than one, with the possible exception of 4 thyroid nodule cases per each core meltdown accident.

Q: How much radioactive waste is produced in a nuclear power plant?

A: A 1,000 MWe nuclear power plant produces about one ton of high level fission product waste per year. However, this amount of waste is contained in the fuel assemblies. Only when the spent fuel is reprocessed do these high level radioactive wastes need to be disposed.

Q: What is the major health hazard of the high level radioactive waste when spent fuel is being reprocessed?

A: If plutonium in this waste is reduced to a level of about 0.005%, the hazard of this waste is comparable to a uranium mine. Techniques to reclaim this valuable plutonium are available. The problem is their economic feasibility.

Q: What does a nuclear plant do to prevent sabotage?

A: Absolute prevention of organized and skilled sabotage is impossible. However, plant security systems, reactor containment building design, and emergency procedures can reduce the probability of sabotage and the corresponding damage to the reactor.

Q: Could the fuel (fresh or spent) shipment be sabotaged?

A: Yes.

Q: Nuclear fuel is stored in swimming pool type facilities at the plant site or other places. Is this dangerous?

A: No. All fuel storage pools are carefully designed and their characteristics analyzed. In fact, the Nuclear Regulatory Commission has to grant a license for using or changing such a facility.

Q: If enough fuel is stolen would it be easy for someone to make an atomic bomb?

A: No. It takes a minimum of 4 kilograms of plutonium or 20 kilograms of enriched (90% or higher U-235 content) uranium to form a critical assembly. Furthermore, these materials must be brought together quickly and held together long enough for the nuclear fission reaction rate to reach a violent explosion state. Nuclear power reactor fuel just does not have such purity.

Q: If there is no absolute guarantee that a nuclear power reactor accident is avoidable, why should we use nuclear energy?

A: The judgment as to the degree of risk our society is willing to accept rests on the public who uses the electricity generated from nuclear power reactors. The major responsibility of nuclear workers is to analyze the problems and to communicate the results to the public so that intelligent decisions can be made.

12.4 Nuclear Power Plant Accidents

The safety record for nuclear power plant operation is far better than for other types of power plants. There was no casualty directly involved with nuclear power reactor operations since the first commercial reactor. Among the several dozen of nuclear power reactors in operation today in the U.S.A. there have been two accidents that have attracted wide attention from the public. They are the a) Browns Ferry Nuclear Plant fire and the b) Three Mile Island accident.

1) Browns Ferry Nuclear Plant (BFNP) accident.

TVA's BFNP has three operating BWR units with a total capacity of approximately 3000 MWe. Units 1 and 2 share a common control room and some instrumentation (on-line process computer). Around noon on March 22, 1975, electrical inspectors were working in the spreading room which is directly below the control room, Figure 12.7. Their assignment was to inspect the air leaks. Foam rubber was used to seal any leaks. They were using candles to determine the directions of the leaks and the air draft. They discovered a 2

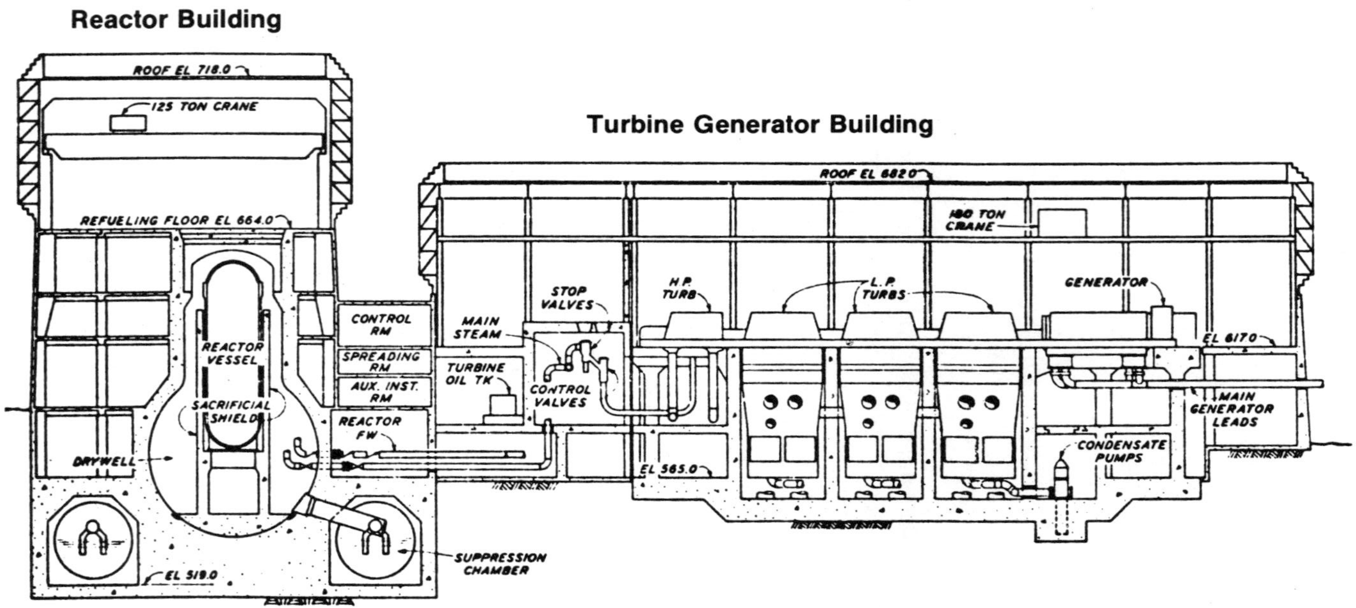

Fig. 12.7: Section of reactor and turbine-generator buildings of BFNP

by 4-inch opening in a penetration window in a cable tray containing several electric cables connected to the control room. This opening was filled with foam rubber. When the candle was brought near the opening to check the leakage, the candle flame was sucked into the tray compartment by the strong draft. As a result the foam rubber caught fire. The time was then 1220.

An initial attempt to extinquish the fire using the flashlight, rags, a CO_2 extinguisher, and a dry chemical extinguisher failed. The flames spread rapidly. The air draft pulled the flaming material from the spreading room into the reactor building. Fire alarms were eventually sounded approximately 15 minutes after the fire started. By this time the indicating lights were flashing on the control console of unit 1. Numerous alarms sounded. By 1240 the emergency core cooling system (ECCS) was activated. The reactor power of unit 1 began to drop. After the failure of the recirculation pumps at 1251, this unit was shutdown by the operator by inserting the control rods. A similar sequence of events occured at the unit 2 control console and this unit was also shutdown, at around 1300.

In the meantime, two major activities went on. The first was the attempt to keep the reactor cores covered with water. To maintain the water levels in these reactors, the operators had to use the condensate booster pumps because all other pumps were inoperable by that time. The second activity was the fire fighting. This effort was not going as well. Due to the fact that CO_2 and dry chemicals could not cool the hot wires, the fire kept on burning. Around 1800, water was used and the fire was finally extinguished 20 minutes later.

The lessons learned from this accident are:

a) fire hazard prevention and drill need to be strengthened,

b) each unit should have its own separate control systems, and

c) reliable and prompt communication systems must be maintained at all times.

2) Three Mile Island Nuclear Plant accident.

The Three Mile Island Nuclear Power Plant, operated by Metropolitan Edison Co., a subsidiary of General Public Utilities Corp. (GPU), is located ten miles southeast of Harrisburg, Pennsylvania. This plant has two PWR units, TMI-1 and TMI-2. On March 28, 1979, several water pumps in the feedwater supply circuit of the TMI-2 stopped working (feedwater pump trips) at 0436 hour. When these feedwater pumps failed, three emergency feedwater pumps were automatically started. However, a valve in each of two of the emergency lines was closed while the state of the third channel was uncertain. Thus the water flow to the steam generator was stopped. Two seconds later the steam turbine and the electric generator were automatically tripped by the plant's safety system. Due to the feedwater stoppage, the reactor coolant temperature increased, expanding the water

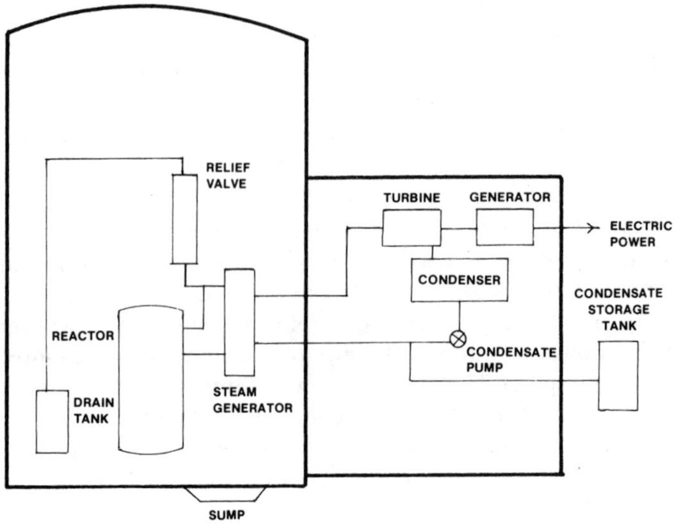

Fig. 12.8: Section of reactor and turbine buildings of TMI

level and compressing the steam in the pressurizer tank. When the steam pressure increased to 2255 lb/in^2 (100 lb/in^2 over the normal operating pressure) the relief valve on the tank opened. As a result, steam and water flew out of the pressurizer into a tank on the containment building floor. Eight seconds into the first pump failure, the reactor was scrammed due to high pressurizer tank pressure. However, after the reactor scram, the heat due to fission products continued to build up. At the same time the pressurizer relief valve failed to close as designed (for five seconds) after the pressure was reduced to 2205 psi. The indicator in the control room did not show this continued open mode of the relief valve. Thirteen seconds into the accident, an operator noticed the low pressurizer water level and turned on a pump to add water to the pressurizer. Thereafter, the water level started to rise (with the relief valve still open)! About 105 seconds into the accident the steam generator had boiled dry causing the reactor coolant temperature to increase. At two minutes after the accident, the reactor pressure dropped sharply. Two large pumps (high pressure injection pumps — HPI)were then turned on automatically, feeding more water into the reactor. An operator who did not know that the relief valve of the pressurizer was open shutdown one of the HPI pumps because of the increase of the water level in the pressurizer tank. Five and a half minutes into the accident, steam bubbles formed in the reactor coolant system, displacing more water to the pressurizer. All this time, water and steam continued to pour into the containment building sumps. At 0415, the drain tank burst, sending slightly radioactive water onto the floor. Such an open "circuit" overworked the HPI pumps causing vibrations. At 0541, due to this vibration, all HPI pumps were shutdown. As a result, by 0600, some fuel rod rupture ocurred, sending fission products into the coolant and onto the containment building floor. Hydrogen and gaseous fission product bubbles were formed. The radiation level in the containment building continued to increase to a level of about 800 rem/hr at around 0720. Thereafter, a general emergency was declared and the containment building was isolated. Later, cleanup work was obstructed due to the explosion of the hydrogen bubbles and the corresponding fire hazard. The cooling of the damaged core was finally achieved in the evening by activating one of the primary pumps.

The radioactive effluent ejected into the environment during the accident was found to be well within the maximum permissible concentration established in 10 CFR 20.

This accident has attracted wide public attention. As a result a presidential commission was appointed to investigate and to study this accident. This commission, chaired by President Kemeny of Dartmouth College, found that the accident was caused by a series of human, institutional, and mechanical failures. A series of recommendations for minimizing later possible safety problems was proposed. Emphases were placed on improved institutional organization, both for the governmental agencies and the utility companies, and operating personnel training. This commission also recommended an evacuation plan as a precondition for the granting for a plant construction permit. In general, improved reactor performance, particularly with regard to safety, can be expected as a consequence of the TMI accident.

Example 12.1

A thermal reactor is cooled by water having an impurity of 5 ppm of sodium. The reactor is operating at an average thermal and fast neutron flux density of 10^{14}n/cm^{2}s. If the once-through coolant incore residence time, or exposure time, is two seconds, what is the induced activity per unit volume of the cooling water a) immediately after it leaves the reactor core and b) two minutes afterwards.

Solution

The neutron activation of water occurs in the following manner:

a) About 0.2% is due to $^{18}O(n,\gamma)^{19}O$ with thermal neutrons. The microscopic activation cross section is 0.2 mb.

b) About 0.04% is due to $^{17}O(n, p)^{17}N$ with fast neutrons. The microscopic activation cross section is 0.007 mb.
c) The rest, 99.76%, is due to the fast neutron absorption in $^{16}O(n, p)^{16}N$. The average microscopic activation cross section is ~ 0.02 mb.

When some impurity, such as sodium, is present in water, the induced activity of the impurity must also be considered. In the case of sodium, the activation is mainly due to $^{23}Na(n, \gamma)^{24}Na$ with an activation cross section of 0.5 b.

Consider the neutron activation equation

$$\frac{dN}{dt} = N_0\sigma_{ac}\,\phi - \lambda N,$$

where N and N_0 are the activated and original nuclei densities, respectively. ϕ is the average neutron flux density. σ_{ac} is the microscopic activation cross section and λ is the decay constant of the induced radioisotope. The initial condition is N = 0 when t = 0.

$$\therefore\ A = \lambda N = N_0\sigma_{ac}\,\phi(1 - e^{-\lambda T_1}),$$

where T_1 is the exposure time.

Calculate the induced activities as follows:

a) $N_0 = \dfrac{6.023 \times 10^{23}}{18} \times 0.2\% \cong 6.7 \times 10^{19}$ atoms/cm^3 of O-18.

$$\lambda = \frac{0.693}{T_{1/2}} = \frac{0.693}{29\text{ s}} = 0.025\text{ s}^{-1}.$$

$$\therefore\ A = (6.7 \times 10^{19}\text{ cm}^{-3})(0.2 \times 10^{-3} \times 10^{-24}\text{cm}^2)(10^{14}\text{n/cm}^2\text{s}) \cdot [1 - e^{-(0.025\text{ s}^{-1})(2\text{s})}] = 65400\text{ dis/cm}^3\text{s}.$$

Two minutes later,

$$A = (65400\text{ dis/cm}^3\text{s})e^{-(0.025\text{ s}^{-1})(120\text{s})} = 3260\text{ dis/cm}^3\text{s}.$$

b) $A \cong 2400$ dis/cm^{3}s. $A(2\text{ m}) \to 0$.

c) $A \cong 1.7 \times 10^7$ dis/cm^{3}s. $A(2\text{ m}) \cong 215$ dis/cm^{3}s.

$A \cong 170$ dis/cm^{3}s. $A(2\text{ m}) \cong 170$ dis/cm^{3}s.

Thus the induced sodium activity remains about the same.

Problems

12.1 Why is the spent fuel of a power reactor stored for a cooling period of 100 days after it is withdrawn from the reactor?

12.2 Radioactive waste is produced in a reactor. What are the factors influencing the amount of its production?

12.3 Describe the chemical procedure of spent fuel processing and the radioactive waste production.

12.4 Construct a flow chart to describe a complete radioactive waste management system of a nuclear power plant.

12.5 What are the methods of disposing high level liquid radioactive waste?

12.6 An underground radioactive waste storage tank is described in Figure 12.6. Discuss the advantages and disadvantages of this system.

12.7 A reactor containing 40,000 kg of 2% enriched uranium operates at a power of 1000 MWe. The spent fuel is removed and cooled for 100 days. Assuming that the burnup of the fuel is one atom percent, estimate the activities per kg of fuel.

12.8 Calculate the reduction of thermal discharge of a 1300 MWe nuclear power plant, if the overall plant efficiency is increased from 28% to 38%.

12.9 Calculate the total surface area of a cooling pond required to dissipate the thermal discharge from a 1300 MWe nuclear power plant. The heat dissipation rate is 1200 kW/km^2. Discuss your result. The plant efficiency is 30%.

Ans: $\sim$ 2538 km^2.

12.10 During an earthquake, is it mandatory to shutdown an operating nuclear reactor? Why?

12.11 In an operating nuclear reactor most of the radioactivity is associated with the fission products. Describe the measures taken to reduce the chance of these products leaking into the environment and causing radiation hazard.

CHAPTER 13

POWER PLANT REACTOR ENGINEERING

13.1 Introduction

In a nuclear power plant, nuclear engineers play key roles in overseeing the performance of the reactor core and its associated equipment. Due to the complexity of the system and safety considerations, strict compliance to the recommended operational procedures is required. Detailed instructions on startup, normal operation, surveillance, refueling, and shutdown are available from the vendors.

All judgements and recommendations pertaining to the reactor operation are subject to the constraints of a) the safety of operation and b) obtaining the maximum amount of electric power. Therefore, it is the duty of the station and shift nuclear engineers to push the reactor operation to as high an output as possible within the regulatory and functional operating limits.

A typical administrative organization of a nuclear power plant is illustrated in Figure 13.1. The actual power production operation is carried out by the shift engineer assisted by an assistant shift engineer; a unit operator; several assistant unit operators; health physics, mechanical, electrical, and instrument technicians; chemical technician; and nuclear engineer. The shift engineer and the nuclear engineer work closely to monitor and to control the reactor operation. Usually, the nuclear engineer, after analyzing the core performance, advises the shift engineer on the proper course of action to ensure the safe and economical running of the reactor for maximum possible power production.

The major tasks of plant nuclear engineers are to

1. perform various reactor tests,
2. determine reactor control requirements and methods, and
3. assist in assuring compliance with NRC and other safety requirements.

Some plant nuclear enginners might be assigned to shift duties. One shift rotation scheme is shown in Figure 13.2.

In the following sections, typical BWR power plant reactor engineering tasks are illustrated.

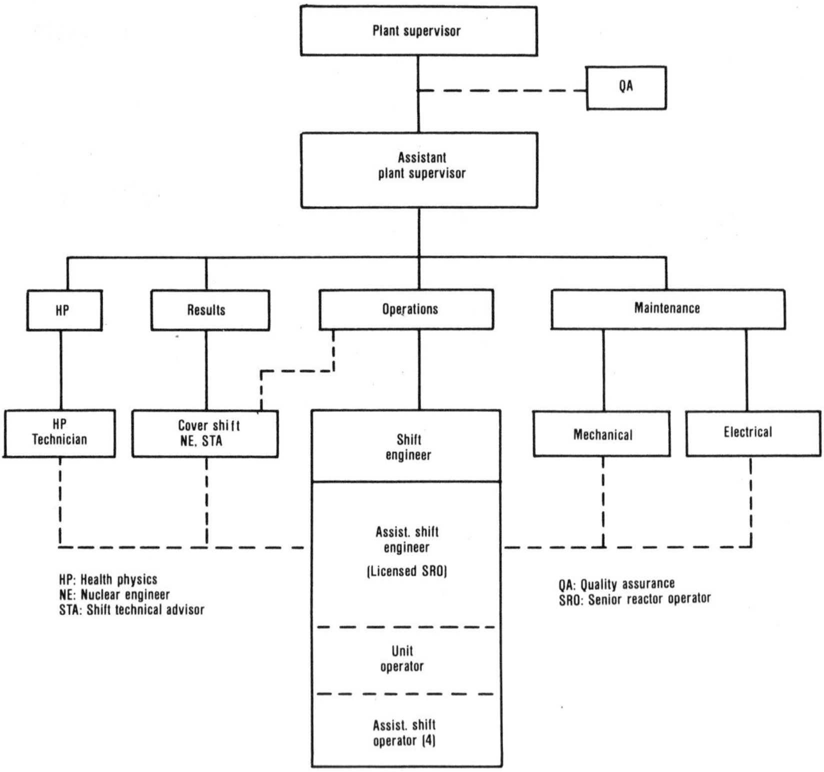

Fig. 13.1: Nuclear power plant organization

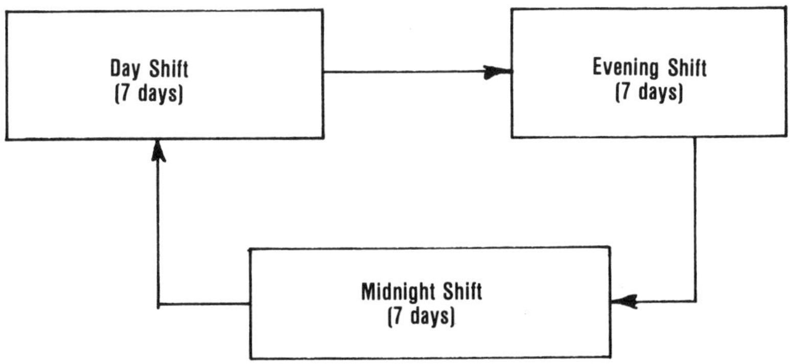

Fig. 13.2: Shift Rotation Scheme

Sometimes shift nuclear engineers are called unit cognizant nuclear engineers. The general responsibilities of the unit cognizant nuclear engineer are to implement the incore fuel management strategy developed by the vendors, to review the daily activities of the unit, to schedule and coordinate major events, and to maintain records of the operating history for the unit. Specifically, the unit cognizant nuclear engineer should:

a. Follow core management closely including review and development of rod patterns for optimum power shape and core thermal limits.

b. Review the unit core performance.

c. Review the daily log. Comment on any unusual results — especially where the computer has erroneous input or runs assymetric mode.

d. Review the collected semi-monthly fuel warranty data for correctness and completeness. Follow up to ensure timely data transmittal.

e. Prepare nuclear performance statements for each monthly report. Review the power history chart for the month.

f. Closely follow and prepare schedules for major events such as rod sequence exchanges, LPRM calibrations, control rod scram time tests, and power shaping to ensure compliance with technical specifications and fuel warranty.

g. Maintain a record of core average exposure.

h. Prepare guideline instructions for accomplishing major events such as rod exchanges and power shaping.

i. Provide guidelines for the unit operator to monitor core thermal limits when engineering shift coverage is not required.

j. Review the unit log book and unit instruction book and ensure they are properly maintained.

k. Establish startup rod sequences and ensure that adequate copies are available in the control room.

As a result of the TMI accidents, the NRC has ordered all nuclear reactor operating utilities to establish shift technical advisor (STA) duties. The STA is responsible for supplying technical advice to the shift supervisor on nuclear safety matters and measures in cases of abnormal plant conditions. The duties of a STA are distinctively different from a shift NE. The former is concerned mainly with safety in operation, such as transient conditions, while a shift NE is mainly concerned with the normal operation of a nuclear reactor.

13.2 Shift Nuclear Engineering

13.2.1 Nuclear Engineering Tasks

A shift engineer and his operators depend on the nuclear engineer's advice as to specific steps to be taken for reactor operation. Thus, to be able to supply such information and proposals, a nuclear engineer covering a shift must regularly perform core performance analysis. The major items in such an analysis are:

a. reactor thermal limits.
b. traversing-incore-probe (TIP) machines.
c. obtaining and updating preconditioning envelope.
d. control rod pattern adjustment.
e. nuclear instrument checkout and operation.
f. corrective analysis for core limit violations.
g. miscellaneous tests.

The nuclear engineer covering a shift is required to keep records on various items on reactor core and flow performance. His most valuable tool is the on-line computer. From the printout of this computer the nuclear engineer can keep an accurate record of the reactor operation. A sample logbook record is shown in Fig. 13.3. The meanings of the entries in this figure are:

CMWT = core thermal power in MW.
CMWE = gross electric power in MW.
CMFLPD = core maximum fraction of limiting power density.
CMPF = core maximum peaking factor.
CPRRAT = ratio of the core critical power ratio.
EFF = thermal to electric conversion efficiency.
WT = coolant flow rate (core) in M lb/hr.
MAPRAT = maximum average power ratio.
CAVEX = core average exposure in MWD/T.

Some of these entries can be obtained directly from the on-line computer output while others must be calculated.

a. The load factor is calculated as follows:

$$\% = \frac{\text{CMWT}}{\text{Rated Thermal Power}} = \frac{2885}{3293} \rightarrow 87.6\%.$$

OPERATING CONDITION Holding

DATE 7-19-78 TIME 1900
CMWT 2885(86.6%) WT 78.20 (76.7%)
GMWE 904 CPRRAT .742(31-30)
CMFLPD 0.691 MAPRAT .703(43,48,4
CMPF 1.93 CAVEX= 10564
EFF 31.3%

Fig. 13.3: Sample of logbook entries

b. The efficiency is

$$\text{Eff.} = \frac{\text{Electric Power}}{\text{Thermal Power}} = \frac{904}{2885} \rightarrow 31.3\%$$

c. The computer printout corresponding to CAVEX is the present exposure rate. This value can be estimated from the rated thermal power and the loaded fuel. If these are 3,000 MWth and 150 tons,

$$\frac{3000 \text{ MW}(1 \text{ da})}{150 \text{ ton}} = 20 \text{ MWD/T}$$

at full rated power. When the reactor is not operating at full power, the exposure rate should be proportionately less. Thus, for the case of an 87.6% load factor, the hourly exposure should be

$$(87.6\%) \times (20 \text{ MWD/T}) \times \frac{1}{24 \text{ hr}} \cong 0.73(\text{MWD/T})/\text{hr}.$$

13.2.2 Reactor Thermal Limits

One of the major responsibilities of a cover shift nuclear engineer is to monitor reactor operations to see if they are within the vendor's recommended thermal limits. In case one or more of these limits are violated, he must advise the shift engineer to take remedial actions. Some of the methods are obvious while others need detailed analysis.

When analyzing core performance, the calculated or recorded data must be compared with the reactor design technical specifications (Tech. Spec.).

a. If the rated power is 3000 MWth, it is obvious that the core thermal power must be less than or equal to 3000 MWth.
b. CMFLPD $\leqslant$ 1.00.
c. MLHGR $\leqslant$ vender's recommended limit.
d. The calculated exposure rate must be compared with the target exposure.
e. MCPR $\leqslant$ vender's (Tech Spec) recommended limit.

Under normal operating conditions, the reactor must be within the above-mentioned limits. If the core thermal power exceeds the rated power, the power must be reduced to a level below this limit. This is usually accomplished by reducing the flow rate. If the reduction of the flow rate is not sufficient to bring down the power to a permissible level, control rod insertion may be used for this purpose.

Perhaps, the most frequent thermal limit violation is that the CMFLPD value is greater than unity. When this happens, the Tech. Spec. might limit power operation to 25% or less (to avoid excessive fuel power peaking damage). In order to operate the reactor at a power above 25%, the shift nuclear engineer has to bring the reactor within limit. This can be accomplished by adjusting the control rod positions. The computer output can be used to find out the most limiting fuel rod(s) and to check the corresponding power peaking factors and locations. Based on this data, the shift nuclear engineer can then make a decision as to which group of control rods are to be moved. It is to be noted that, after a control rod position change, it will take a few minutes (3-5 min) for the reactor to reach equilibrium. Only then can monitoring data be taken as reliable.

13.3 In-core Neutron Monitoring Systems

To cover the whole range of reactor operations from pre-fuel loading to full power, different types of specially designed neutron detectors must be used. Source range monitor (SRM), intermediate range monitor (IRM), and power range monitor (PRM) are used to monitor neutron levels at different power ranges, Figure 13.4.

The block diagram of a typical neutron monitoring system is shown in Figure 13.5. The only difference between this system and a standard laboratory neutron detection system is the addition of a driver to move the detector. Depending on the type of monitors, the recording devices are different. These will be discussed in each of the appropriate sections.

Each monitoring system must be able to satisfy the general criteria of

1. signal/noise — 2:1.
2. long lifetime under normal operating conditions.
3. in-core, shutdown counting rate $\geqslant$ 3 cps.

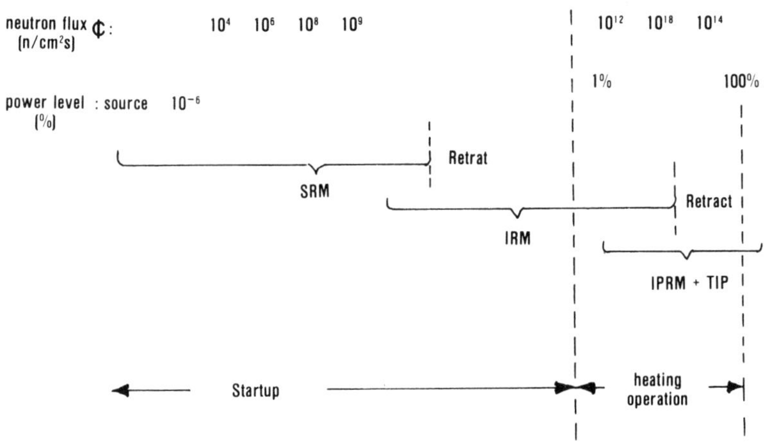

Fig. 13.4: Monitoring ranges

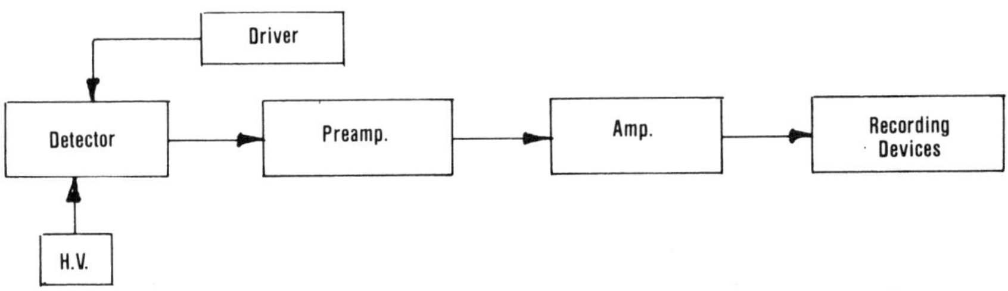

Fig 13.5: A neutron detection system

4. trip set point at $\sim 10^5$ cps with operating limit $\sim 10^7$ cps (counting rate saturation!) for FLC and SRM.

The signal-to-noise ratio can be obtained by taking counts with the detector in and out of the reactor. If the in-core and out-of-core counting rates are C_i and C_o, respectively, the S/N ratio is then given by

$$S/N = \frac{C_i - C_o}{C_o} = \frac{C_i}{C_o} - 1.$$

In the following sections, each of these neutron monitoring systems will be discussed separately.

13.3.1 Source Range Monitor (SRM)

The source range monitor system consists of a miniature fission chamber, with a U_3O_8 (>90% enrichment of U-235) coating on its central electrode, filled with argon in a case of titanium. The fission chamber is operated with a H.V. of a ~200 V - 700 V plateau and is attached to a low loss quartz fiber insulated transmission cable. This assembly is designed to operate in a maximum temperature of ~300° C with a neutron exposure limit of 10^{19}. The detector sensitivity is ~2×10^{-3} cps/nv in a gamma field of up to 10^5R/hr and 10^{-4}cps/nv at 2.5×10^7R/hr. Due to the large neutron flux increment during the startup stages, the SRM channel recording must utilize logarithmic scales since the SRM will experience several decades of neutron

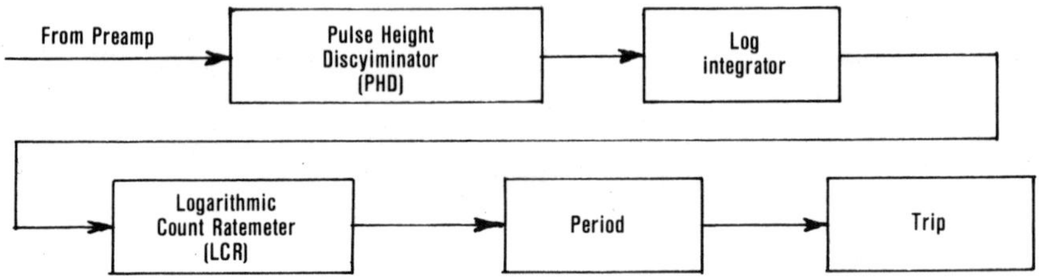

Fig. 13.6: SRM

flux increment during the startup. The SRM is not connected to the reactor protection system. However, it does provide the following rod motion blocking interlock signals:

1. SRM detectors not fully inserted with < 100 cps.
2. SRM counting rates greater than 5×10^5cps.
3. SRM channel inoperative.

13.3.2 Intermediate Range Monitors (IRM)

IRM detectors are miniature fission chambers with an enriched U_3O_8 electrode coating, similar to SRM detectors. They are, however, designed and tested to be operated under a maximum temperature of ~540° C. It has to operate so as to overlap with the neutron flux and/or power indications of the SRM and PRM systems. In a large BWR such as the BFNP units 1, 2, or 3, there are 8 such channels. IRM channels are capable of generating trip signals to block rod withdrawal and/or scram as well as feed the reactor protection system.

The recording part of an IRM is a multi-ranged (5 decades) direct reading meter and recorder calibrated proportionately to the reactor power. It has a calibration and diode logic unit that includes a triangular wave generator of adjustable amplitude to provide a means for full scale calibration of the power meter. The IRMs are calibrated against APRMs on controlled shutdowns with a maximum frequency of once per week. During the IRM calibration the reactor power should be ~25-30%. The calibration involves comparing the IRM readings of % core power to the readings of the APRMs which can be obtained either from panel meters or the on-line computer.

13.3.3 Power Range Monitors (PRM)

Power range monitors consist of the local power range monitors (LPRM), the transversing in-core probes (TIP), the average power range monitors (APRM), and the rod block monitor (RBM). Strictly speaking the rod block monitor and the APRMs are not neutron monitoring devices but a logic assembly for executing actions when certain LPRM signals are fed into the RBM and APRM input selectors.

LPRM detectors are also miniature fission chambers. Since they must reflect accurate reactor power, their calibrations are much more involved than SRMs and IRMs. In order to use the LPRM readings for power estimation, all LPRMs must be properly calibrated. There are two methods to perform this task: on-line computer and hand calculation. These are described following the discussion of TIP systems.

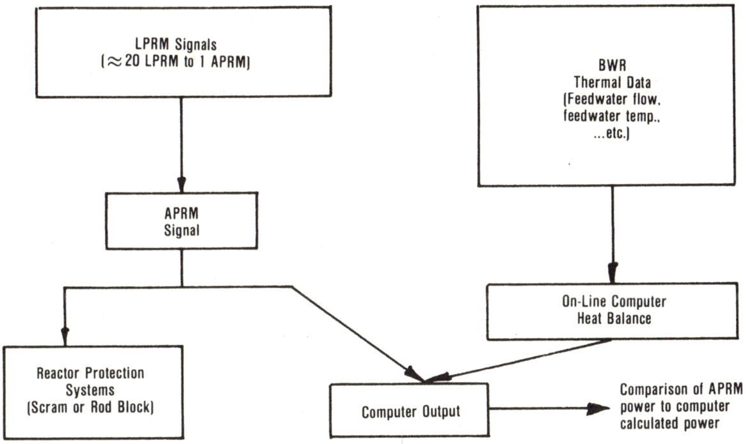

Fig. 13.7: APRM flow diagram

The APRM readings are averaged to give the core thermal power. The on-line computer calculates the actual core thermal power and the gain of the APRMs is adjusted accordingly. The reactor protection system's rod block and scram set point depend upon the APRM power level. With time the APRMs will require adjustment.This is determined by the on-line computer. The APRM gain is defined as:

$$\text{APRM gain} = \frac{\text{actual core thermal power}}{\text{APRM power}}.$$

The technical specifications require that the scram set point be 120% of full power. Assume that the APRM gain is 1.05. It is evident from the equation above that this would allow the actual core thermal power to be greater than 120%. For example, when the APRM gain = 1.05 and APRM power is = 120%, the actual core thermal power = 1.05 × 120% = 126%. This is a tech spec violation.

13.3.4 TIP Systems

The traversing in-core probe (TIP) system measures the axial neutron flux distribution by having a miniature neutron detector (a fission chamber) traversing in a guide tube inserted in certain locations of the core. The DC current output from the detector is converted into a voltage output plotted on the y-axis of an x-y recorder. Thus the x-y recorder printout is proportional to the axial neutron flux distribution and therefore the axial power distribution.

For the convenience of comparing different TIP readings and for TIP intercalibration, a common location (guide tube) is provided for all 5 TIP subsystems. Correction factors can be obtained by comparing the TIP readings of each TIP subsystem when the TIP probe is traversing through this common guide tube. A TIP guide tube is inserted with an LPRM in an instrument tube for TIP calibration. Normally calibration calculations are per-

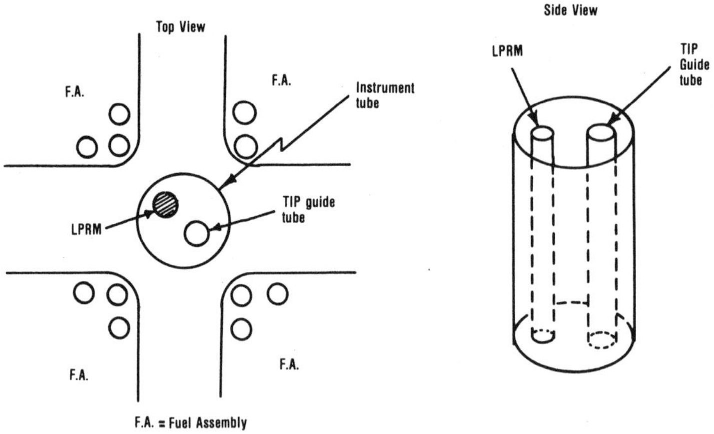

Fig. 13.8: LPRM and TIP in-Core arrangement

formed by the on-line computer based on the TIP data. In case the process computer is inoperable, a simple and approximate method to hand calculate the LPRM calibration parameters can be made.

13.3.5 LPRM Calibration

LPRMs are calibrated using a TIP trace set; this should be performed every 1000 fullpower hours, ~42 days at rated condition. A full set of TIP traces is obtained and fed into the on-line computer. This updates the base distribution and enables the computer to use accurate calibrated LPRM readings to calculate core performance parameters. Any change made to the LPRM amplifier gains must be followed immediately with complete TIP set measurements in order to update the computer.

The neutron sensitivity of the local power range monitors decreases with accumulated reactor exposure caused by the burnup of U-235. This results in a decrease in the neutron induced current of the detector. By periodically increasing the gains of the individual LPRM amplifiers in the control room, the detectors can be made to read the value of the heat flux at the detector locations. Since the LPRM amplifiers are linear, a test input current necessary to produce a standard reading can be divided by the gain adjustment factor to find a new input current. Using this current, the amplifier gain is adjusted until the standard reading is obtained on the meter. Increasing the gain on an LPRM amplifier also increases the reading on the associated APRM channel. This requires a readjustment of the APRM amplifier gain to provide meter indication. The LPRM gain adjustment factor (GAF) is normally obtained as an output of the process computer.

A hand calculation method used by GE to calibrate an LPRM and a TIP reading is the TBAR method. This is based on the fact that both the LPRM and TIP readings should represent the same power, therefore the heat flux, at the nodal point in question. If, taking an idealized situation for an example, the radial and axial neutron distributions are all uniform, then all the LPRM and TIP readings will be the same, separately of course. Thus only a constant conversion factor needs to be used for each set of data to convert the respective readings to represent the power production.

Problems

The problems included in this chapter can be perused to prepare for shift nuclear engineer qualification. As a consequence additional references such as the *GE Nuclear Engineering Training Manual* and plant operation instructions are needed to solve some of these problems.

13.1 What are the functions of water flowing through the reactor core?

13.2 Where does the energy come from when a uranium nucleus is fissioned?

13.3 Draw simplified diagrams of BWR and PWR plants.

13.4 Briefly explain why inserting control rods in the core causes the power to go down.

13.5 Why is it necessary to employ a neutron source in a reactor during startup? What would happen to the reactor power if the source is left in a critical (initially) reactor?

13.6 Define reactivity in terms of the effective multiplication factor.

13.7 Name three reactivity coefficients of a BWR and their effects on reactivity.

13.8 Graphically illustrate the xenon concentration variation as a function of power.

13.9 Name four factors that effect the power distribution in a reactor.

13.10 A reactor is being started from a cold condition (initial xenon concentration ~0). Describe the reactivity variation as the reactor goes from cold to hot to exposure conditions.

13.11 Briefly explain why the injection of water into the vessel via the HPCI system will cause a power spike.

13.12 What effect do voids have on the reactor power shape in a BWR and how is this compensated for?

13.13 Why is there Gadolinia in fuel assemblies?

13.14 Draw a graph to show the K_e variation as a function of exposure.

13.15 From what in-core monitoring devices can a new base distribution be obtained?

13.16 When is an APRM calibration required (pertain to its gain setting or drift)? Why?

13.17 The technical specifications limit the peak heat flux of a BWR to certain specified values. This limit is neccessary in order to prevent two things from occurring. What are they?

13.18 To prevent clad melt, the Critical Power Ratio must be kept greater than 1.05. Why is the technical specification limit 1.25?

13.19 Fuel preconditioning from low power to full power at a slow rate is performed to prevent fuel failures. What happens to the fuel and the clad during heatup that causes this problem (i.e., why is preconditioning necessary)?

13.20 During a scram recovery from 100% power where a preconditioned envelope and rod pattern have been established, is it possible to return to power using the exact same rod pull sequence as during the initial startup? Explain.

13.21 At what power level does the process computer program P-1 become operable? What is the minimum power level at which P-1 can be used with a reasonable assurance of accuracy?

13.22 Will MAPLHGR be limiting during the low power (0 - 40%) portion of a startup? What about CMFLPD? Why?

13.23 At what power level does the APRM flow bias system come into effect?

13.24 Why is it important to obtain a strongly bottom-peaked power shape during a startup? Why is this especially important at low core flows?

13.25 What accidents are the RWM and RSCS designed to mitigate? How?

13.26 What is the main function of a recirculation pump?

13.27 How does the RSCS limit rod move between 100% and 50% rod density? What about the RWM?

13.28 What alternative is available when the RWM is found to be inoperable during pre-startup surveillance testing? What about the RSCS under the same circumstances?

13.29 What precautions must be taken to bypass the Rod Worth Minimizer when operating at approximately 22% core thermal power?

13.30 Discuss the group notch control mode of the RSCS.

13.31 What is normally the initial Sm at the beginning of a startup? Why?

13.32 At what point during a startup would the rod withdrawal sequence be changed from the typed sequence sheets?

13.33 When is a substitute core flow required? Explain the steps necessary for inserting a substitute core flow.

13.34 When are the RWM and RSCS required to be operable?

13.35 Consider a reactor startup in rod sequence A.

a. Will the RSCS allow the selection of an A34 rod if some, but not all, of the A12 rods are full out?

b. If all the A12 rods are full out and the dim lights on the rod select panel do not extinguish, what problem is suspected and how can it be corrected?

13.36 At what power level (approximately) does the RSCS automatically bypass?

13.37 Consider a power reduction from 100% to 70% for the weekly control valve test. Why is it important to minimize the time spent at the reduced power level?

13.38 Which technical specification core thermal limit is most strongly exposure dependent? Does this dependence vary with fuel type? Does the limit increase or decrease with exposure?

13.39 Which of the core thermal limits are actual fuel damage limits? (i.e., if these limits are violated fuel damage may occur) Which are only derived limits? (i.e., if these limits are violated fuel damage is not expected to occur)

13.40 What are the consequences of a P-1 run in the asymmetric mode?

13.41 What would be the effect on core thermal limits of a feedwater temperature sensor reading higher than the actual feedwater temperature?

13.42 The limiting CMPF occurs at node 8 off of a rod positioned at 40. The diagonally adjacent rod is at 04 as shown: (48 = rod out, 00 = rod in)

04	48
48	40

PCIOMR is not a consideration since the power level is low and all 4 rods have been preconditioned out to position 48. What steps may be taken to improve CMPF? Explain. How do the proposed changes affect CPR in this location?

13.43 What precautions must be taken to monitor the core when an LPRM is bypassed?

13.44 Give an explanation of the core thermal limit MAPLHGR.

13.45 Does the local peaking factor vary with exposure? Explain.

13.46 If a PCIOMR violation is observed should you reduce power and correct the problem or hold steady and assume that the damage (if any) is already done?

13.47 Which technical specification core thermal limit is most strongly flow dependent? Explain.

13.48 Discuss the advantages of fission chambers over boron-lined proportional counters for in-core neutron monitoring.

13.49 Discuss the advantages and disadvantages of control rods in terms of their effectiveness in controlling the radial and axial power distribution.

13.50 Why is chemical shim more effective than control rods for control of power distribution?

13.51 The trace obtained on a startup power monitoring channel employing a B-10 proportional counter and a logarithmic rate meter with a maximum counting rate capability of 10^5 cps is shown below. Discuss the reason for the behavior shown at points A and B.

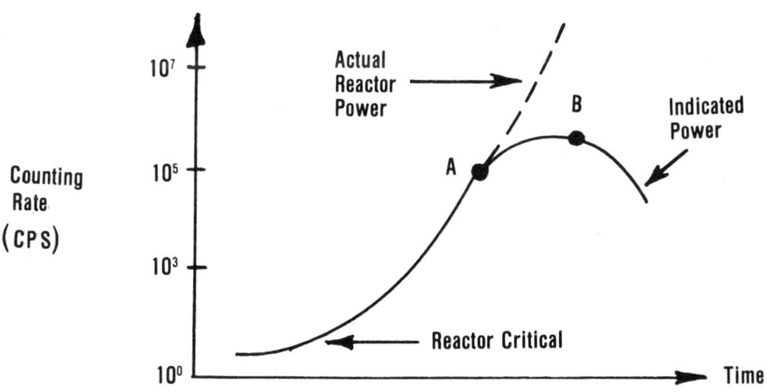

13.52 Consider a B-10 proportional counter connected to a ratemeter with an integral discriminator. The detector is placed in a field of thermal neutrons and gamma rays. The gamma ray energies are uniformly distributed up to about 1 MeV. Assuming the voltage is set at the constant operating voltage, plot a curve to show the counting rate as a function of the discriminator settings and point out the proper discriminator setting for neutron counting.

13.53 Describe plant response and operator actions which would accompany the following:
a) Severe load rejection.
b) Reactor coolant pump trip.
c) Sudden increase in a fuel bundle exit temperature.
d) Sudden increase in condenser off-gas activity.
e) Reactor scram due to high neutron flux.

13.54 Assume two similar fuel bundles are operating with the same radial peaking factor and the same coolant flow rate. Both bundles have a maximum axial peaking factor of 1.35, but in bundle A the axial peak occurs 1 foot below the core centerline and in bundle B the axial peak occurs 1 foot above the core centerline. Which bundle will probably have the lowest burnout ratio? Why? If the radial power distribution looked as shown below, what would it indicate about the control rod pattern?

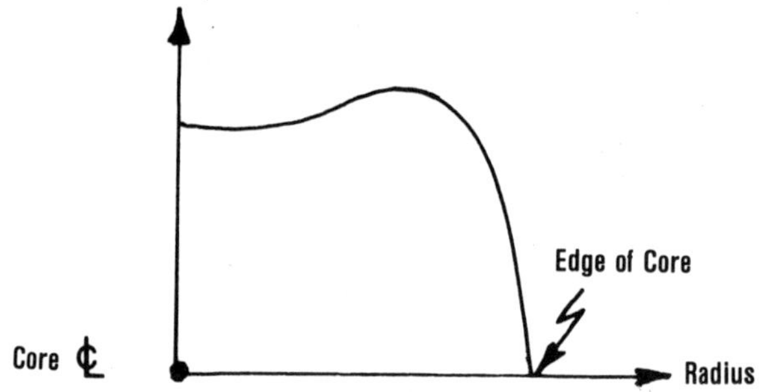

13.55 Consider a 300 bundle BWR operating at 600 MWTh (or 2.05×10^9 BTU/hr). Each bundle has a total of 60 ft^2 of heat transfer surface area. At full power, the "hottest" bundle is operating at a radial peaking factor of 1.35. The corner rod peaking factor is 1.33. At a point 10 inches below the core centerline, the axial peaking factor is 1.43 and the burnout heat flux is 750,000 BTU/hr ft^2. At a point 12 in above the core centerline (℄), the axial peaking factor is 1.24 and the burnout heat flux is 600,000 BTU/hr ft^2. What is the burnout ratio at each location at full power? The operating license specifies that the minimum burnout ratio (BOR) must be greater than 1.5 at 125% power. Is this limit being met?

13.56 Suppose we have a BWR fueled with 2.2% enriched UO_2 rod type fuel and the strength of the control rod system is 17% $\Delta K/K$. At a later refueling, new fuel of 2.5% enrichment and a corresponding higher K_∞ is placed in the core in order to allow greater burnup between refueling outages. No changes are made to the control rod system, but several sheets of burnable poison are placed in the core to help control the more reactive fuel. With the newly constituted core, would you expect the control rod strength to be greater or less than 17%$\Delta K/K$? Why?

13.57 The K_e of a fully controlled reactor at ambient temperature, xenon free, is 0.955. What is the shutdown margin? If the operating license for this facility states that the reactor must be at least 1% $\Delta K/K$ shutdown with the strongest single control rod withdrawn, what is the maximum allowable worth of this rod under these conditions?

13.58 With no fuel in a core, the neutron source gives a reading of 200 cps on a detector. After 10 fuel assemblies have been added to the core the counting rate is 28000 cps. Assuming that the detector has responded perfectly to the increase in K_e, what is the K_e of the core now?

13.59 Why does it take longer for the neutron population to equilibrate when K_e is changed from 0.9991 to 0.9998 than it does to equilibrate when K_e is changed from 0.969 to 0.970?

13.60 Power goes from 60 MWT to 82 MWT in 30s. What is the period?

13.61 A reactor is started up and operated for 50 hours at which time a scram occurs. Eight hours later the reactor is brought to 50% power and is operated there for 40 hours. The reactor is then taken to full power and operated for 50 more hours. Draw a curve showing the xenon reactivity worth throughout this period.

13.62 How many fissions per second are taking place in a reactor which is producing 200 Megawatts of power?

13.63 Explain the fuel densification phenomenon and its possible effects on reactor operations.

13.64 What is the range of the nodal power, as compared with the preconditioned envelope, within which the TIP base distribution has to be updated?

13.65 What is the basic information furnished by a TIP trace?

13.66 Find the fuel element meltdown time of a fuel element in a reactor which has suddenly lost the coolant without a scram or the emergency coolant in operation. Assume that the UO_2 fuel element has a diameter of 3/4 in with a heat generation rate of 5×10^6 BTU/hr ft^3 and is cooled by water at 500° F with h= 200 BTU/hr ft^2 °F. The fuel element melting temperature is ~5000° F. Also, assume that k = 1 BTU/hr ft^{2}° F, c = 0.06 BTU/lb °F, ρ = 530 lb/ft^3, and that the cladding has no effect on heat transfer.

13.67 Reactor operation modes may be classified according to the fuel temperature, T_f, and the moderator temperature, T_m. When $T_m \cong T_f \leqslant$ 200° F, the reactor is said to be in a cold condition. A typical hot condition might have T_f = 900° F and T_m = 600° F. Prescribe a reactor operation schedule in terms of the exposure (fuel burnup) and time, labeling the proper modes of operation (hot or cold).

CHAPTER 14

FUEL MANAGEMENT IN POWER REACTORS

14.1 Fuel Cycle

Fuel management for nuclear reactors may be divided into two parts, namely out-core fuel management and in-core fuel management. Out-core fuel management includes the details of prospecting, mining, refining, enrichment, conversion, fabrication, reprocessing and management of radioactive wastes, while the in-core fuel management generally means the fuel management in reactors, fuel composition as a function of burnup, state of spent fuel elements, etc.

The entire nuclear fuel process, starting from the mining of the uranium ore, through the stages of the fuel element preparation and depletion to the recycling of the spent fuel, is called the nuclear fuel cycle. A detailed analysis of this path furnishes the reactor fuel cost which is an important factor in the selection of a reactor type. A typical uranium fuel cycle is illustrated in Figure 14.1.

At current uranium prices, a deposit of uranium is economically feasible to be mined if it contains at least 1% uranium in the form of xUO_2, yUO_3. The ratio of y/x varies from zero to two. The various forms of uranium

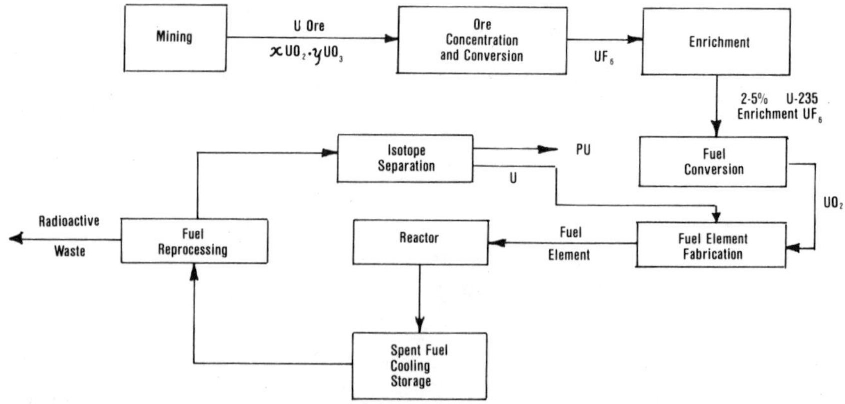

Figure 14.1: Uranium fuel cycle

oxides, UO_2 to U_3O_8, in the ore are concentrated by leaching about 80% of U_3O_8. For isotopic enrichment, U_3O_8 is converted to UF_6 by the following reactions:

$$U_3O_8 + 2H_2 \rightarrow 3UO_2 + 2H_2O, \; UO_2 + 4HF \rightarrow 2H_2O + UF_4,$$

and

$$UF_4 + F_2 \rightarrow UF_6. \tag{14.1}$$

UF_6 vapor is then sent through a multiple stage diffusion chamber to increase the $^{235}UF_6$ contents. Normally 2 to 5 percent enrichment of U-235 is required for light water reactors. It is then converted into UO_2 for fuel fabrication. A typical 1000 MWe light water power reactor normally has some 800 fuel assemblies in each of which there are some 60 fuel elements. One typical fuel element contains about 200 fuel pellets. For such a reactor the total UO_2 mass is approximately 180 metric tons, or a net uranium content of 160 metric tons. The fuel depletion due to burnup in such a power reactor amounts to approximately 30 metric tons of UO_2 per year with a 1% burnup limit for replacement.

14.2 Out-core Fuel Management

The radiation emitted by uranium and thorium makes their prospecting relatively simple. The detectors generally used for this purpose are Geiger-Muller tubes and scintillation counters. A well planned exploration program would include preliminary reconnaissance of detailed geologic structures and the study of aerial photographs.

a) Mining

Uranium mining methods are similar to those of other minerals. For shallow deposits open pit mining is used, while for greater depths underground mining techniques are employed. The maximum depth for the open pit mines is around 100 meters. In addition to the usual mining hazards of contamination of dust and gases resulting from drilling and blasting, emission of strata gases, mine fires, etc., uranium miners face the additional problem of radioactivity. The radioactive decay of uranium leads to the production of radon gas, which decays into a series of short lived products. These products tend to deposit in the respiratory systems of miners. Special precautions such as the channelling of fresh air and the sealing of the areas in which mining is discontinued are taken in order to minimize the radiation hazard to the workers.

b) Milling

Uranium ores found in various uranium mines contain different concentrations of uranium. Concentrations ranging from 0.1 to 1 percent are quite common. Ores having concentrations much lower than 0.1 percent are not economical at present. The milling of uranium employs the extraction of uranium from a large body of the ore. The ore is first crushed, then finely ground and leached by percolating chemicals through it. Diluted acids and alkaline carbonates are used for the ore processing. Hundreds of tons of ore can be processed daily in a uranium mill extracting almost all the uranium from various precipitates and solutions. Ion exchange or solvent extraction techniques are used for separating uranium from the numerous impurities present in the solutions. These processes yield relatively good quality uranium in the form of U_3O_8, commonly known as yellow cake.

c) Refining

The yellow cake obtained from uranium mills still contains numerous impurities which cannot be tolerated in nuclear fuels. The uranium is therefore purified in refining plants to standards which are more characteristic for the pharmaceutical industry than the chemical industry. Uranium is first dissolved in nitric acid, resulting in the production of an impure uranyl nitrate solution which is then mixed with a solvent, such as tributyl phosphate dissolved in an inert hydrocarbon, such as kerosene oil. The solvent extracts pure uranyl nitrate from the solution leaving impurities in the waste nitric acid solution. The solvent containing uranium is next mixed with pure water and the uranyl nitrate is transferred back from the solvent to the water. The uranium, now available in pure form, is converted to uranium hexafluoride if enrichment of the fuel is desired.

d) Enrichment

Most of the presently available thermal reactors use enriched uranium fuel since natural uranium places severe limits on the choice of structural materials, fuel burnup and core size. Although numerous methods, i.e. diffusion technique, centrifuge technique, nozzle method and laser techniques are used for uranium enrichment, gaseous diffusion technique is the chief method used for the large scale production of enriched uranium at present. A gaseous diffusion plant generally consists of hundreds of stages commonly known as diffusion cascades. The degree of separation of a single stage is a function of the square root of the ratio of the masses of $^{235}UF_6$ and $^{238}UF_6$. A diffusion process is carried out at pressures much lower than atmospheric pressure in order to increase the mean free path of the gaseous molecules. The entire diffusion plant is very complex and requires huge quantities of electrical energy. The three diffusion plants in the USA consume over 50 billion kilowatt-hours of electricity annually for operating the gas circulating pumps, maintaining desired pressures, etc.

e) Chemical Conversion and Fabrication

The uranium from the refinery or enrichment plant is chemically converted from the metallic or hexafluoride form to a form in which it is to be used for power production. The most commonly used form of fuel in commercial power reactors is UO_2. Various stages in the fuel fabrication are: compaction of small cylindrical pellets from uranium dioxide powder; loading of the pellets in cladding tubes and sealing the tubes; clustering of a number of such fuel tubes by means of spacer devices, resulting in fuel bundles. A large number of such fuel bundles held together in a definite pattern in the reactor vessel constitute the reactor core.

f) Fuel Reprocessing and Reconversion

The average lifetime of nuclear fuel in a typical enriched uranium reactor is about three years. Although the fuel still contains large quantities of fissionable materials at the end of this period, the physical distortions of the fuel elements due to radiation damage and the buildup of fission product poisons require the fuel to be removed and replaced. Because of the residual fissionable material, the spent fuel is too valuable to be discarded. The spent fuel is therefore shipped to a reprocessing plant where it is dissolved and chemical operations are performed to recover and purify the residual uranium and plutonium. A single reprocessing plant can handle the spent fuel from many reactors. The separated fuel solutions from reprocessing plants are finally converted to the specific fuel form desired for use in nuclear power reactors or uranium enrichment plants.

14.3 In-core Fuel Management

a) Fuel Composition

In order to carry out accurate calculations regarding the fuel composition as a function of time, computer calculations are required. An approximation of how the fuel composition changes with burnup can be obtained from simplified calculations for the uranium and plutonium isotopes. Consider a reactor initially charged with uranium, having the U-235 and U-238 concentrations equal to $N^{235}(0)$ and $N^{238}(0)$ per unit volume. The reactor is then operated at a neutron flux density ϕ. The equations for the change of fuel composition with time can be formulated as follows:

i) Uranium-235

$$\frac{dN^{235}}{dt} = -N^{235}\sigma_{235}(a)\phi(t).$$

$$\therefore\ N^{235}(t) = N^{235}(0)e^{-\sigma_{235}(a)\int_0^t \phi(t')dt'}$$

$$= N^{235}(0)e^{-\sigma_{235}(a)\theta}, \qquad (14.2)$$

where $N^{235}(0)$ and $N^{235}(t)$ are the number of atoms of U-235 per unit volume initially and at time t, respectively; $\theta \equiv \int_0^t \phi(t')dt'$ is called the integrated neutron flux density or flux-time; $\sigma_{235}(a)$ is the microscopic absorption cross section of U-235.

ii) Uranium-236

$$\frac{dN^{236}}{dt} = N^{235}\sigma_{235}(c)\phi(t) - N^{236}\sigma_{236}(a)\phi$$

$$= \frac{N^{235}\sigma_{235}(a)\alpha_{235}\phi(t)}{1 + \alpha_{235}} - N^{236}\sigma_{236}(a)\phi(t), \qquad (14.3)$$

where

$$\alpha_{235} = \sigma_{235}(c)/\ \sigma_{235}(f).$$

If $N^{236} = 0$ at $t = 0$, the solution of the above equation is

$$N^{236}(t) = \frac{N^{235}(0)\sigma_{235}(a)\alpha_{235}}{[\sigma_{235}(a) - \sigma_{236}(a)](1 + \alpha_{235})}[e^{-\sigma_{236}(a)\theta} - e^{-\sigma_{235}(a)\theta}]. \qquad (14.4)$$

iii) Uranium-238

$$\frac{dN^{238}}{dt} = -N^{238}\sigma_{238}(a)\phi(t)$$

$$\text{or } N^{238}(t) = N^{238}(0)e^{-\sigma_{238}(a)\phi t}.$$

As the change in the U-238 concentration during the practical irradiation period is only a few percent, an approximation is

$$N^{238}(t) \cong N^{238}(0)[1 - \phi\sigma_{238}(a)t]. \qquad (14.5)$$

iv) Plutonium-239

The differential equation governing the net formation of Pu-239 may be written as

$$\frac{dN^{239}}{dt} = N^{238}\sigma_{238}(c)\phi(t) + \eta_{235}N^{235}\sigma_{235}(a)\phi(t)\epsilon p_1(1 - p)$$

$$+ \eta_{239}N^{239}\sigma_{239}(a)\phi(t)\epsilon p_1(1 - p) - N^{239}\sigma_{239}(a)\phi(t), \qquad (14.6)$$

where p = resonance escape probability
and p_1 = fraction of neutrons which do not escape the reactor while slowing down from fission to resonance energy.

$$\eta_{235}N^{235}\sigma_{235}(a)\phi(t)p_1(1 - p) \text{ and } \eta_{239}N^{239}\sigma_{239}(a)\phi(t)\epsilon p_1(1 - p)$$

represent the formation of Pu-239 by the absorption of resonance neutrons produced respectively by U-235 and Pu-239 fissions in U-238. The solution of equation (14.4) subject to the condition $N^{239} = 0$ initially is

$$N^{239}(t) = C_1 + C_2e^{-\sigma_{235}(a)\theta} - (C_1 + C_2)e^{-\sigma_{239}(a)\gamma\theta}, \qquad (14.7)$$

where

$$C_1 = \frac{N^{238}(a)\sigma_{238}(a)}{\sigma_{239}(a)\gamma},$$

$$C_2 = \frac{N^{235}(0)\sigma_{235}(a)\eta_{235}\epsilon p_1(1 - p)}{\sigma_{239}(a)\gamma - \sigma_{235}(a)},$$

and

$$\gamma = 1 - \eta_{239}\epsilon p_1(1 - p).$$

The concentration of higher isotopes of plutonium i.e. Pu-240, Pu-241, etc. can be worked out by solving the differential equations

$$\frac{dN^{240}}{dt} = N^{239}\sigma_{239}(c)\phi(t) - N^{240}\sigma_{240}(a)\phi(t)$$

and

$$\frac{dN^{241}}{dt} = N^{240}\sigma_{240}(c)\phi(t) - N^{241}\sigma_{241}(a)\phi(t).$$

The rates of formation of the fission products from U-235, Pu-239 and Pu-241 are given by

$$\frac{dN_f^{235}}{dt} = N^{235}\sigma_{235}(f)\phi,$$

$$\frac{dN_f^{239}}{dt} = N^{239}\sigma_{239}(f)\phi,$$

and

$$\frac{dN_f^{241}}{dt.} = N^{241}\sigma_{241}(f)\phi. \tag{14.8}$$

The total increase in the concentration of the total fission products due to all these fissionable materials may be written as

$$\frac{dN_f}{dt} = \left(\frac{N^{235}\sigma_{235}(a)}{1 + \alpha_{235}} + \frac{N^{239}\sigma_{239}(a)}{1 + \alpha_{239}} + \frac{N^{241}\sigma_{241}(a)}{1 + \alpha_{241}}\right). \tag{14.9}$$

This could again be solved by applying the appropriate conditions. Note that no burnout term is included and the products of the fast fission effects of U-238 are ignored.

b) Handling of Spent Fuel Elements

At the time of removal from a reactor core the spent fuel elements are very hot and at the same time highly radioactive. The separation and chemical reprocessing plants designed for the treatment of such elements must take into consideration the handling and treatment of hot and intensely radioactive fuel elements. The temperature and level of radioactivity of the fuel elements are functions of the fuel burnup and the cooling time.

The fission products produced in the fuel elements cover such a wide range in the periodic table that it would be extremely difficult to analyze their formation and decay schemes in the reactor. Approximate statistical relations provide simple equations sufficient for engineering applications. These relations giving the rates of emission of beta and gamma rays, i.e. N_β and N_γ, can be used from a time of 10 s up to a time of several weeks after the fission reaction has ceased to exist.

$$N_\beta = 3.8 \times 10^{-6} t^{-1.2} \text{ particles/s fission} \qquad (14.10a)$$

and

$$N_\gamma = 1.9 \times 10^{-6} t^{-1.2} \text{ photons/s fission,} \qquad (14.10b)$$

where t is the time in days after the fission reaction has stopped. As the mean energy of the beta particles and gamma rays from the fission products is around 0.4 and 0.7 MeV respectively, the rate of emission of the total energy due to the release of beta and gamma radiations is given by

$$E_{\beta,\gamma} = 2.8 \times 10^{-6} t^{-1.2} \text{ MeV/s fission.} \qquad (14.11)$$

Now consider a more practical case in which a reactor has been operating at a steady power level for a time T_1. During the steady power operation assume that the fission rate was R_f. The reactor is shutdown at the time T_1 and it is desired to evaluate the rate of energy emission T_2 days after the reactor shutdown. In order to calculate $E_{\beta,\gamma}$, consider Figure 14.2. As the number of

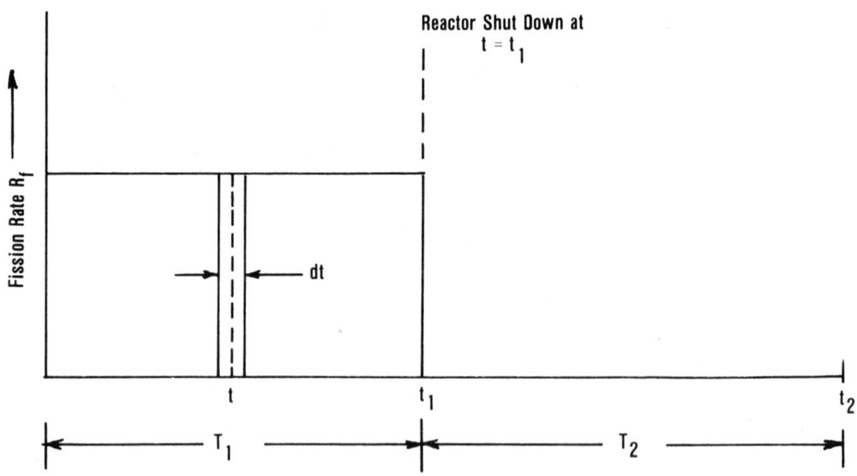

Fig. 14.2: Energy release as a function of reactor operation and shutdown time.

fissions produced in an infinitesimal time interval dt at any time t is $R_f dt$, the beta-gamma energy released by all the fission products formed in the time dt at time t_2 is given by

$$dE_{\beta,\gamma} = 2.8 \times 10^{-6}(t_2 - t)^{-1.2}R_f dt. \qquad (14.12)$$

The total rate of energy emission per unit time due to reactor operation from $t = 0$ to $t = t_1$, observed after a cooling period of T_2 days can therefore be written as

$$E_{\beta,\gamma} = \int_0^{t_1} dE = 2.8 \times 10^{-6} \times Rf \int_0^{t_1} (t_2 - t)^{-1.2}dt. \qquad (14.13)$$

Suppose the reactor has been operating at a steady power level of P_0 watts during the time intervai dt. Since 2.68×10^{15} fissions per day yield one watt of power, the fission rate R_f is given by $2.68 \times 10^{15}P_0$. The expression for $E_{\beta,\gamma}$ hence becomes

$$E_{\beta,\gamma} = 7.5 \times 10^9 P_0 \int_0^{t_1} (t_2 - t)^{-1.2}dt$$

or

$$E_{\beta,\gamma} = 3.8 \times 10^{10}P_0 [T_2^{-0.2} - (T_1 + T_2)^{-0.2}] \text{ MeV/s}. \qquad (14.14)$$

Some other relations for handling spent fuel elements are given here:

$$\frac{P}{P_0} = 6.1 \times 10^{-3}[T_2^{-0.2} - (T_1 + T_2)^{-0.2}], \qquad (14.15)$$

where P is the power at time t after the reactor shutdown.

$$A_\beta = 1.4 P_0 [T_2^{-0.2} - (T_1 + T_2)^{-0.2}] \text{ Curies}, \qquad (14.16)$$

where A_β is the activity due to the beta particles. This is based on the assumption that one beta particle corresponds to one nuclear disintegration.

By comparing the expressions for the rates of emission of beta and gamma rays, it is clear that the gamma activity is half that of the beta activity. It is obvious from the relations of beta and gamma activities that if a reactor has been operating at a power level of the order of Megawatts, the total activity will be of the order of Megacuries soon after the reactor is shut down. For this very reason, the spent fuel from nuclear power reactors is cooled in shielded storage facilities for 60 to 120 days to allow the short lived radioactive fission products to die out.

14.4 In-core Analysis Plant Support

Detailed reactor analysis for fuel cycle management, safety analysis, quality assurance, fuel reload specification, and licensing is usually a major in-house task of a utility away from the nuclear plant. Computer programs are organized to simulate a reactor operation. This in-house reactor simulation capability, in parallel with vendor design and recommendation, is established by

1. the selection of suitable computer codes for calculating group constants, reactor power and reactivity distributions, fuel exposure, and other parameters,
2. the determination of appropriate input and output mechanisms and parameters for the chosen codes, and
3. the final benchmarking of the reactor simulation model.

Figure 14.3 illustrates the modeling calculations required to benchmark the system for the reactor design review and operation support. There are many computer codes, transport theory and diffusion theory codes, such as PDQ, CPM, CASMO, SIMULATE, RECORD, and PRESTO etc. which are capable of calculating reactor power and reactivity distributions. However, for modeling a specific reactor, detailed developmental work as shown in Figure 14.3 must be conducted to organize the chosen codes for benchmarking.

Specific reactor model capabilities must be checked against available data. Axial and radial power distributions, reactivity (K_e and K_∞), thermal limits, and exposure are some of the comparison items. In general, a set of design tolerances is established to judge the modeling capability. One example of a set of design tolerances is:

power miscalculation $\leqslant$ 10%.
K_e miscalculation $\leqslant$ 0.3%.

Example 14.1

The fuel burnup of a power reactor can be expressed in Megawatt days per metric ton of fuel (MWD/T), percent atom burnup, or fissions per cubic centimeter. If natural uranium is used as fuel, find the conversion factors for these units.

Solution

Assuming that 200 MeV is released per fission,

$$1 \text{ MWD/T} = (10^6\text{w})(24 \times 60 \times 60 \text{ s})/\, 10^6\text{g of fuel}$$
$$= 8.64 \times 10^4\text{J/g of fuel} = 2.7 \times 10^{15} \text{ fissions/g of fuel}$$
$$= 0.45 \times 10^{-6} \text{ atom \%.}$$

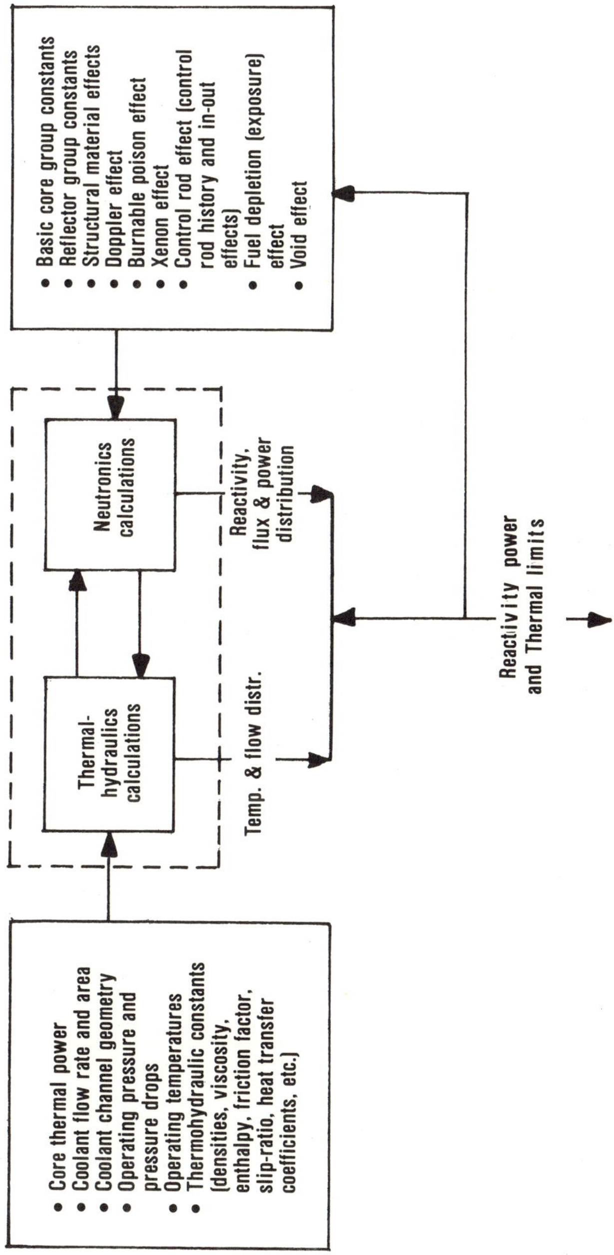

Fig. 14.3: Reactor modeling

$$= \frac{8.64 \times 10^4}{200 \times 10^6 \times 1.6 \times 10^{-19}} \text{ fissions/g of fuel}$$

$$= 2.7 \times 10^{15} \text{ fissions/g of fuel.}$$

One gram of fuel contains $\frac{6.023 \times 10^{23}}{A}$ atoms, where A is the atomic mass,

$$\therefore 1 \text{ MWD/T} = \frac{2.7 \times 10^{15} A}{6.023 \times 10^{23}} \times 100 \text{ atom \%}$$

$$= 0.45 \times 10^{-6} \text{ atom \%.}$$

Also, one gram of fuel occupies a volume of $1/\rho \text{ cm}^3$ if the density is in g/cm^3.

$$\therefore 1 \text{ MWD/T} = 2.7 \times 10^{15} \text{ fissions}/\ 1/\rho \text{ cm}^3 = 2.7 \times 10^{15} \rho \text{ fissions/cm}^3.$$

Example 14.2

Calculate the energy released in MeV/s in a reactor which has been operating continuously for 100 days at 100 MW(th) power, 10 days after the reactor shutdown.

Solution

$$E = 3.8 \times 10^{10} P_0 [T_2^{-0.2} - (T_1 + T_2)^{-0.2}] \text{ MeV/s.}$$

Now $P_0 = 10^8$ W.

$T_1 = 100$ d.

$T_2 = 10$ d.

$$\therefore E_{\beta,\gamma} = 3.8 \times 10^{10} \times 10^8 [(10)^{-0.2} - (110)^{-0.2}] \text{ MeV/s}$$

$$= 9.1 \times 10^{17} \text{ MeV/s.}$$

Example 14.3

A reactor is fueled with slightly enriched uranium with 2 percent U-235 and 98 percent U-238. Assuming a constant flux of 5×10^{13} neutrons/cm^2s, calculate the following:

a) How many days are required to bring the U-235 concentration down to 1.5 percent?

b) At the end of this interval what is the plutonium concentration of the fuel?

Solution

a) $N^{235}(t) = N_0^{235} e^{-\phi\sigma_{235}(a)\,t}$

$0.015 = 0.02 \text{ Exp } (-5 \times 10^{13} \times 590 \times 10^{-24} t).$

$\therefore t \cong 113$ d.

b) The plutonium concentration of the fuel is governed by the following relationship:

$$\frac{dN^{239}}{dt} = \sigma_{238}(a)\phi N^{238} - \sigma_{239}(a)\phi N^{239},$$

where $N^{238}(t)$ and $N^{239}(t)$ are the U-238 and Pu-239 concentrations at any time t.

$$N^{238}(t) = N_0^{238} e^{-\phi\sigma_{238}t}.$$

$$\therefore N^{239}(t) = \frac{\sigma_{238}\, N_0^{238}}{\sigma_{239} - \sigma_{238}} (e^{-\phi\sigma_{238}t} - e^{-\phi\sigma_{239}t}).$$

When t= 113 d, it can be shown that

$N^{239}(113 \text{ d}) \cong 0.00136.$

Problems

14.1 A reactor has a steady state neutron flux density $\phi = 10^{15} n/cm^2 s$. Calculate the fractional fuel atom burnup as a function of time. If the fuel is 2 percent enriched U-235, examine the fuel composition after the reactor has operated for 100 days.

14.2 Calculate the rate of the beta particle emission from fission products due to a single fission process ten seconds after the fission. What is this rate ten hours later?

14.3 Calculate the power from fission products due to beta and gamma emissions per fission one hour after the occurrence of one fission process.

14.4 What is the total $(\beta + \gamma)$ energy release from a fission process, if equation (14.11) is used for the calculation? Discuss your result.

14.5 Calculate the energy release due to the fission of one gram of U-235. Express your answer in MWD/g of U.

Ans: 0.95 MWD/g of U.

14.6 The KANUPP power reactor (CANDU-type), which uses natural uranium fuel with D_2O as moderator, has been operating for 100 days at an average neutron flux density $\phi = 3 \times 10^{14} n/cm^2 s$. The fuel burnup is 3700 MWD/T. What is the fuel burnup in atom %. Explain your result.

14.7 Calculate the burnup of the U-235 mass per day required for producing one MWth power.

Ans: 1 g/day.

14.8 In power reactor operations, an upper limit of one percent atom burnup is frequently used for fuel discharge. This criterion is based on actual operational experience which indicates that this value corresponds to the point at which fuel material failure might occur. Calculate the annual fuel discharge of a 100 MWe reactor with 150 tons of UO_2 fuel.

14.9 A three percent enriched UO_2 pellet 1.5 cm in thickness and 1 cm in diameter was placed in an average thermal flux of 10^{14} $n/cm^2 s$ for a period of 6 months. Calculate the following, giving all the simplifying assumptions used:

a) Burnup of the sample in MWD/T

b) Amount of U-235 left in the sample, in grams, after irradiation

c) Amount of Pu-239 produced in the sample after irradiation.

14.10 The Department of transportation has established a limit of 30 mrem as the maximum permissible radiation dose received by a truck driver per spent fuel or waste shipment. Is this limit consistent with the general guideline described by the formula DE = 5(A - 18)?

CHAPTER 15

POWER PLANT PLANNING

15.1 Planning and Licensing

The demand for electric power increases with time; an average increase of about 5% per year in electric power production has been maintained in recent years. Utility companies must take steps to meet such increases. At present, they can choose between fossil or nuclear power plants. After a detailed analysis of economical, environmental, social and political, as well as other factors, if the decision is to go nuclear, the utility company has to apply for construction and operation permits from the Nuclear Regulatory Commission. Figure 15.1 described the major steps to be taken to reach the final stage of the power operation of a nuclear plant.

In the USA the U.S. Atomic Energy Commission was responsible for promoting the development of atomic energy as well as for regulating it until 1975. In January of that year the USAEC was disbanded, the development and promotional activities being transferred to the new Energy Research and Development Agency (ERDA), while an independent Nuclear Regulatory Commission (NRC) was established to carry out the regulatory and licensing functions of the federal government in nuclear and radioisotope applications. ERDA has since been reorganized into the Department of Energy.

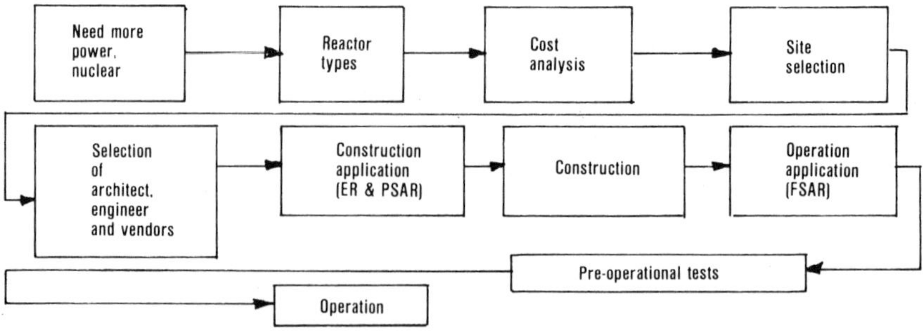

Fig. 15.1: Nuclear Power Plant Planning.

The procedure for the licensing of a nuclear power plant may be divided into two stages. The first stage involves the collection of adequate technical and environmental data, carrying out economic and social analysis by the utility company, and ends with the issuing of a construction permit by the NRC. In the second stage of licensing, the applicant starts the construction of the plant, and prepares to secure an operating license.

An applicant for a power plant construction permit must file applications to the NRC under the specifications of a class 103 license which is issued to those who qualify to transfer or receive in interstate commerce, manufacture, produce, transfer, acquire, process, use, import, or export under the terms of an agreement for cooperation, a production or utilization facility for industrial or commercial purposes (as stated in the 10CFR-energy 1974). Nine to 36 months prior to the application of the construction permit, the applicant must present detailed analysis on power requirements, capacities and other pertinent data and information to the NRC and the U.S. Attorney General's Office for anti-trust review and hearings. The application for a class 103 license must be followed, within six months, by the submission of a preliminary safety analysis report (PSAR). In the PSAR, technical information on site description and safety analysis, facility description, facility design, quality control and assurance programs, and emergency measures are presented. A detailed list of the contents can be found in part 50.34 of 10 CRF-energy 1974. Furthermore, an environmental report (ER) must also be submitted. The application form, the PSAR and the ER, complete the requirements for a construction permit application. The NRC then proceeds with the staff review and public hearings which are conducted by the Atomic Safety and Licensing Board (ASLB). At the same time, the Advisory Committee on Reactor Safeguards (ACRS) carries out its own initial review of applications. With the successful conclusion of these applications, review, and hearings, the NRC then issues a construction permit to the applicant after a further review by the Atomic Safety and Licensing Appeal Board (ASLAB).

Upon completion of the power plant construction and component testing, the applicant must file the application for an operating license. The first step is the submission of a final safety analysis report (FSAR) as prescribed in part 50.34 (b) of 10 CFR-energy. The only difference between the FSAR and the PSAR is that in the FSAR all descriptions of the facilities and the components pertain to what they actually will be when in operation instead of the original designs. The NRC reviews and hearings follow. The operating license is issued when the Commission finds that:

(1) Construction of the facility has been substantially completed in conformity with the contruction permit and the application as amended.

(2) The facility will operate in conformity with the amended application.

(3) There is reasonable assurance that the activities authorized by the operating license can be conducted without endangering the health and safety of the public.

(4) The applicant is technically and financially qualified to engage in the activities authorized by the operating license.

(5) The provisions of Part 140 of 10-CFR pertaining to the applicant have been satisfied.

(6) The issuance of the license will not be inimical to the common defense and security.

15.2 Power Reactor Types

A nuclear reactor may be defined as an assembly of nuclear fuel and other materials in which a chain reaction can be maintained. Adequate controls and safeguards are provided in a reactor to protect the operators and workers as well as the surrounding population against any kind of accident and subsequent spread of radioactivity. To extract the fission energy for the large scale production of thermal and electrical energy, reactors are provided with liquid or gaseous coolants.

A nuclear reactor designed for large scale electric power production is equivalent to the coal or oil burning furnace of existing conventional power plants. There are differences between nuclear power plants and conventional plants in fuel, basic reactions involved, and control mechanism design and safety considerations, but the function of a nuclear reactor is exactly the same as that of the coal or oil fired furnace. The essential features of conventional and nuclear power plants are shown in Figures 15.2 and 15.3.

Nuclear reactors may be classified according to their applications as (a) central station power reactors, (b) package power reactors (with power in the range of 20 kWe to 10 MWe), (c) production reactors (for isotope productions), (d) test reactors, (e) desalination reactors, and (f) research reactors.

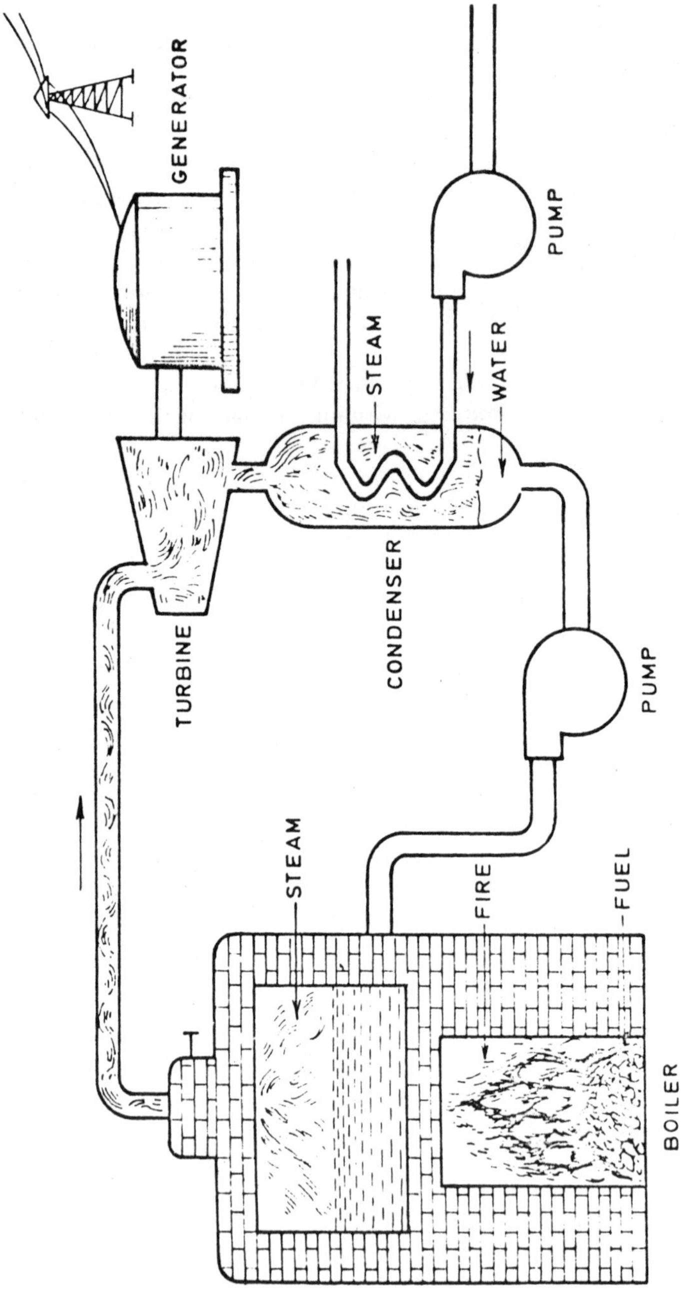

Fig. 15.2: Coal-fired power plant.

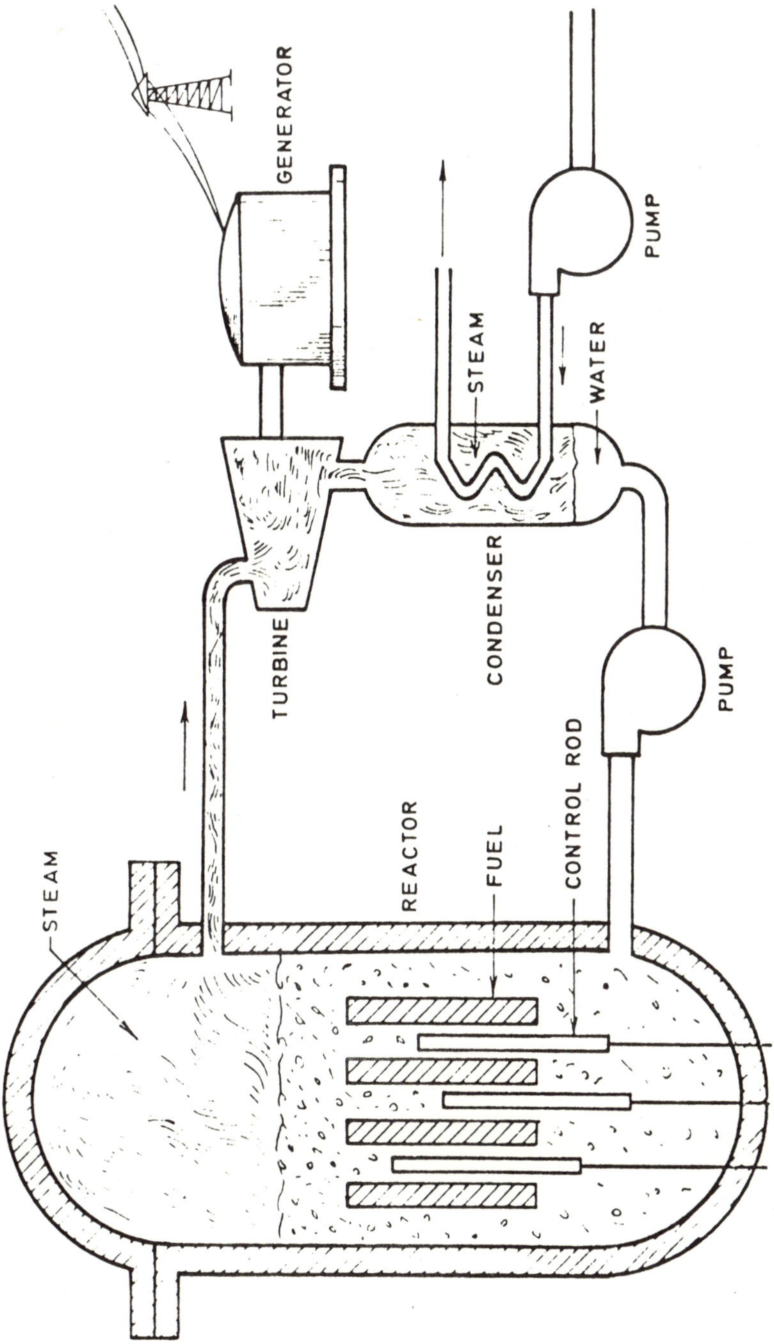

Fig. 15.3: Nuclear power plant.

Central station power reactors are generally designed to supply electrical energy to the general public. The size of the reactor depends upon the local requirements and is used to supply high quality steam or gas to turbines for the generation of electricity. A detailed description of the various types of commercially available nuclear reactors used for this purpose is presented in the following sections. At present there are several types of power reactors in commercial operation: high temperature gas-cooled reactors (HTGR), pressurized water reactors (PWR), boiling water reactos (BWR), and Candu-type reactors.

The HTGR is a further evolution of the gas-cooled reactor concept, yielding higher temperatures and very high burnup values. Helium is used as a coolant and the fuel is a mixture of highly enriched coated uranium dicarbide and thorium dicarbide particles dispersed uniformly in a graphite matrix. The fuel in Fort St. Vrain HTGR in the USA contains 100 micron UC_2 particles and 400 micron ThC_2 particles with four coatings of porous carbon, pyrolytic carbon, silicon carbide and pyrolytic carbon, successively. The function of this multilayer particle coating is to reduce the fission gas release and the retention of metallic fission products. These particles are dispersed in graphite and formed into two inch long cylindrical rods of about 0.5 inches in diameter. These rods are inserted into holes drilled in the hexagonal moderator blocks.

Due to the special fuel design, extremely high burnup values are predicted for this class of gas-cooled reactors. The predicted values are around 100,000 MWD per ton (70,000 MWD per ton in fully enriched uranium and 40,000 MWD per ton in the fertile fuel particles). Such values of burnup have made this type of reactor very attractive and numerous countries are now developing such systems. Two types of fuel cycles, known as high enrichment and low enrichment schemes, are possible in HTGR's. In the first case, highly enriched U-235 fuel is used along with thorium dicarbide particles while the second type utilizes 4 to 5 percent enrished U-235 particles. In these fuel cycles, Th-232 or U-238 convert into U-233 and Pu-239 respectively and contribute very significantly towards power production along with the initially present fissionable material. In Figure 15.4 a simple diagram of a typical HTGR is given.

In a PWR the entire core is enclosed in a vessel capable of withstanding the full system pressure. This results in the requirement of large, very thick-walled reactor vessels. As a matter of fact, these pressure vessels have become the principal distinguishing feature of such reactors. Fabrication of a large thickness steel pressure vessel is a very complex problem. The quality control problems of these vesels are such that fuel fabrication is difficult, and so far only shop-fabrication has been attempted. The overland transportation of large pressure vessels is quite difficult. A PWR of around 600 MWe would require a pressure vessel weighing around 500 tons and ranging from 15 to 20

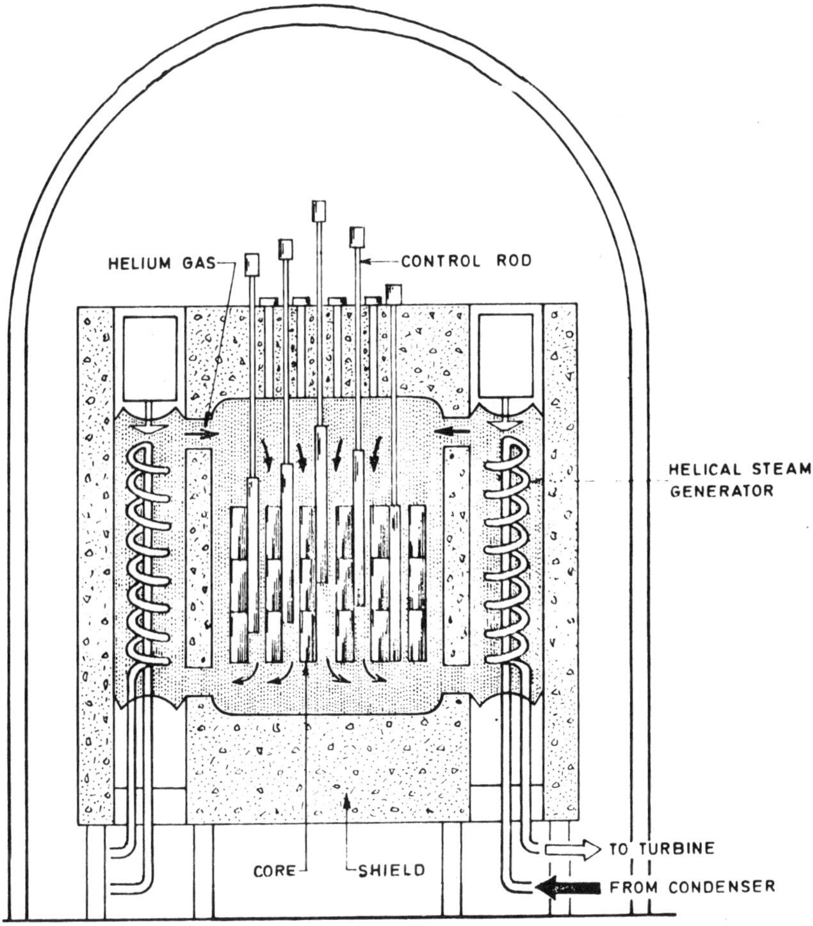

Fig. 15.4: A high temperature gas-cooled reactor

feet in diameter with wall thicknesses from 8 to 15 inches!

The fuel used in PWRs is uranium-235 having an enrichment of 2 to 4 percent. The material is compacted into short cylindrical pellets around one-half inch in diameter and are loaded into close fitting thin-walled cladding tubes. The cladding material in early PWRs was stainless steel but now zircaloy is generally used. The zircaloy tubes are bound together in bundles. A typical bundle contains 150 to 200 fuel rods generally arranged in a square pattern. The rods in each bundle are held apart so that cooling water can flow along them and remove the heat produced due to the fission reactions. All of these fuel bundles, ranging generally from 10 to 12 feet in height, are positioned so that they form a right cylinder having a diameter matching its height. Spaces are provided for the insertion of control rods of any desired shape, i.e., cruciform, slab or pencil. The commonly used control materials are hafnium, silver-indium-cadmium alloy or boron compounds. Chemical shim (i.e. soluble poison) and burnable poison techniques are used in most of the PWRs in order to obtain a longer core life and smaller excess reactivity in the fresh core.

The coolant is circulated through a number of primary loops containing steam generators and pumps. A pressurizer is used to maintain the system pressure by the electrical heating or spray cooling of water in a small pressure chamber. The steam generators commonly used in PWRs are of the U-tube type with an arrangement for feed water recirculation. Steam separators and dryers are situated on the top of each unit. Fueling of the reactor is carried out at an interval of 12 to 18 months, including the reactor downtime. In order to attain uniform and higher values of fuel burnup, multi-region loading has been introduced in PWRs. The core is generally divided into three regions, each carrying an equal number of fuel bundles but different enrichments. This arrangement also results in reducing the hot channel factors. The fuel burnup in these reactors is generally over 30,000 MWD per ton. Figure 15.5 shows the main features of PWRs with steam generator and safety containment barriers.

The initial concept of BWR was to allow boiling to take place in the core region of water cooled reactors. It was found that controlled boiling could be achieved in a self-stabilizing condition at around half the system pressure of a PWR. The lower operating pressure for a given steam temperature, higher attainable thermal efficiency, decrease in capital cost due to a less elaborate pressure vessel and elimination of a separate steam generator system are attractive features that have resulted in the adoption of BWRs for large scale commercial power production in many countries.

The fuel technology of BWRs is similar to that of the PWRs. Reactor control is achieved with cruciform rods inserted hydraulically between the fuel assemblies from below the core and also by variation of the recirculation flow rate. The control problems of BWRs are, however, comparatively more difficult. The soluble poison method is not used because it is hard to control the concentration of the poison in a coolant which is undergoing a continuous change of phase. This necessitates a large number of control rods to suppress the excess reactivity. The core of a BWR is not as compact as that of a PWR due to jet pumps and steam dryers within the core; this results in a larger but thinner pressure vessel. The main difference is that the bulk boiling of water is allowed in the BWRs. The reactor acts as a recirculation boiler with steam separators and dryers situated in the top section of the reactor pressure vessel. The recirculation flow of water from the steam separators and feedwater returning from the turbine condenser is usually achieved with jet pumps situated around the reactor core and driven by smaller external pumps. In BWRs, a special type of pressure suppression containment is generally

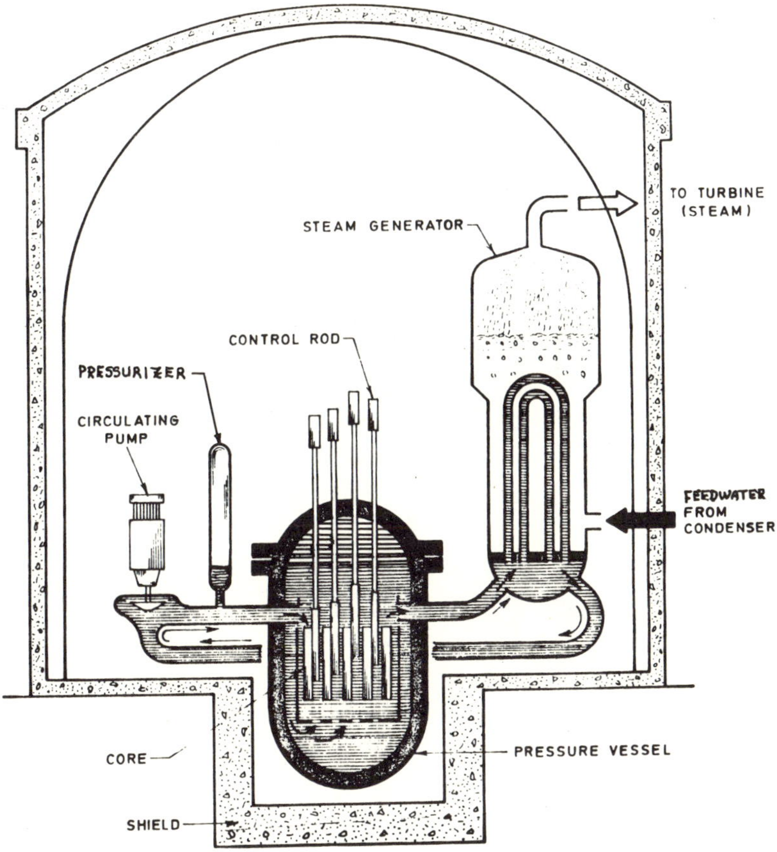

Fig. 15.5: A pressurized water reactor showing one of several (usually three) sets of steam generator, pressurizer, and coolant circulating pump.

adopted in order to deal with the consequences of a primary circuit rupture. A pressure relief system is arranged around the reactor to vent steam in a reservoir of water which acts as an emergency condenser. Figure 15.6 shows the overall features of a BWR.

Numerous types of heavy water moderated reactor systems have been proposed and several different concepts have been developed to the prototype or demonstration plant stage. The only heavy water reactor operating on a commercial scale is, however, the CANDU type, developed and brought to the present state of technology by Canada. Heavy water reactors derive their

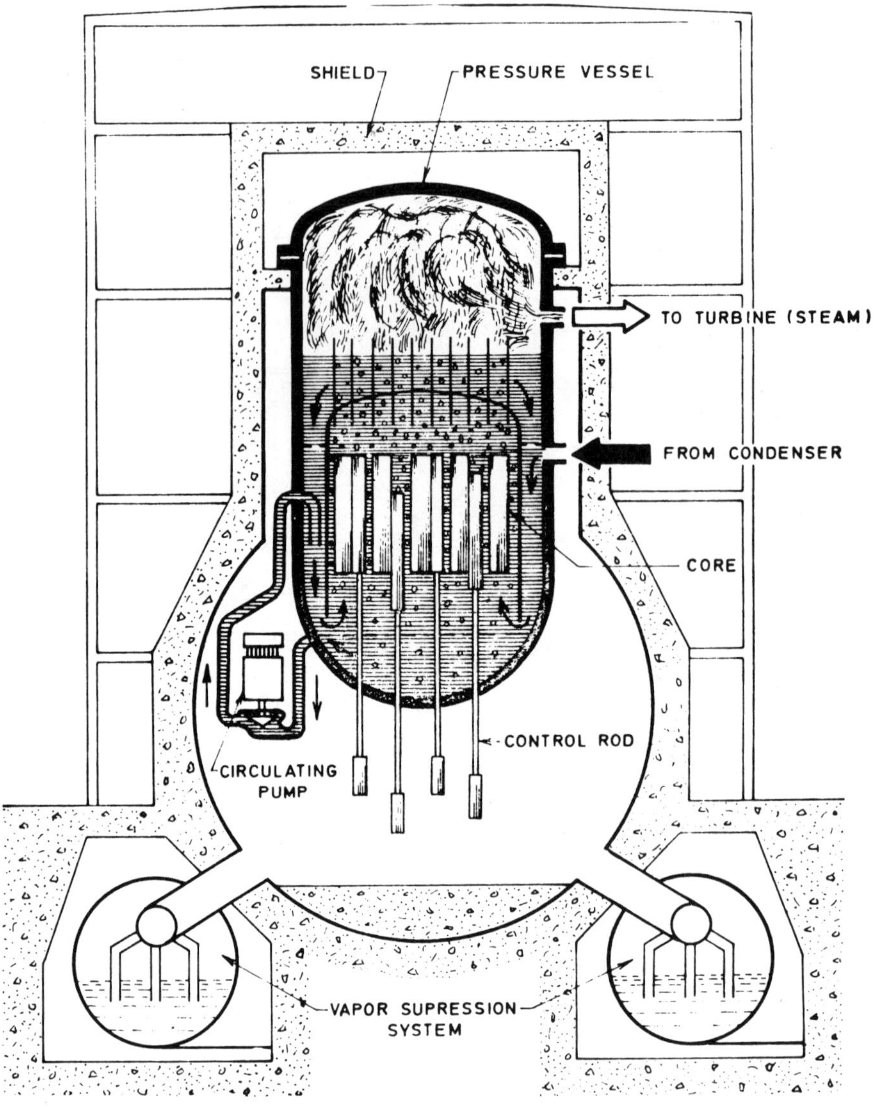

Fig. 15.6: A boiling water reactor.

main advantage over other systems from the excellent fuel economy of the heavy water moderator. Since only a small percentage of neutrons is lost during moderation, considerably less of the uranium-235 isotope is needed per unit of power output. Besides, the use of natural uranium fuel allows a simple fuel cycle as reprocessing of the spent fuel is not really essential.

The fuel rods used in these reactors consist of natural UO_2 pellets in zircaloy tubes. Relatively short fuel bundles, consisting of 19 to 28 fuel rods, are used. The fuel is loaded into horizontal pressure tubes which pass through a large horizontal cylinder known as the Calandria. Heavy water is used as the moderator and coolant. The heavy water moderator is maintained at a lower temperature relative to that of the coolant to improve the neutron economy. The use of pressure tubes for the fuel in the reactor core allows the coolant system to be pressurized without the need for massive pressure vessels required by PWRs. The steam generators are generally of the U-tube type, similar to those used in PWRs. Use of natural uranium requires on-power fueling. This is done with the help of two remotely controlled machines at each end of the Calandria to insert and withdraw the fuel bundles. Control is accomplished with control rods and by varying the moderator level. An additional method of achieving quick shutdown is provided by a dump tank for the moderator below the Calandria. Enriched uranium booster rods are generally provided to override the xenon-poisoning effects and restart the reactor soon after shutdown. Figure 15.7 shows a typical CANDU reactor system.

15.3 Nuclear Power Cost Analysis

For about 15 years following the first nuclear chain reaction nuclear reactors were considered to be too expensive to play any significant role in the production of electric energy. This viewpoint changed after the mid-fifties when successful operations of power reactors in the UK, USA and USSR demonstrated the scope of nuclear power as a reliable and large scale alternate source of energy. After the mid-sixties, when large-sized nuclear power reactors became available, the balance further shifted in favor of nuclear energy. After the energy crisis in the early-seventies, the picture changed altogether and nuclear power became economical in practically all the regions of the world. Many countries that have abundant fossil fuel reserves have now initiated large scale nuclear power programs in order to save their reserves as raw materials for petrochemical industries.

Cost estimates for a nuclear power plant in the design stage or under construction inevitably involve many approximations and uncertainties. Only after the plant has been commissioned and has been in operation for a few years is it possible to accurately calculate the power costs based on actual data. Every cost estimate should therefore indicate the reference data upon which it is based. Preliminary estimates during the design and/or construction stage, though not very accurate, are useful in project planning.

The calculation of the cost of generating electrical energy from nuclear

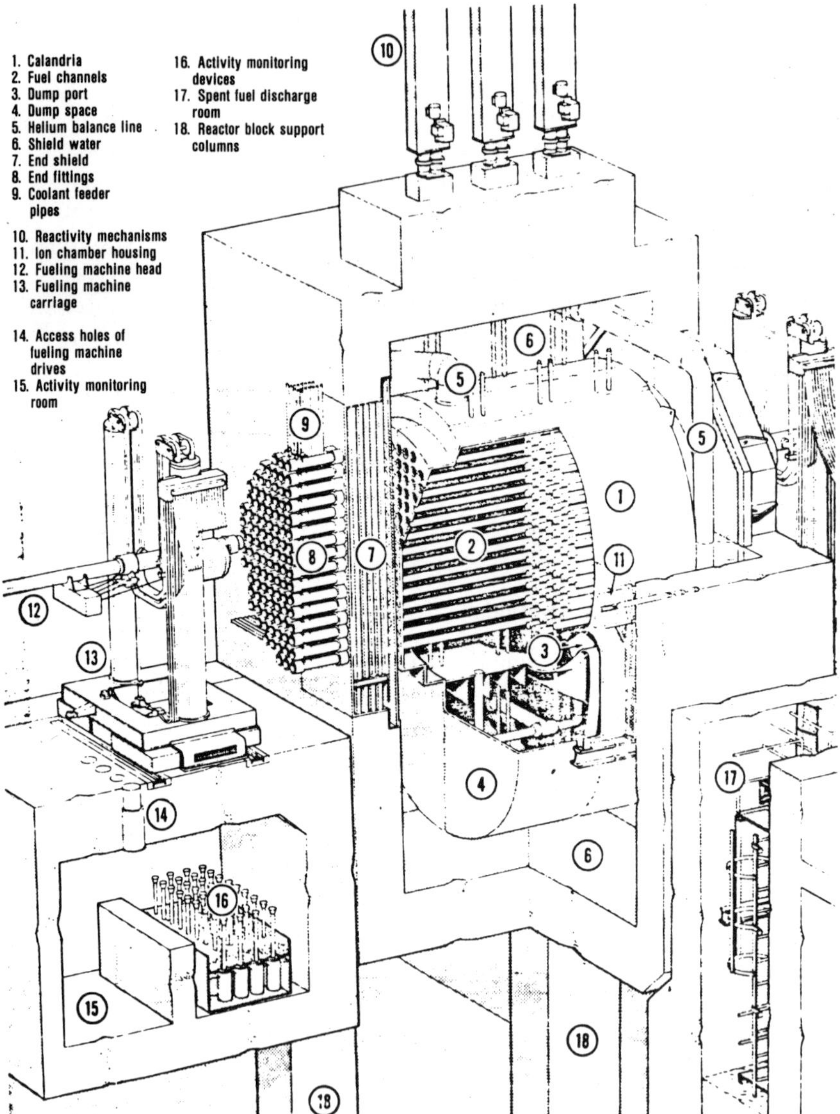

Fig. 15.7: Cutaway of CANDU reactor.

reactors is quite similar to that of conventional power plants. In both cases the unit electrical energy cost is determined on the basis of costs for the construction, fueling, and operation of the plant. Although the cost in both types of power plants is grouped under similar broad headings, the unique features of nuclear fuel, complexity of reactor types and uncertainty in the reactor life make nuclear power cost estimation a relatively complex problem. To some extent the estimates are affected not only by physical and economic factors but also by the degree of optimism or pessimism of the person estimating the nuclear power costs. While comparing the nuclear energy costs with conventional energy costs, the area of uncertainty further widens as the assumed prices for conventional and nuclear fuels may not hold for many years ahead. In spite of all these uncertainties in the estimation of nuclear energy costs, it is now well established that in practically all the regions of the world nuclear power plants are economically attractive. The low fuel cycle cost for nuclear power has now more than compensated for its drawbacks of higher capital cost.

15.4 Factors Constituting Nuclear Power Costs

The unit energy of electricity cost from a power plant may be broken up into three components: (1) the fixed or capital cost; (2) the fuel cycle cost and (3) the operation and maintenance cost.

(a) Fixed or Capital Cost

The fixed cost generally constitutes the major item in the cost of nuclear energy generation. For a given type of nuclear power plant the capital cost varies significantly with the nature and location of the site. There are great differences among various countries with respect to labor rates, labor efficiencies, equipment costs and building materials. The cost of land and the safety requirements of exclusion areas also vary from one region to another. The varieties of exchange rates in the controlled currency countries, various design practices, and numerous other factors can have a major effect on the cost of building a reactor in one country as compared to another. The various factors which constitute and influence the capital cost are as follows:

i) Cost of preliminary studies and project report: This involves various factors such as the economics of various sources of energy, engineering design, availability of sufficient coolng water, survey of possible sites for reactor installation, facilities of waste management, etc. This cost is relatively high for the first nuclear station in any country, particularly for those countries where local trained manpower is limited.

ii) Cost of land and site development: Included in this item are the cost of the land required to accomodate the nuclear power plant and the fuel storage facilities as well as the cost of drainage, leveling, site improvement, electrical and water supply during construction, access roads to sites, and temporary buildings needed for site improvement works.

iii) Cost of installation of the nuclear power plant and engineering supervision: These costs are normally covered by a contract for construction by some experienced firm and may be divided into three main categories, called the nuclear island or the nuclear steam supply system (NSSS), the turbine-generator, and engineering design and supervision.

iv) Miscellaneous contingencies: Many other expenses may arise during the course of construction of a nuclear power plant due to extra work and unforeseen difficulties. Delays in the arrival of equipment, discovery of new subsoil conditions and the corresponding modifications of the civil engineering problems associated with the foundation work and the replacement of some of the equipment during transportation and storage are some of the factors which should be taken into consideration.

v) Interest rate and interest during construction: Since the capital investments in nuclear power plants are higher than those for coal or oil-fired power plants, a higher interest rate affects the cost economics of nuclear power much more adversely than conventional power plants. Besides, as a large amount of capital is unproductively tied-up during the construction period of a nuclear plant, it is important that this period is not very prolonged. The interest during this period can be calculated using a representative expenditure model.

vi) Depreciation rates: Depreciation is a cost item that takes into account the gradual decline in the remaining physical life of the plant. The plant physical life ends at the time when the operation of the plant is no longer economically or technically justified. Presently, there is insufficient experience to determine the precise useful life of a nuclear power plant, but the value adopted by most of the utility companies is between 25 to 30 years. The method universally used in economic studies to calculate plant depreciation is called the sinking fund method. In this method, a fixed amount of capital is invested every year in securities which yield interest and can be readily converted into cash. The amount of cash accumulated in this way is just sufficient to replace the plant value completely when its life is over. Some details of this method are given later in this chapter.

vii) Stretch capability: A factor affecting the fixed cost in a nuclear power plant is the kilowatt capacity of the plant and the possibility of increasing it after some running experience through what is generally known as the 'stretch capability'. This capability can significantly reduce the nuclear power cost. It is estimated that the maximum potential stretch of a nuclear boiler is around 15 percent of its rated capability. If the rest of the plant was initially sized to accomodate the potential stretch, then the capability of the whole power plant would evidently become about 15 percent greater.

viii) Plant factor: This is the ratio of actual kilowatt-hours of electricity produced to the kilowatt-hours that could be produced if the plant ran all the time at full rated power. As the capital or the fixed cost component for nuclear energy is much higher as compared to an equivalent fossil fuel-fired plant, the nuclear energy cost is very sensitive to the plant factor. In order to estimate this factor accurately, numerous studies on the technical capability of the plant for energy production, the demand of electrical energy in that region and the relative operating and transmission costs of nuclear and conventional power plants in the same region need to be carried out. It is difficult to predict precisely the variation of the plant factor during the assumed life of the plant. The value of the plant factor also depends on the nature of the future energy demands as well as the nature of additional power plants which might be added into the already existing grid system. While any new plant in an interconnected system should operate at the base load at the beginning of its life, a nuclear power plant having relatively higher capital cost and much lower production cost, should remain on the base load for a longer period than a conventional plant.

ix) Escalation: Due to extremely high inflation rates, costs of all items are changing rapidly. As the time required for designing, licensing and the construction of a nuclear plant runs into many years, an escalation factor should be included for all cost evaluations.

(b) Fuel Cycle Cost

Nuclear power plants have an advantage over the conventional power plants because of the relatively low fuel cycle cost. Whenever the savings in the fuel cycle cost of a nuclear power plant become large enough to offset its higher capital cost, nuclear energy becomes competitive with other energy sources. The fuel cycle cost has two components: (1) the variable cost resulting from the consumption of the fuel, and (2) the fixed cost resulting from the immobilization of the capital necessary for acquiring the fuel.

Preparation of fabricated fuel elements ready for use in nuclear power plants involves many cost items. These include the cost conversion of uranium to uranium hexafluoride or uranium dioxide, enrichment of fuel, fabrication of these materials into the desired form of fuel elements, losses of materials during various processes, etc. The fuel element fabrication includes chemical, metallurgical, and mechanical processing, inspection, testing, scrap recovery, etc. The cost of fabrication per unit mass of fissionable material is affected by design, size, dimensional tolerances, alloying material, the number of fuel elements fabricated, enrichment of the fuel elements, etc. The fuel in the form of fabricated elements may either be purchased or leased in the case of enriched uranium. When the fuel is leased the payment for the use of the fuel is computed from the loss of enrichment suffered during the power production. The reprocessing cost of the irradiated fuel and credit for the new fissionable materials produced in the reactor constitute other cost elements especially for enriched uranium fuel. For the natural uranium systems the situation is slightly different. Generally speaking, the reprocessing of irradiated fuel in such systems is not very attractive economically.

Some other items of the fuel cycle cost are fixed charges on fuel inventory when the fuel is in transit, cooling and reserve, as well as the transport of the fuel to the reator and the fuel reprocessing plants. Any escalation of the uranium prices would result in increased fuel cycle cost. However, the generating costs of enriched and natural uranium reactors would be affected differently due to different changes in the fuel cycle costs of these reactor systems. A nuclear plant having higher thermal efficiency results in a lower fuel cycle cost.

(c) Operation and Maintenance

These costs include the operation and maintenance expenses for the plant as well as the charges for the make-up of special material and spare parts, insurance, taxes, and all other charges not included under the capital and fueling costs. Sometimes taxes and plant insurance are handled as a part of the capital charges. Although operation and maintenance constitute only about 10 to 15 percent of the total cost of electrical power generation, savings, if any, in the cost of power generated can be affected only in this area and in fueling once the capital has been invested in the construction of a nuclear power plant.

The operation and maintenance charges for nuclear plants are relatively higher than those of the equivalent conventional power plants. This is mainly because of a larger number of diverse skills required in fields like nuclear engineering and health physics as well as the higher salaries of the experienced and qualified workers. Besides, the professional staff for a nuclear power plant has to be appointed several years before the nuclear power plant goes critical. All these costs should be considered for the accurate determination of nuclear power generating costs.

Maintenance of nuclear power reactors requires much higher standards than those of a conventional plant. Various materials which are required for maintenance are lubricating oils, make-up coolant (if other than light water), chemicals, and spare parts for replacement. An annual operating allowance should be made for the replacement of spare parts such as pumps, valves, control rod tips, electronic equipment, etc.

Sample cost calculations for a 500 MWe enriched uranium reactor with a capital cost of 400 million dollars at 10% yearly interest are presented here. The plant life and plant factors are assumed to be 30 years and 80% respectively. In order to calculate the cost per kilowatt-hour of energy, it is necessary to relate the fixed cost, fuel cycle cost, and operation and maintenance cost to the total kilowatt-hours produced by the plant.

1. Capital Cost Component

For calculating the capital cost component of electricity, the sinking fund method is generally used. If R is the annual amount which accumulates yearly to replace the plant value, i is the annual compound interest rate, n is the useful life of the plant in years and C_t is the total capital cost after adjustment for interest during construction,

$$C_t = R[(1 + i)^{n-1} + (1 + i)^{n-2} + (1 + i)^{n-3} + ... + 1]$$

$$= R\left[\frac{(1 + i)^n - 1}{i}\right].$$

$$\therefore R = C_t\left[\frac{i}{(1 + i)^n - 1}\right]. \quad (15.1)$$

The quantity within the brackets in the last equation is called the sinking fund rate, s. In countries where the power supply system is government-owned, fixed charges on new investments do not usually include taxes or insurance and are given by what is known as the annuity rate or the capital recovery factor (CRF). This is defined as the fraction of the initial loan to be paid per period and is defined as

$$CRF = i + \frac{i}{(1 + i)^n - 1} = i + s. \quad (15.2)$$

If C_a represents the total annual cost,

$$C_a = C_t(i + s) = C_t(CRF). \quad (15.3)$$

For privately owned projects, additional expenditure incurred due to state and local taxes, insurance, etc., must be taken into account. The sum of the interest rate, sinking fund rate and all other expenditures mentioned above is generally known as total capital charge rate or more commonly as fixed charge rate (FCR) and sometimes as cost of money. The difference between FCR and CRF depends on the taxes and insurance rates.

The contribution of capital investment on the unit energy generation cost, U_c is given by:

$$U_c = C_t(i + s)/E, \quad (15.4)$$

where E is the total energy produced by the plant annually. If the total capital cost C_t is taken in mills and the energy E in kilowatt hours, then U_c is given in mills per kilowatt-hour. The unit cost of electricity in terms of capital cost per kilowatt installed, C_n, is given by

$$U_c = C_n(i + s)/h, \tag{15.5}$$

where h is the annual hours of full operation. The energy cost in mills per kilowatt-hour (kWH) can hence be written as

$$\text{mills/kWH} = \frac{(C_n\$/\text{kWe})(\text{FCR or CRF})(10^3\ \text{mills}/\$)}{(8760\ \text{hr/yr})(\text{annual plant factor})} \tag{15.6}$$

Now the capital cost in mills/kWH can be calculated as follows:

$$C_n = \frac{400 \times 10^6\$}{500 \times 10^3\ \text{kWe}} = 800\ \frac{\$}{\text{kWe}}\ .$$

$$\text{CRF} = 0.1 + \frac{0.1}{(1.1)^{30} - 1} = 0.1061$$

$$\therefore \text{Cost in mills/kWH} = \frac{800 \times 0.1061 \times 10^3}{8760 \times 0.8} = 12.1\ \text{mills/kWH}.$$

2. Fuel Cycle Cost

For calculating the fuel cycle cost it is usually convenient to first evaluate the cost in terms of dollars per kilogram of uranium and then convert it to mills per kilowatt-hour by the relation:

$$\text{Fuel cycle cost (mill/kWH)} = \frac{(\$/\text{kgU})(10^3\ \text{mills}/\$)}{(\text{MWD/kgU})(24000\ \text{kWH/MWD})\eta},$$

where η = plant efficiency = $\frac{\text{kWH (e)}}{\text{kWH (th)}}$ and $\frac{\text{MWD}}{\text{T}} = 10^3 \times \frac{\text{MWD}}{\text{kgU}}$.

The calculation of the cost of uranium in dollars per kilogram is quite complicated. The various factors that must be considered for evaluating this accurately include the costs of yellow cake, fuel enrichment, fuel fabrication, fuel reprocessing and transportation, cask charges for the spent fuel and credit for plutonium as well as unused uranium. Assume that for a given fuel the allowed burnup is 30,000 MWD/T and the fuel cost is 600 dollars per kilogram,

$$\frac{(600\ \$/\text{kgU}) \times (10^3\ \text{mills}/\$)}{(30\ \text{MWD/kgU})(24000\ \text{kWH/MWD})(0.3)} = 2.76\ \text{mills/kWH}.$$

A more detailed calculation involving several main components of the fuel cycle cost such as yellow cake, fuel enrichment and fuel fabrication is illustrated below. The following data is assumed:

U_3O_8 (yellow cake) cost	= 40 \$/kg
Enrichment cost	= 50 \$ per kg of SWU
Fuel fabrication cost	= 150 \$ per kg
Plant factor	= 80 percent
Fuel enrichment	= 3 percent
Plant efficiency (η)	= 30 percent
Fuel burnup	= 30,000 MWD/T

The requirement of enriched uranium per year is given by

$$\frac{\text{Power (MWe)} \times \text{Plant factor} \times \text{days/yr}}{\eta \times \text{Burnup (MWD/T)}}$$

$$= \frac{500 \times 0.8 \times 365}{0.3 \times 30{,}000} = 16.22 \text{ tons/yr.}$$

The amount of natural uranium required for an enrichment plant needs to be computed. For 3.0 percent, the equivalent kilograms of natural uranium are 5.479. Therefore, the mass of natural uranium is equal to 16.22 × 5.479 or 88.88 tons. Neglecting the losses during fabrication, conversion, etc. the quantity of yellow cake required is given by

$$88.88 \times \frac{238 \times 3 + 16 \times 8}{238 \times 3} = 104.9 \text{ tons.}$$

In order to calculate the contribution of the various fuel cycle components in mills per kWH, it is necessary to first calculate the total quantity of energy produced annually in kWH. This is given by

$$500 \text{ MWe} \times 10^3 \text{ kWe/MWe} \times 365 \times 24 \text{ hr/yr} \times 0.8 = 3.5 \times 10^9 \text{kWH.}$$

∴ The contribution of the cost of yellow cake per kWH is:

$$\frac{104.88 \text{ tons} \times 10^3 \text{kg/T} \times 40 \text{ \$/kg} \times 10^3 \text{mills/\$}}{3.5 \times 10^9 \text{kWH}} = 1.2 \text{ mills/kWH}$$

The contribution of fuel enrichment cost per kWH of electricity is

$$\frac{16.22 \text{ tons} \times 10^3 \text{kg/T} \times 50 \text{ \$/kgSWU} \times 4.306 \text{ SWU} \times 10^3 \text{mills/\$}}{3.5 \times 10^9 \text{ kWH}}$$

$= 1$ mill/kWH.

The contribution of the fuel fabrication cost per kWH of energy is

$$\frac{16.22 \text{ tons} \times 10^3 \text{kg/T} \times 150 \text{ \$/kg} \times 10^3 \text{mills/\$}}{3.5 \times 10^9 \text{ kWH}} = 0.7 \text{ mills/kWH}.$$

The fuel cycle cost component can now be obtained by adding up the three individual fuel cycle cost items and is equal to 2.9 mills/kWH.

3. Operation and Maintenance Cost Component

The operation and maintenance cost per kWH can be calculated by dividing the total annual expenditure on operation, maintenance, insurance, etc. by the total quantity of energy in kWH produced per year. For example, if the annual operation and maintenance expenditure of the 500 MWE reactor operating at an average plant factor of 80 percent is 6 million dollars, this cost component is then given by

$$\frac{6 \times 10^6 \times 10^3 \text{mills/yr}}{500 \times 365 \times 24 \times 0.8 \times 10^3 \text{kWH/yr}} = 1.7 \text{ mills/kWH}.$$

Summing up all the component costs from these calculations, the total cost is

$$(12.1 + 2.9 + 1.7) \text{ mills/kWH} = 16.7 \text{ mills/kWH}.$$

Problems

15.1 Draft a construction cost breakdown for a BWR power plant having a capacity of 1300 MWe.

15.2 For a typical reactor fuel cycle, economic design requires the following information: reactor thermal and electric power, initial fuel loading, fuel enrichment, fuel burnup, fuel discharge loading, fuel discharge enrichment level, fuel region lifetime, total energy produced, and others. Explain how and why each of these parameters is used to carry out the fuel cost analysis.

15.3 Calculate the annual payment to repay the loan of 1,000 million dollars secured to cover the capital cost of a 1300 MWe nuclear power plant, at an annual interest rate of 10%. The plant life and plant factor are assumed to be 30 years and 0.8, respectively.

15.4 The plant factor is defined as the actual kWHt produced/kWHe rated. Discuss the significance of this parameter in power plant cost analysis.

15.5 Describe the series of events leading to the operation of a nuclear powered merchant ship. Specifically, develop the objectives and the design specifications for the reactor.

15.6 A BWR power plant requires a capital investment of one billion dollars. Also $150,000,000 must be spent for the fuel processing and fabrication. The plant rated power is 1000 MWe with an efficiency of 30%. If the operating cost is $15,000,000 per year, what is the cost per kWH. Assume a 75% plant factor.

15.7 Construct a table to show the major differences among a boiling water reactor (BWR), a pressurized water reactor (PWR), and a Candu-type reactor.

15.8 A one billion dollar capital cost is to be repayed according to a sinking fund rate of 10%. What is the annual amount (R) to be accumulated to pay off the capital cost? The plant life is 30 yrs.

15.9 Develop a set of nuclear power plant siting criteria with an emergency evacuation plan. Reference: *Transactions, Conference on Nuclear Power Plant Siting,* TANSAO 19 (Suppl. 1), pages 1-53, ANS, 1974.

15.10 A utility company needs to increase its total electric power output at an annual rate of 5%. The present capacity is 10 Gigawatts. The company has committed itself to build PWRs to satisfy the future power demand. Due to the TMI accident, all scheduls for the following plants have been delayed:

Unit Number	Construction	Operation
1	1970	1980
2	1972	1982
3	1974	1984
4	1976	1986
5	1978	1988

All units have the rated power at 1,000 Megawatts. Develop a contingence plan to lessen the impact due to this delay.

CHAPTER 16

FAST FISSION BREEDER REACTORS

Materials that can be converted into fissionable materials by fast or thermal neutron absorption followed by the subsequent decay into fissile nuclei are called fertile materials. For example, U-238 absorbs a fast neutron to become U-239 which decays into Pu-239 as shown by the equation

$$^{238}U + n \rightarrow {}^{239}U \xrightarrow{\beta^-} {}^{239}Np \xrightarrow{\beta^-} {}^{239}Pu. \tag{16.1}$$

Similarly, another fissile material, U-233, can be produced from Th-232 by the following nuclear reaction

$$^{232}Th + n \rightarrow {}^{233}Th \xrightarrow{\beta^-} {}^{233}Pa \xrightarrow{\beta^-} {}^{233}U. \tag{16.2}$$

Both reactions are caused by fast neutrons. The reactions described here are very important because if all the available U-238 and Th-232 could be converted into Pu-239 and U-233 respectively, the world nuclear fuel reserves would increase by order of magnitude.

Reactors which are capable of converting fertile materials into fissile materials are called convertors. When the production of fissionable material in a reactor exceeds the fuel consumption, it is called a breeder reactor. Generally speaking the natural or enriched uranium reactors do not produce more fissionable material than the amount of U-235 consumed. The most promising breeder reactor uses Pu-239 as the fuel and U-238 or Th-232 as the fertile material for the production of Pu-239 and U-233 respectively. Various factors which make Pu-239 the best nuclear fuel for breeder reactors are now discussed.

In reactor operations the parameter used most often to investigate the conversion of fertile to fissile material is the conversion ration (CR) defined by

$$CR \equiv (\eta - 1) - L, \tag{16.3}$$

where η is the number of fission neutrons per one absorbed neutron and L is the leakage factor. This gives an estimate of the number of new fissile nuclei which could be formed for each nucleus consumed. Table 16.1 gives the nuclear data for some fissile and fertile materials. From this table one can see that the fertile materials U-238 and Th-232 cannot be used alone for producing fissionable nuclei. However if any one of the three fissile materials is used

Table 16.1: Nuclear data for various fertile and fissile materials.

Nuclear data	U-235 with fast n's	Th-232 with fast n's	Pu-239		U-233		U-235	
			Th	fast	Th	fast	Th	fast
σ_f(b)	0.05	0.01	740	2.0	530	2.7	580	1.8
σ_a(b)	0.25	0.25	1000	2.4	580	3.0	680	2.2
ν	2.7	2.54	2.9	2.96	2.5	2.57	2.4	2.5
η - 1	0.46	0.09	1.15	1.47	1.28	1.31	1.05	1.05

as nuclear fuel, some excess neutrons can be made available for breeding purposes in addition to the one neutron required for maintaining the chain reaction. For Pu-239 fast fission $\eta - 1 = 1.47$ and is by far the most favorable material for breeding.

Two other important parameters for judging the breeding properties are the breeding gain (BG) or breeding ratio (BR) and the doubling time. The former is given by the following equation,

$$BR = BG = CR - 1. \tag{16.4}$$

This determines whether or not the material can be used as fuel for a breeder reactor so far as the neutron economy is concerned. For example Pu-239 is the best fuel as it has the highest breeding gain (0.47) for an infinite fast breeder reactor.

The most important quantity for fast reactor considerations is the doubling time (DT), defined as the time for the reactor to produce twice the amount of the initial reactor fuel. If the reactor has a flux density ϕ and if the fertile material nuclei density N is so large that for an operating time t it is considered to be constant, the fissionable nuclei produced during this period can be calculated from

$$n = N\sigma_a \phi t(\eta - 2 - L), \tag{16.5}$$

where n is the produced fissionable nuclei density, σ_a is the microscopic absorption cross section of the fertile material, η is the neutron regeneration factor of the fuel, and L is the neutron leakage factor. Note that this equation is based on the neutron balance concept. The right hand side is the net neutron excess in time t while the left hand side simply means that each excess neutron will lead to the production of a fissionable nucleus. It is implied therefore that the amount of fertile material used for the production of fissionable material is very large. Since in most breeder reactor configurations blanket material surrounds the reactor completely, the leakage term L can be neglected in equation (16.5) to obtain

$$n = N\sigma_a \phi t(\eta - 2). \tag{16.6}$$

The doubling time can now be obtained by setting n = N in this equation.

$$t_D = \frac{1}{\sigma_a \phi(\eta - 2)}. \tag{16.7}$$

In general the doubling time of a breeder reactor is expected to be less than 20 years.

For detailed design considerations equation (16.7) must be drastically revised. The factors that must be considered are fuel-fertile material, plant factor, the microscopic capture to fission cross section ratio, the fuel cycle inventory to core inventory ratio, and breeding gain. The capture to fission cross section ratio is of special interest in the choice between a fast and a thermal breeder reactor. Its value decreases with increasing neutron energy. For Pu-239, $\alpha = \sigma_c/\sigma_f = 0.4$ at thermal energies to 0.2 for fast neutrons (1 MeV or higher energy).

The importance of breeders was emphasized for the first time by Enrico Fermi at Los Alamos in 1945 by the following statement, "The country which first develops a breeder reactor will have a great competitive advantage in atomic energy". It is clear that a breeder will be of the utmost importance for a large scale nuclear power economy. If only U-235, which is directly fissionable, is used then the presently known reserves of uranium will give much less energy than the known fossil fuel reserves. However, if fast breeders are developed, most of the U-238 would be converted into Pu-239. A large scale development of fast breeder reactors will therefore increase the fuel utilization from about one percent usable in the present day reactors to over 60 percent. Figure 16.1 illustrates the utilization of the reserves of uranium using present day thermal reactors, advanced converters and fast breeders. It also indicates the total energy available in U-235 and natural uranium.

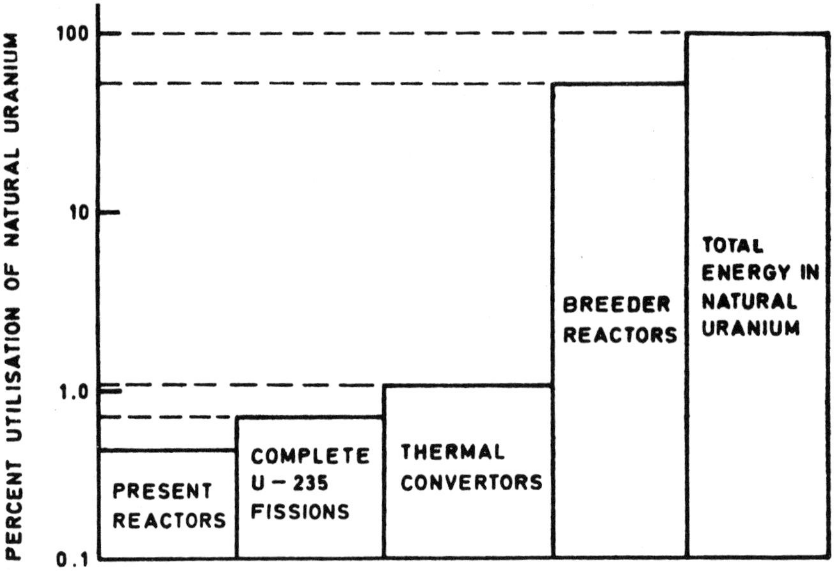

Fig. 16.1: Utilization of the uranium reserves using various types of reactors.

The efficiency of the utilization of uranium and thorium depends greatly on the conversion or breeding ratio of a reactor. This is shown in Figure 16.2. Note that while the fuel utilization changes very slowly for values of a conversion ratio smaller than unity, it increases rapidly for small increments in the value of conversion ratios over unity. The efficiency of fuel utilization also depends on the fuel burnup values as well as the loss of uranium and plutonium per cycle.

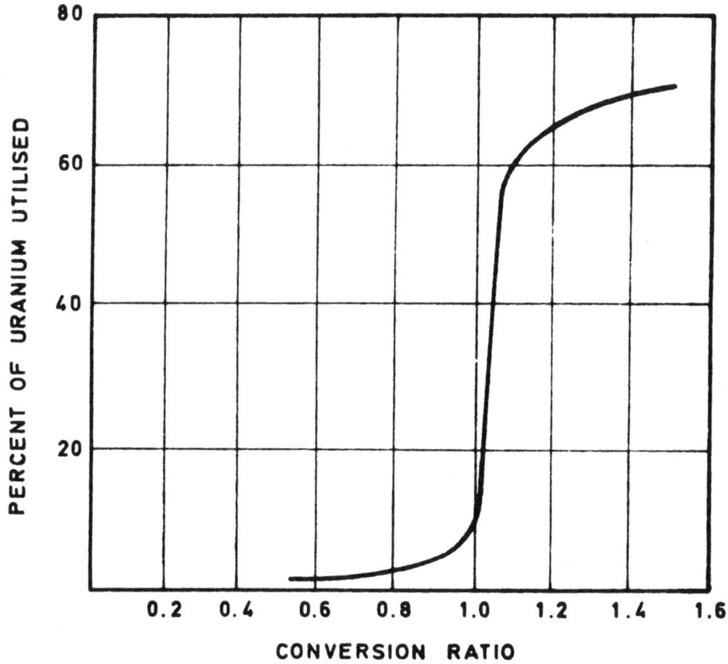

Fig. 16.2: Utilization of uranium as a function of conversion ratio.

The conversion ratios for thermal breeders are smaller than those predicted for fast breeders. Consequently the doubling time for thermal breeders is expected to be from 12 to 18 years as compared to 6 to 10 years for fast breeders. Since the doubling time for electrical energy in most countries is from 8 to 12 years, it is expected that the rate of breeding of fuel beyond the reactor's needs could provide sufficient fuel for the new breeder reactors to keep pace with the increase in the demand for electricity. It is therefore clear that fast breeder reactors would be most desirable if their power generating costs become competitive with other means. There is now little doubt that large fast breeder reactors will be commercially available within the next decade or so. However, a number of technological problems have to be solved before economical fast reactors can be established. Table 16.2 outlines the fuel, structural materials, specific inventory, conversion ratio and doubling times of various types of commercially available reactors, advanced converters and fast breeders.

Table 16.2: Comparison of conversion ratios and doubling times of various reactor types.

Reactor type	Fuel	Cladding	Specific inventory kg/MWe	Conversion ratio	Compound fuel doubling time (yr)
LWR	UO_2	Zr	2.2	0.5	—
HWOCR	UC	Zr	0.9	0.7	—
HWBLW	UO_2	Zr	1.8	0.78	—
HTGR	(U, Th) C	—	1.8	0.80	—
LMFBR	(U, Pu) O_2	SS	2.5	1.3	12.2
MSBR	(U, Pu) F_4	—	1.1	1.07	14.4
LMFBR	(U, Pu) C	SS	3.3	1.45	6.1
GCFR	(U, Pu) O_2	SS	3.5	1.5	10.0
SCFBR	(U, Pu) O_2	SS	4.1	1.4	12.0

Although it is expected that the initial inventory of Pu-239 for the fast breeder reactors would be derived from the presently operating thermal reactors, suitably enriched U-235 might be used for initial inventory in order to avoid waiting for enough plutonium to become available. This would mean that fast reactors are likely to be complimentary to thermal reactors until a large number of fast reactors have come into operation in many countries.

Besides pure neutron economy considerations there are some basic differences between a fast reactor and a thermal reactor. In fast reactor operations one must consider the short neutron lifetime in the light of the fact that it is easier for a fast reactor to become supercritical. A fast reactor has an average neutron lifetime of $l = 10^{-7}$ s as compared with $l = 10^{-4}$ s for a thermal reactor. Due to the small volume of a fast reactor and the small fuel pin dimensions, fuel rearrangement causing a supercritical condition is a very real problem in these reactors. The coupling of these two factors certainly is a serious safety problem in the reactor design, which must contain provisions to prevent the reactor from becoming prompt critical.

As the fast neutron induced fission cross section is very small, a larger quantity and higher concentration of fissionable fuel is needed to achieve criticality. The minimum enrichment on which an infinitely large core could go critical is around 7 percent of U-235 or 5 percent of Pu-239. While a thermal reactor could go critical on a few kilograms of concentrated fissionable fuel, the Enrico Fermi fast breeder reactor has a critical mass of 444 kg of U-235 at 25 percent enrichment! As the cost of enriched fuel is quite high, a fast reactor should have a very high specific power to produce competitive nuclear energy. This requirement introduces the following problems.

1) The requirement of a large heat transfer area necessitates a finely

divided core and hence a complicated core design and a high cost of fabrication of special fuel elements.

2) Coolants possessing low moderating properties and efficient heat transfer characteristics are required. Liquid metals fulfill these requirements but their use introduces additional problems like corrosion, induced radioactivity, mass transfer, etc. Liquid sodium is used as the coolant in most of the operating fast breeder reactors. Other possibilities are mercury, potassium, lithium and sodium-potassium alloys.

3) For better cost economics very high fuel burnup is necessary. As the fuel temperature is very high the fuel pins tend to swell easily due to the internal accumulation of fission gases. The development of a fuel capable of undergoing very large burnup values without swelling is hence extremely important for the economic utilization of fast power reactors.

4) The cladding material choice is very limited as it must be compatible with both the fuel and the coolant high temperatures.

5) The safety and control problems of fast reactors are more complicated as compared with those encountered in the thermal reactors. The fission product heating after shutdown requires cooling for many days. About half the core size of a fast reactor is generally occupied with coolant channels. In case of any loss of coolant, the heat produced is sufficient to melt down the entire core. This could result in a more compact configuration of a very large enriched fissile fuel assembly which may become highly supercritical and destroy the core completely. Special safety considerations are therefore required for the safe operation of fast reactors.

The fuel used in fast reactors may be U-235, Pu-239 or U-233. The use of any lighter material in the core is avoided except when these are incorporated in fissionable materials as carbides or oxides for improving the irradiation stability of the pure metal. Absorbers generally recommended for controlling the fast reactors are B-10 and Li-6. The choice of the coolant and structural materials is generally based more on their compatibility and strength rather than on nuclear grounds. Stainless steel, zirconium, vanadium and niobium are considered to be good structural materials.

Although the critical mass of a fast reactor is much larger than a thermal reactor, its size is smaller. The leakage of fast neutrons is therefore expected to be much higher if a single region fast reactor is designed. A fast reactor, therefore, consists of two regions, i.e. the core and the blanker. While the core contains both fissionable and fertile materials, the blanket consists only of the fertile material such as U-238 or Th-232. The fast neutrons that would have escaped in a single region fast reactor are well utilized in a two region reactor. Prototype fast reactors under construction are rather similar. The main features of a typical fast breeder in the range of 250-300 MWe are outlined here.

a) The fuel used presently in fast reactors is mixed Pu-239 and uranium oxides with fuel enrichment exceeding 20 percent.
b) The volumetric composition of the core is around 40 percent fuel, 35 percent sodium and 25 percent stainless steel as cladding and structural material.
c) As the average specific power is very high, fuel pins with very small diameters (about 6 mm) are used.
d) The sodium outlet temperature exceeds 550° C and the efficiency of the plant is over 40 percent.
e) To flatten the power density the core is divided into concentric zones of equal volumes with an enrichment ratio of outer to inner zone equal to about 1.25.
f) In order to improve the conversion ratio, the core is surrounded by natural or depleted uranium oxide in an axial and a radial blanket about 30 cm thick. As the power density in the blanket is relatively low, the fuel rod diameter is about 15 mm.

The core consists of a mixture of fissile and fertile materials for the following reasons:

i) In the absence of U-238 the core size would become extremely small. This would result in the production of extremely high temperatures and melting.
ii) Without U-238 in the core the inner breeding ratio would be zero, the reactivity would be less and the power shift from core to blanket would be much higher with increasing burnup.
iii) The breeding ratio is improved due to the U-238 fast fission factor.
iv) U-238 in the core leads to a negative Doppler coefficient. This prompt-acting temperature coefficient of reactivity plays an essential role in fast reactor safety.
v) U-238 fission in the core increases the effective delayed neutron fractions since it is higher for U-238 than for Pu-239.

g) In the prototype fast reactors the breeding ratio varies from 0.2 to 0.5. About 45 percent of the plutonium is bred in the core and about 55 percent in the blanket. This corresponds to an internal or core breeding ratio ranging from 0.05 to 0.20 and an external or blanket breeding ratio ranging from 0.15 to 0.30. The Pu-239 that is bred in the core helps to reduce the reactivity loss due to fuel burnup.

h) The blanket not only increases the reactor breeding ratio by capturing the escaping neutrons from the core but it also acts as a reflector and thus reduces the critical mass and flattens the power distribution. Besides, about 10 percent of the reactor power is generated in the blanket. Thus, a blanket performs the following functions:

i) increases the breeding ratio of the reactor,
ii) flattens the power distribution,
iii) reduces the critical mass, and
iv) reduces the core size for a given power output.

The initial power in the blanket is produced through U-235 and U-238 fissions. This keeps on increasing as more and more plutonium is accumulated in the blanket. Sincre coolant flow in the blanket must be designed for the plutonium rich condition it will be overabundant in the clean condition. However, the coolant flow is adjusted as a function of Pu-239 formation.

i) A two region fast reactor is often called a 'washing machine' for plutonium. The core of the fast reactor uses plutonium coming either from thermal or fast reactors. This is not clean and contains numerous higher isotopes of plutonium. Reprocessing of this spent fuel also yields 'dirty' plutonium. As the blanket does not initially contain any plutonium, the plutonium produced in the blanket will be of a very high quality. This is so because at high neutron energies, the ratio of capture cross section of Pu-239 to that of U-238 is quite low. Hence a fast reactor takes only 'dirty' plutonium and produces both clean and dirty plutonium.

j) It is assumed that fast reactors will attain burnup values above 100,000 MWD/T. Initially fuel swelling and fission gas pressure were thought to be the limiting factors, but now it appears that the limitation would come from the swelling of structural material due to the high integrated neutron doses.

k) In order to avoid high initial excess reactivity, distortion of power distribution and neutron losses, it is generally recommended to reload the core in batches.

l) For a typical prototype fast reactor the maximum core flux density is about 10^{15}n/cm^{2}s. The prompt neutron lifetime is about 0.4 μs and β_{eff} is approximately 0.3 percent. Short neutron lifetime and delayed neutron fraction offer numerous control problems.

m) Control is usually achieved by control rods containing a neutron poison. Generally boron carbide is preferred. Tantalum and rare earth elements are also considered. Fuel movement is also used for reactivity control.

Oxide fuel is generally preferred for fast reactors because it allows higher melting points and better burnup values. Although metallic fuel is better from the point of view of breeding ratio because of higher fuel density and harder spectrum, the oxide fuel leads to a softer spectrum and hence a higher Doppler coefficient and slightly less critical mass. Work on carbides and carbo-nitrides as possible alternatives to oxide fuel is in progress for the future. As these fuels have higher densities and better heat conductivities, a higher breeding ratio, smaller critical mass and shorter doubling time is expected. The work on fabrication, compatibility with structural materials, burnup and reprocessing of these fuels will ultimately indicate their usefulness as fast reactor fuels.

Presently the fast reactors can be divided into three main classes, depending upon the nature of the coolant. These are sodium-cooled, gas-cooled, and steam-cooled fast reactors. Sodium is presently the favorite because of its low slowing down capability, low neutron capture, excellent heat transfer properties and high boiling point. However, it has some serious problems like induced radioactivity, opacity, chemical reactivity, positive void coefficient etc. A sodium-potassium alloy was at one time considered as an alternate. The main attractive feature of such an alloy is that it remains liquid at ambient temperatures. Experience with sodium-cooled reactors has shown that, in the event of a sodium leak, freezing of the leaking sodium helps stop the leak.

Steam-cooled fast reactors were initially considered because of the well developed technology of present day light water reactors. Such reactors would result in a lower power density and a softer neutron spectrum. Besides, the breeding ratio of such reactors is expected to be much smaller. Mainly due to these and some other technological problems steam is no longer considered for fast reactors.

Gas(He)-cooled fast reactors seem to offer some advantages over the steam-cooled reactors. The power density of such reactors is lower than sodium-cooled reactors. In the case of helium the spectrum is relatively harder and the breeding gain is higher. Besides, gas-cooled fast breeders can benefit from HTGR technology. However, it appears that the large-scale research and development expenditure required for the gas-cooled fast breeders will be justified only if these reactors showed a decisive edge over the sodium-cooled reactors.

Stainless steel is the most commonly used structural material in fast breeders as it has good compatibility with the sodium metal and oxide fuels. Also, stainless steel has a reasonably low parasitic neutron absorption cross section. Neutron induced swelling is, however, a problem but at present no other material is available to replace stainless steel.

Numerous experimental and prototype fast reactors ranging from a few MW to several hundred MW are presently operating, under construction, or in the planning stages in various countries. The competitive fast reactors are expected to be in the range of 1000 MWe. Reactors as large as 2000 MWe are also being discussed. The countries that are actively engaged in the fast reactor program are the USSR, USA, France, U.K., and Federal Republic of Germany. Numerous other countries have initiated large scale research and development work in this direction. At present there are many fast breeder reactors in operation in various countries. Table 16.3 gives a list of some of these reactors.

Table 16.3: Fast breeder reactors

Reactor	Country	Power (MWth)	Fuel	Coolant
Clementine	USA	0.025	Pu metal	Hg
EBR-I	USA	1.2	U metal	NaK
EBR-II	USA	62.5	U metal	Na
Enrico Fermi	USA	430	U metal	Na
Br-2	USSR	0.1	Pu metal	Hg
BR-5	USSR	5	Pu Oxide	Na
Dounreay	UK	72	U metal	NaK
Rapsodie	France	20	U and Pu Oxide	Na

The components of a liquid metal fast breeder reactor may be arranged in either of the following two ways: All the primary heat transfer components and the reactor core are contained in a large liquid metal filled tank, Figure 16.3. This is the so-called pot system. Another arrangement is the pipe system. This system is similar to a thermal reactor arrangement in that the coolant is pumped through a small reactor containment vessel and the reactor core as illustrated by Figure 16.4.

The latest development in the field of fast breeder reactors is the Fast Flux Test Facility (FFTF) at Richland, Washington and the Clinch River Demonstration Plant. The FFTF is a 400 MWth plant design for fast flux testing and irradiation. The Clinch River Demonstration Plant near Oak Ridge, Tennessee, is the first full scale fast breeder reactor for power production. It is designed to have a capacity of up to 500 MWth to be operated by TVA when it is completed. This project is at present in a stagnant stage due to concerns of plutonium production associated with proliferation considerations.

Problems

16.1 Based on the consideration of the fission neutron mean lifetime and

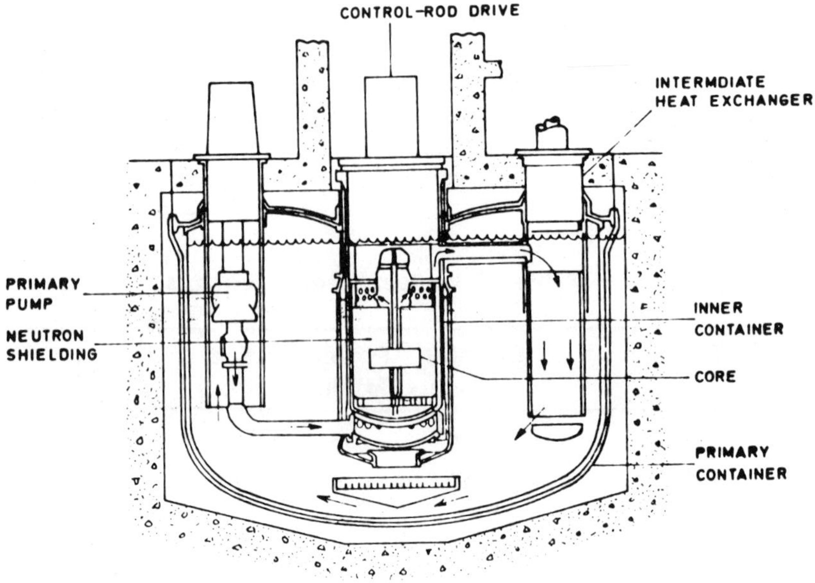

Fig. 16.3: Liquid metal fast breeder — pot system

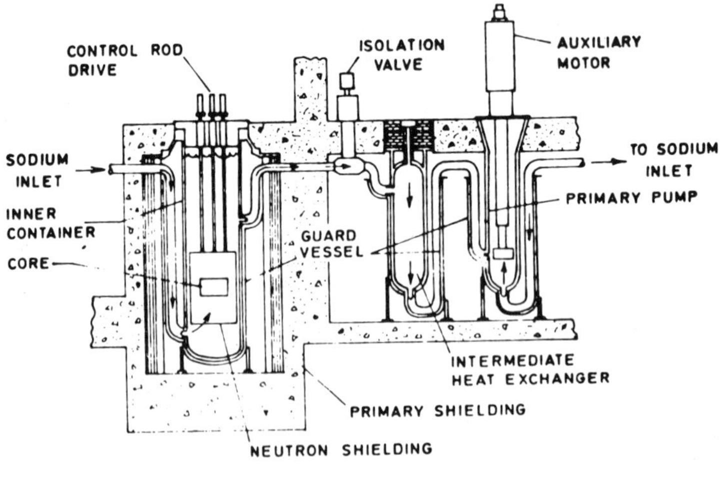

Fig. 16.4: Liquid metal fast breeder — pipe system.

delayed neutron fraction, discuss the fast reactor stability problems.

16.2 Liquid metal is frequently proposed and tested as a coolant of fast breeder reactors. Discuss the advantages and disadvantages of this choice.

16.3 What is the difference between a converter reactor and a breeder reactor?

16.4 In fast breeder reactor criticality calculations, the quantity α is defined as the ratio of the neutron capture cross section to that for fission. $\alpha \equiv \sigma_c / \sigma_f$. Discuss how α influences the values of η and DT.

16.5 Explain why positive values of (η - 2) are a prerequisite for a breeder reactor.

16.6 Consider and discuss circulating fuel as a means of varying the reactivity of a fast reactor.

16.7 The total amount of fuel in a fast breeder reactor fuel cycle is 50 kg. The breeding gain is 0.65. If the fuel consumption is ten grams per day, what is the doubling time?

Ans: 21.1 yr.

16.8 In the last problem, if the fuel microscopic absorption cross section is 2 barns, the neutron flux density is 10^{15} n/cm^{2}s, and the neutron reproduction factor $\eta = 2.65$, what is the doubling time? Consider incore factors only.

Ans: 24 yr.

16.9 Compare the advantages and disadvantages of a fast breeder reactor with a thermal breeder reactor.

16.10 Calculate the infinite multiplication factor of a Pu-239 dioxide and molten sodium mixture with volume fractions (v/o) of 40 and 60 respectively. $\Sigma_{af} = 0.06$ cm^{-1} and $\Sigma_{am} = 4 \times 10^{-5}$ cm^{-1}.

16.11 In the last problem the core composition is changed to 4 v/o Pu-239 dioxide, 25 v/o U-238 dioxide, 25 v/o stainless steel (SS), and 46 v/o sodium. $\Sigma_{ass} = 1.5 \times 10^{-3}$ cm^{-1}. $\Sigma_{au} = 0.008$ cm^{-1}. What is the new infinite multiplication factor?

16.12 If fertile materials are installed as a fast reactor core blanket, will this system be more efficient than a homogeneous mixture of fuel, coolant, and fertile material?

CHAPTER 17

THERMONUCLEAR ENGINEERING

17.1 Introduction — Thermonuclear Engineering

Enormous amounts of energy are released from the reduction of the total mass in a fusion reaction. Controlled nuclear fusion reactions could provide unlimited sources of energy. The research for exploring means of extracting this energy has been going on in many countries for many years. This research is basically aimed at achieving a net energy gain from a system in which a fusion reaction is maintained. Controlling it would have profound consequences on energy cost, and hence on the living standard of the world. The basic fuel for a fusion reactor is deuterium. For every 6,500 atoms of light hydrogen in water there is one atom of deuterium and the cost of extracting deuterium from water is quite small. This means that fuel for fusion reactors would be available in an almost unlimited quantity.

Because of the very high temperature needed for a fusion reaction to proceed, it is also called a thermonuclear reaction. Recently, with the ever increasing promise of the realization of controlled fusion reaction, a new field known as thermonuclear engineering has emerged. It is defined as:

> THERMONUCLEAR ENGINEERING IS THE BRANCH OF NUCLEAR ENGINEERING THAT DEALS WITH THE DESIGN, CONTROL, AND UTILIZATION OF THERMONUCLEAR (FUSION) REACTORS.

Since there is no fusion device or experiment that would produce a net energy output, no nuclear fusion reactor or power plant has been engineered so far. However, due to energy problems, increased emphasis has been placed on thermonuclear research recently. To shorten the time needed to carry out a demonstration experiment on a practical nuclear fusion power plant, it is necessary that the engineering design of fusion reactors take place simultaneously with the research. An unlimited and almost free fuel supply combined with a much smaller environmental hazard make thermonuclear reactors extremely attractive. If the intensive research and development work is continued in various aspects of thermonuclear engineering, it is expected that full scale fusion reactor will be in operation in the early part of the next century.

17.2 Plasma and Fusion Reactions

A nuclear fusion reaction involves the fusing of two light nuclei to form a heavier nucleus, releasing energetic subatomic particles at the same time. The positively charged light nuclei can be made to fuse in spite of the Coulomb repulsion between them. Figure 17.1 illustrates the electric potential V(r) as a function of the distance r between the two nuclei. Up to a certain minimum distance R_0 similarly charged nuclei experience a force of repulsion the magnitude of which is given by Coulomb's law. The height of the potential barrier is equal to their mutual electrostatic potential energy and is given by

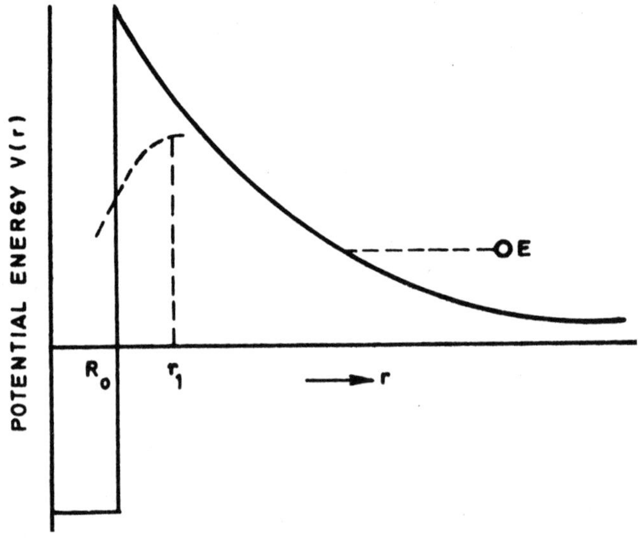

Fig. 17.1: Potential barrier.

$$V(r) = \frac{1}{4\pi\epsilon_0} \cdot \frac{Z_1 Z_2 e^2}{r}, \tag{17.1}$$

where Z_1 and Z_2 are charges of the two light nuclei, r is the distance between the nuclei, e is the electric charge and ϵ_0 is the electric permittivity of vacuum. For distances greater than R_0, as the distance decreases the height of the barrier increases. However at the critical distance R_0 the nuclear attractive forces become dominant. Therefore, if two nuclei could be brought close enough to one another, fusion would occur.

Since the height of the potential barrier is directly proportional to the nuclear charges of two fusing nuclei, its height increases with the increasing atomic numbers of the nuclei. It is therefore obvious that more energy will be needed to overcome a higher potential barrier if heavier nuclei are to be fused. Naturally, the first reactions to be considered for the release of useful energy by fusion reactions involve the isotopes of hydrogen. Unfortunately the three reactions of the lighter hydrogen isotopes have very small fusion cross sections and are therefore of no practical use. These reactions are

$$H + H \rightarrow D + e^+. \tag{17.2}$$

$$H + D \rightarrow {}^3He + \gamma.$$

$$H + T \rightarrow {}^4He + \gamma.$$

Some more useful fusion reactions are given here.

$$D + D \rightarrow {}^3He + n.$$

$$D + D \rightarrow T + H. \tag{17.3}$$

$$D + T \rightarrow {}^4He + n.$$

Upon examining the nuclear masses of the elements in all of these reactions, it is obvious that the masses of the reaction products are slightly less than those of the reacting nuclides. This results in a net release of energy. Although there are hundreds of fusion reactions that can produce energy, the main emphasis is placed on those that are possible at relatively low energies of the colliding particles and, thus, involve nuclides of very low atomic numbers. A list of important fusion reactions along with the amounts of energy released in MeV is given in table 17.1.

Table 17.1: Important fusion reactions.

Fusion reactions	Q-values (MeV)
$D + D = n + {}^3He$	+ 3.25
$D + D = p + T$	+ 4.00
$T + p = \gamma + {}^4He$	+ 19.70
$T + D = n + {}^4He$	+ 17.60
$T + D = 2n + {}^4He$	+ 11.30
${}^6Li + n = T + {}^4He$	+ 4.80
${}^6Li + p = {}^3He^3 + {}^4He$	+ 4.00
${}^6Li + D = {}^4He^4 + {}^4He$	+ 22.40
${}^7Li + p = {}^4He^4 + {}^4He$	+ 17.30
${}^7Li + p = \gamma + {}^8Be$	+ 17.20
${}^7Li + D = {}^4He^4 + {}^5He$	+ 14.20
${}^7Li + D = n + {}^8Be$	+ 15.00

Some practical problems found in thermonuclear reactor development are now discussed.

a) High Temperature and Self Sustaining Reaction

As pointed out earlier, the fusion nuclei must come sufficiently close together so that the short range attractive nuclear forces may come into play. Because of electrostatic repulsion between similar charges, close encounters can be achieved only when the reacting nuclides have very high kinetic energies. Even then it is far more probable that the nuclei will scatter rather than undergo fusion. The nuclei must be confined in a region where they can approach each other consistently until fusion eventually occurs. The kinetic energies necessary for the nuclei to overcome the forces of repulsion and begin to fuse together depend on the atomic numbers of the nuclei. For example, for two touching deuterons, the height of the potential barrier is around 500 keV. Fortunately, barrier penetration allows these reactions to occur well below this energy. But energies greater than 10 keV are generally required for fusion reaction to occcur with sufficient efficiency. In order to convert this kinetic energy into equivalent temperature, consider the energy due to the thermal motion of particles. This energy is given by (3/2)KT, where K is the Boltzmann constant and T is the absolute temperature. Thus,

$$\frac{3}{2}\ KT = 10^4\ eV$$

and

$$T = \frac{2 \times 10^4 \text{ eV}}{3 \times 8.62 \times 10^{-5} \text{ eV}/K^\circ} = 10^8 \,^\circ K.$$

Temperatures of the order of 100 million degrees correspond to those in the interiors of stars; the confinement of particles at these temperatures is one of the major problems in building a practical thermonuclear reactor. It may be added here that at very high temperatures, atoms are completely stripped of their orbital electrons and become an assembly of bare nuclei and electrons known as plasma. An important requirement of any useful fusion device is that the reaction be self sustaining. This means that once the plasma temperature has been increased to the point at which nuclear fusion occurs at an appreciable rate, the energy released by the plasma must at least be sufficient to maintain the temperature.

Figure 17.2 shows a plot of power generated as a function of plasma temperature. Also shown is the power radiated per cubic centimeter for both D-D and D-T reactions. For a given reaction to be self-sustaining and useful the power generated must be greater than the power radiated. Thus the reaction will be useful only above a certain critical temperature. The temperature at which the power produced in the plasma from thermonuclear reactions is equal to that radiated away is called the ignition temperature. It is seen that the D-T reaction has a lower ignition temperature than a D-D reaction. Fusion cross sections of some useful reactions as a function of energy are shown in Figure 17.3.

b) Plasma Confinement

Ionized gases containing light nuclei are said to be in their plasma state. A major requirement for a sustained fusion reaction is the suitable confinement of the plasma. No material walls can withstand the enormous fusion temperatures. The bare nuclei may travel thousands of miles before undergoing fusion. They cannot be permitted to rebound back and forth from the walls of the container, not only because the walls would vaporize, but also because the plasma would become cooled as a result of the energy transfer and the reaction would stop. It is therefore necessary to confine the plasma to the interior of the container. Various methods have been tried for plasma confinement; most use strong magnetic fields.

When a plasma is confined by a magnetic field, the inward pressure exerted by the magnetic field and the outward pressure of the plasma particles must follow the relationship

$$\beta = \frac{\text{plasma pressure}}{\text{confinement pressure}} \leqslant 1. \tag{17.4}$$

When $\beta = 1$, this corresponds to a bursting condition. However, with the

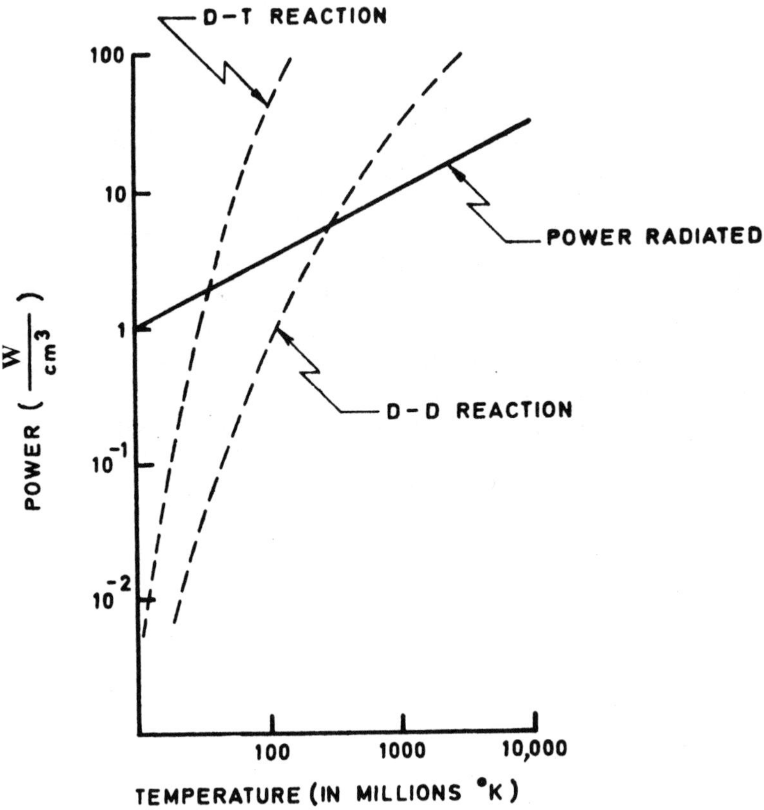

Fig. 17.2: Power generated vs. plasma temperature (plasma densities of about 10^{15} particles per c.c. are assumed).

development of superconducting magnets, sufficiently high magnetic fields would be available to avoid this.

c) Radiation Losses

A plasma consists of bare nuclei and free electrons. Any deflection of the charged particles in a Coulomb field gives rise to the continuous emission of radiation. This phenomenon is called Bremsstrahlung. Although losses from a plasma at temperatures of thermonuclear significance are rather high they are not disastrous because as the rate of this energy loss increases as $T^{1/2}$, the rate of thermonuclear energy production roughly varies as $T^{5/2}$.

In addition to Bremsstrahlung from high temperature plasma there is another mechanism which also results in energy loss. It is known as cyclotron

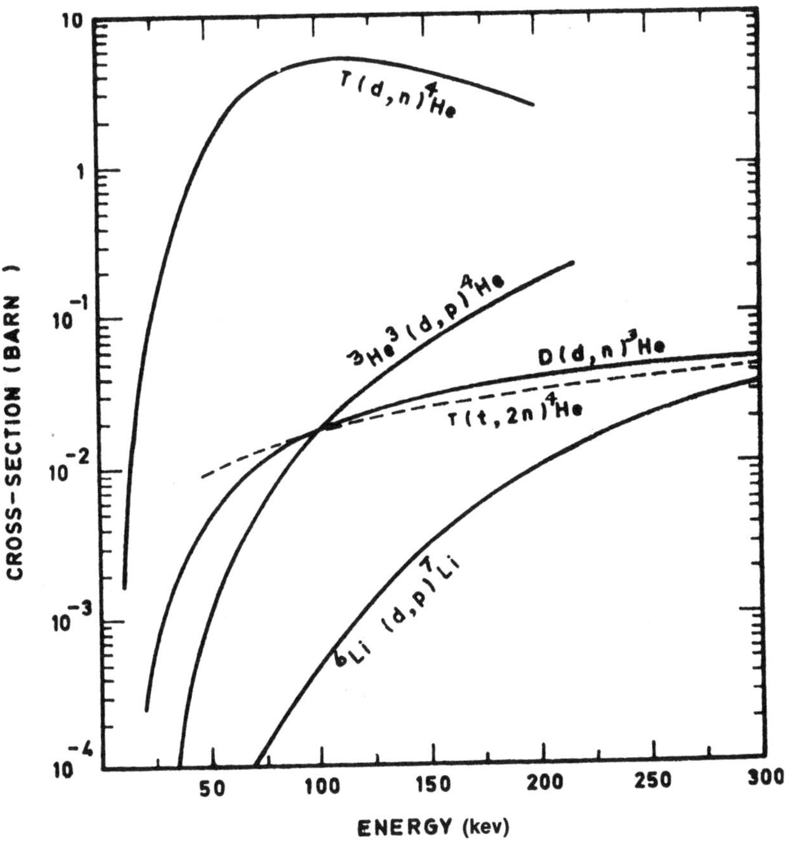

Fig. 17.3: Some useful fusion reaction cross sections.

radiation. In the presence of huge magnetic fields, positively charged bare nuclei and electrons spiral around with very high frequencies. The circular acceleration of the charged particles results in the emission of radiation known as cyclotron radiation. The rate of energy loss from electrons is much higher than that from heavy bare nuclei.

Consider a plasma confined in a space for a confinement time τ. If the plasma ion density is n which has an equal atom density of deuterium and tritium, the balancing equation for the generated power and the radiated power loss is expressed by

$$\eta \left\{ \frac{1}{4} n^2 < \sigma v > Q_T + \frac{3nT}{\tau} + P_b \right\} = \frac{3nT}{\tau} + P_b, \tag{17.5}$$

where

η = thermal to electric efficiency in %,
n = sum of D and T ion density per cc,
$<\sigma v>$ = the average value of the product of reaction cross section and particle speed, in cm^3/s,
Q_T = Q-value of the fusion reaction in keV,
T = plasma temperature in keV,
τ = particle confinement time in s, and
P_b = $3.36 \times 10^{-15} T^{1/2} n^2$ = radiation power loss of a unit volume of plasma in $keV/cm^3 s$.

From the above equation, one can obtain the relationship betweel the plasma parameters $n\tau$ and T. This is the Lawson criterion,

$$n\tau = 12(1 - \eta)T/[\eta\,(\sigma v)Q_T - (1 - \eta)\ 1.3 \times 10^{-14} T^{1/2}]. \qquad (17.6)$$

Note that $<\sigma v>$ is also a function of T. When $\eta = 33\%$, $n\tau$ as a function of T is illustrated by Figure 17.4. A rule of thumb numerical value attributed to Lawson for the D-T fusion breakeven state is $n\tau \cong 10^{14} s/cm^3$.

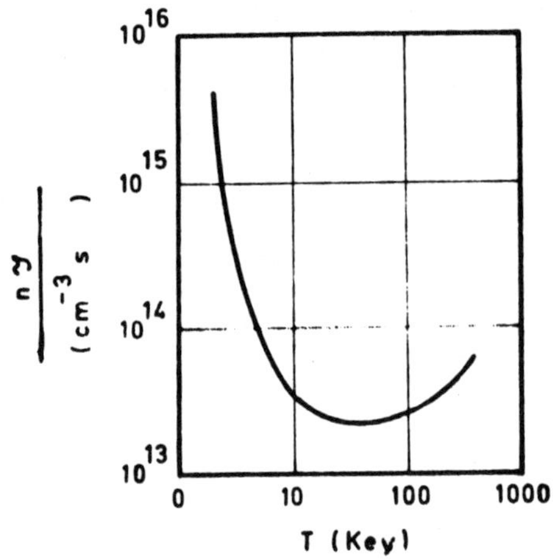

Fig. 17.4: $n\tau$ versus T relation based on the Lawson criterion.

d) Effect of Impurities

Impurities of any type have very adverse effects on ignition temperatures for fusion reactions. It has been estimated that the presence of 1 atomic percent of oxygen impurities will increase the rate of energy loss by Bremsstrahlung by about 80 percent! In the case of D-D reactions such an impurity will almost double the ignition temperature. This is due to the fact that the rate of energy loss as Bremsstrahlung varies as the square of the atomic number in a completely ionized gas. Besides, the impurity nuclei increase the scattering to a considerable extent and cause the plasma to escape.

e) Plasma Density

If the plasma density is too low, the rate of energy generation per unit volume from fusion reactions ceases to have any practical significance. On the other hand, a very high density might exert enormous outward pressures against the confining walls. Therefore, a compromise has to be made between very low and very high plasma densities. In general, the density can vary from 10^{14} to 10^{18} particles per cm^3.

f) Development of Useful Fusion Power

Finally the question arises that even if all the above difficulties and limitations are overcome, how would the energy be extracted and harnessed as useful power? In a D-T reaction most of the energy is carried by the neutrons. To recover this energy, the neutrons must be slowed down in some suitable moderator outside the discharge chamber. The heat thus produced in the moderator may be removed by a suitable coolant and used to generate electrical energy. In addition these neutrons could be used to produce T-3, Pu-239 and U-233.

17.3 Nuclear Fusion Reactors

There are several basic criteria for a controlled fusion reactor: the temperature T, density n, and confinement time τ of the plasma, and the β value. For deuterium and tritium specifically, the requirements are

Plasma temperature: $T = 5 \times 10^7\ ^\circ K$

The product of plasma density and confinement time =
$$n\tau = 10^{14} s/cm^3. \tag{17.7}$$

A nuclear fusion reactor would be a practical devise for the continuous execution of controlled thermonuclear reactions and the conversion of fusion energy into electric. This reactor would certainly be more bulky and complicated than nuclear fission reactors. The main design features of a thermonuclear reactor are outlined here:

1. In the core of the fusion reactor, there must be a stable and controllable region for thermonuclear reactions. The present experimental devices are just a partial model of such a reactor core. According to the methods of plasma confinement and production, the nuclear fusion reactors may be divided into the following representative types:

1) tokamak, 2) magnetic mirror, 3) theta pinch, and 4) laser triggered. The plasma production methods are all different. The first three types use magnetic fields for confinement, while the last type does not. For the first two types the plasma movements are stable but for the last two these are pulsed.

2. A blanket structure must be designed to surround the core for
   a) transferring fusion and plasma radiation energy,
   b) producing tritium from neutrons, and for
   c) neutron and other radiation shielding.

3. The experience gained in nuclear fission can be used in heat transfer and converter design. However, if direct conversion of fusion energy to electricity is desired a completely new approach must be initiated.

4. Auxiliary systems such as the superconductor magnetic field system, vacuum system, plasma heating system, tritium separator, and fuel handling and control systems are essential parts of the reactor. Due to the special requirements, some difficulties in the design of these auxiliary systems must be anticipated.

A typical nuclear fusion power plant is illustrated in Figure 17.5. This is a tokamak type. There are several blankets sandwiched between the vacuum vessel and the magnetic field coils as shown by the cross sectional view in the left side of the figure.

Besides the shells and blankets, and the fuel injection and waste disposal systems, a tritium separation system must be used. The entire tokamak housing is supported by the iron core of the transformer. It is estimated that for a 5000 Megawatt tokamak-type fusion reactor this circular body would be about 9.5 meters high with a length and width of about 21 meters.

17.4 Functions and requirements of the main parts

(1) Plasma — This is the core of the reactor in which the $D + T \rightarrow \alpha + n$ reaction occurs. The energy release per reaction is 17.6 MeV. The kinetic energy of the alpha particle is 3.5 MeV and that of the neutron is 14.1 MeV. Neutrons are slowed-down by the blanket material. Alpha particles, however, reduce their kinetic energies by interacting with electrons as well as deuterons and the tritium nuclei, resulting in plasma heating. The accelerated ions emit radiation; the energy may be absorbed by the vacuum vessel wall. In principle, all fusion energy can be converted into heat and electric energy.

(2) Vacuum Vessel — its function is to 'contain' plasma in a prescribed degree of vacuum and to separate the heat transfer material and plasma. A key problem here is the structural materials, which should

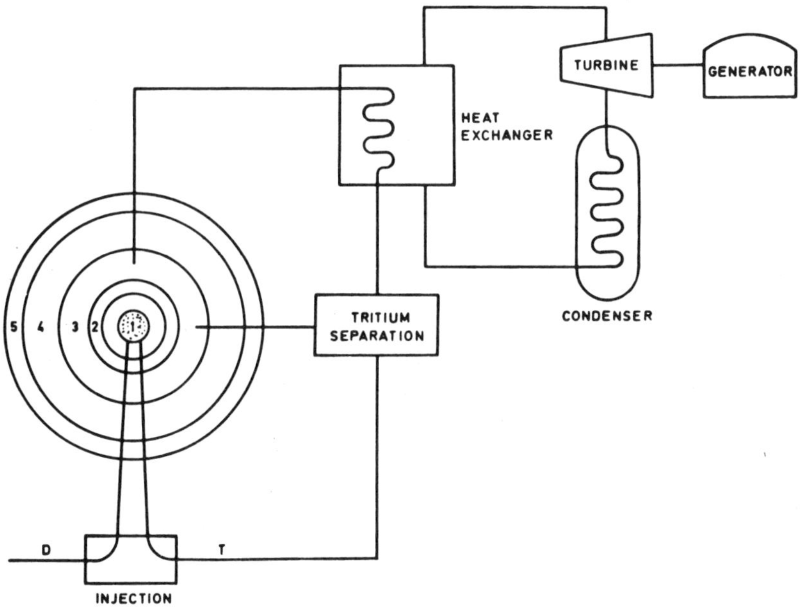

1. Plasma
2. Vacuum
3. Heat transfer agent and tritium breeding
4. Radiation shield
5. Superconducting coil.

Fig. 17.5: A fusion power plant.

have high heat resistance, low diffusive reflectivity, be radiation-proof, and anti-corrosive. Presently, tantalum, niobium, and vanadium might be used to build the wall which is about 1 to 3 centimeters thick. Assuming the maximum permissible thermal power received per unit area of the wall to be $P_W(MW/m^2)$, and the total plasma thermal power to be $P_T(MW)$, one can determine the tokamak vacuum vessel radius R (in meters) from the following relation:

$$R = \left(\frac{AYP_T}{4\pi^2 P_w} \right)^{1/2}, \tag{17.8}$$

where $A = R/r_p$ with r_p as the cross sectional radius of the plasma, $Y = r_p/r$, r being the chamber cross sectional radius.

(3) Blanket and tritium breeding layer — the requirements of the blanket are thermal compatibility with the structural materials and the ability to produce tritium through neutron-lithium reactions. Molten lithium or lithium floride-beryllium floride are two possibilities.

A 14.1 MeV neutron loses about 10% of its kinetic energy as it passes through the vacuum chamber. Once it reaches the blanket, it is slowed-down and interacts with Li-6 or Li-7 according to either of the following reactions:

$$^{6}Li + n \rightarrow {}^{4}He + T + 4.8 \text{ MeV}.$$

$$^{7}Li + n + 2.8 \text{ MeV} \rightarrow {}^{4}He + T + n - 2.5 \text{ MeV}. \quad (17.9)$$

Tritium is produced in both reactions so that its consumption is compensated. Note that 4.8 MeV of energy is released in the first reaction. Although the second is an endoergic reaction, it releases a new neutron besides a tritium atom.

(4) Radiation shield — a superconducting coil for the magnetic field operated at $\sim 4°K$ while the temperature of the vacuum chamber is about 600°-800°C or higher. Therefore, the radiation and thermal shield must be used to reduce the heat transfer and activation to acceptable levels. Lead-boron-water-lithium hydride may be used as the radiation shield. The approximate thickness of the shield is 30-50 cm.

(5) Superconducting magnetic field — the maximum magnetic flux density, B_{max} is $\sim$ 100 kilogauss. The superconducting coil radius is about 4-5 m. It is installed at the extreme external position to the core because of radiation and heat problems. Due to the large radius of this coil, the cost and the inhomogeneity of the magnetic field causing plasma instability are unavoidable difficulties.

(6) Heat transfer loop — Figure 17.5 illustrates the direct heat transfer loop. In practical design however it is always desirable to have indirect heat transfer, from lithium to potassium to water. This method has a higher thermonuclear efficiency.

There are many problems in the process of carrying the tokamak experiment to a practical fusion reactor. Some of the obvious problems are summarized here.

(1) New plasma heating methods must be found.
(2) Large capacity energy storage devices for magnetic fields are not very economical.

(3) It must be continuously operable.

(4) Due to the inherent complexity of its structure, a fusion reactor must accomodate many equipments in a small space.

These problems are not limited to tokamak-type machines. Any fusion reactor concept that uses the magnetic field confinement method has the same problems.

New possibilities of attaining nuclear fusion are raised by the development of laser technology. Due to the rapid progress in the field of laser triggered nuclear fusion, many laser fusion reactor design concepts are proposed. Figure 17.6 illustrates one of these concepts. Basically, this design

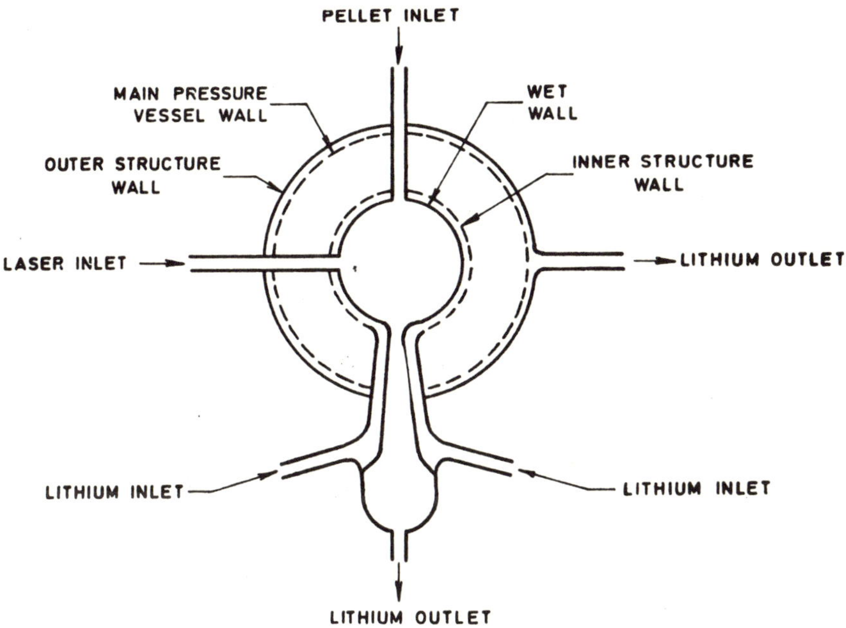

Fig. 17.6: Illustration of a laser triggered fusion reactor model.

uses a high power pulsed laser beam to strike solid deuterium-tritium pellets. These pellets thus acquire a very high temperature which is sufficient for the fusion reaction to start. Magnetic confinement is not used. Instead the inertia confinement of the pellets themselves is used. Recently a method of utilizing multilaser beams has been proposed. Pellets are illuminated simultaneously from different directions realizing the so-called centripetal implosion. This process occurs in the space of the spherical cavity illustrated in Figure 17.6. The empty cavity is surrounded by a layer of about one meter molten lithium for heat transfer and neutron absorption.

Solid deuteron-tritium pellets are injected into the cavity center from the inlet. At the same time, single or multiple beams of laser strike the pellets to ignite the fusion explosion. The thermal radiation and ions are absorbed by the inner side of the wet wall and the lithium layer. After some operating time the lithium layer is completely evaporated. The lithium vapor is ejected, cooled, and recycled. The pressure wave of the fusion explosion is mainly absorbed by the main pressure vessel wall. Fusion neutrons which carry off about 80% of the fusion energy are slowed-down by the surrounding lithium layer. The energy loss by these neutrons causes the temperature of lithium to increase rapidly. At the same time neutrons interact with lithium to produce tritium. The lithium flow is sufficient for heat removal from the reactor. The thermal energy so generated is used in the turbogenerator for the production of electricity.

The laser triggering of the fusion reaction is pulsed and is executed every few seconds. The main design consideration is to have the system recover its original condition before the next explosion is induced. For example, the original condition requires the re-establishment of the thin layer of lithium and the recovery of the cavity vapor pressure to 10^{-4} mm of mercury. Although the explosions are discontinuous, the lithium loop can be operated to keep a constant heat flux output.

A detailed introductory discussion on the various aspects of thermonuclear engineering can be found in *Fusion Energy Conversion* by George H. Miley, ANS Monograph, 1976.

Problems

17.1 A man consumes about 1.4×10^{12} kWH of energy in one year. What is the mass equivalent of this energy? Calculate the amount of coal, uranium, deuterium necessary to produce the same amount of energy. Given:

i) 7000 cal/gram for coal;
ii) 200 MeV/fission for U-235;
iii) the thermonuclear reaction under consideration is $D + D \rightarrow {}^4He$.

17.2 Calculate the height of the potential barrier for D-D and D-T reactions in keV. What should be the deuteron temperature so that the energy is sufficient to surmount the barrier? How is it possible that the fusion reaction does in fact take place even at lower temperatures?

17.3 Show that the approximate expression for the height of any nuclear potential barrier is $B^2 = 0.76\, zZ^{2/3}$. Assuming $r_o = 1.5 \times 10^{-13}$ cm, where z = Atomic number of incident particle, and Z = Atomic number of target nucleus.

17.4 Ordinary water contains one atom of deuterium for every 6500 atoms of ordinary hydrogen. Determine the energy released in Joules per gram of water by the following reactions:

i) $4H \rightarrow {}^4He + 2e^-$.

ii) $2D \rightarrow {}^4He$.

17.5 Calculate the Q-value of the reaction
$D + D \rightarrow {}^3H + H$.

17.6 The minimum $n\tau$ value can be obtained from equation (17.6). Assuming $\eta = 33\%$, $Q_T = 10$ keV, and $\langle\sigma v\rangle = 4 \times 10^{-16}$ cm^3s, find this minimum $n\tau$ value and the corresponding temperature.

17.7 For a tokamak using the D-T-Li fuel cycle, tritium can be produced in a lithium blanket. Analyze this process based on the natural abundances of Li-6 and Li-7 and equations (17.9).

17.8 Explain the concept of a hybrid reactor. Examine the possibility of using a hybrid reactor for the dual purpose of producing electric power and disposing high level radioactive waste from nuclear fission reactor fuel cycles.

17.9 What are the environmental effects of a fusion reactor?

17.10 Draw a diagram to illustrate a tokamak, with a clear indication of the major radius R, the chamber cross sectional radius (minor radius) r, and the plasma radius r_p.

PART III

RADIATION ENGINEERING

CHAPTER 18

INTRODUCTION

18.1 Radioisotopes

All radioisotopes possess the same physical and chemical properties as their stable isotopes. However, unlike stable isotopes, their numbers of nuclei decrease according to their decay properties. A radioactive decay process can be represented by:

$$ {}^{A}_{Z}X \rightarrow {}^{A_1}_{Z_1}Y + {}^{A_2}_{Z_2}b + Q, \qquad (18.1) $$

where A, A_1 and A_2 are the mass numbers of the radionuclide X, the product nuclide Y, and the subatomic particle b, respectively, Z, Z_1 and Z_2 are the respective atomic numbers. The conservation of charge, mass and energy of this reaction can be expressed by three equations.

$$ A = A_1 + A_2, \qquad (18.2) $$

$$ Z = Z_1 + Z_2, \qquad (18.3) $$

$$ Q = m_X - m_Y - m_b = E_Y + E_b - E_X. \qquad (18.4) $$

The product nuclide Y may be stable or radioactive. If it is radioactive, radiation will be emitted from this nuclide until eventually a stable isotope is formed. The energies of the emitted particles depend on the masses and energetics of the nuclides. For example, for electron emission

$$ {}^{A}_{Z}X \rightarrow {}^{A}_{Z+1}Y + {}^{0}_{-1}e + Q_e. \tag{18.5} $$

In this case

$$ Q_e = M(A,Z) - M(A,Z+1) - m_e, \tag{18.6} $$

where M(A, Z) and M(A, Z + 1) are the nuclear masses of X and Y, respectively, while m_e is the rest mass of an electron. If A = 28, Z = 13,

$$ {}^{28}_{13}Al \rightarrow {}^{28}_{14}Si + {}^{0}_{-1}e + Q_e. \tag{18.7} $$

To determine if Al-28 can indeed emit electrons, calculate

$$ Q_e = M(28, 13) - M(28, 14) - m_e $$

$$ = 27.9908 - 27.9858 - 0.0005 $$

$$ = 0.0045 \text{ amu} > 0. $$

Therefore, Al-28 is radioactive and emits an electron to become Si-28.

There are naturally existing and artificially made radioisotopes. The naturally existing radioisotopes are mainly derived from the decay chains of uranium and thorium starting from U-235, U-238 and Th-232 as described in Figures 18.1, 18.2 and 18.3.

It is believed that a violent cosmological event occurred about 5 billion years ago, at the time of formation of the universe, in which all nuclides, stable or unstable were formed in varying amounts. Those unstable nuclides with half-lives much smaller than 5×10^9 years have long since decayed into stable nuclides; however there are 14 unstable nuclides (given in table 18.1) the half-lives of which are comparable to or longer than the age of the universe. These nuclides are still found in measurable amounts in nature.

Table 18.1: List of long lived radioisotopes.

Element	Decay mode	Half-life (yr)	
K-40	β^-, EC	1.2×10^9	
V-50	EC	4×10^{14}	
Rb-87	β^-	6.2×10^{10}	
In-115	β^-	6×10^{14}	
La-138	β^-, EC	1.0×10^{11}	
Ce-142	α	5×10^{15}	All elements in this
Nb-142	α	3×10^{15}	table decay into
			stable daughters

Element	Decay mode	Half-life (yr)	
Sm-147	α	1.2×10^{11}	
Lu-176	β^-	5×10^{10}	
Re-187	β^-	4×10^{12}	
Pt-192	α	10^{15}	
Th-232	α	1.39×10^{10}	10 Radioactive generations
U-232	α	7.07×10^{9}	11 Radioactive generations
U-238	α	4.51×10^{9}	14 Radioactive generations

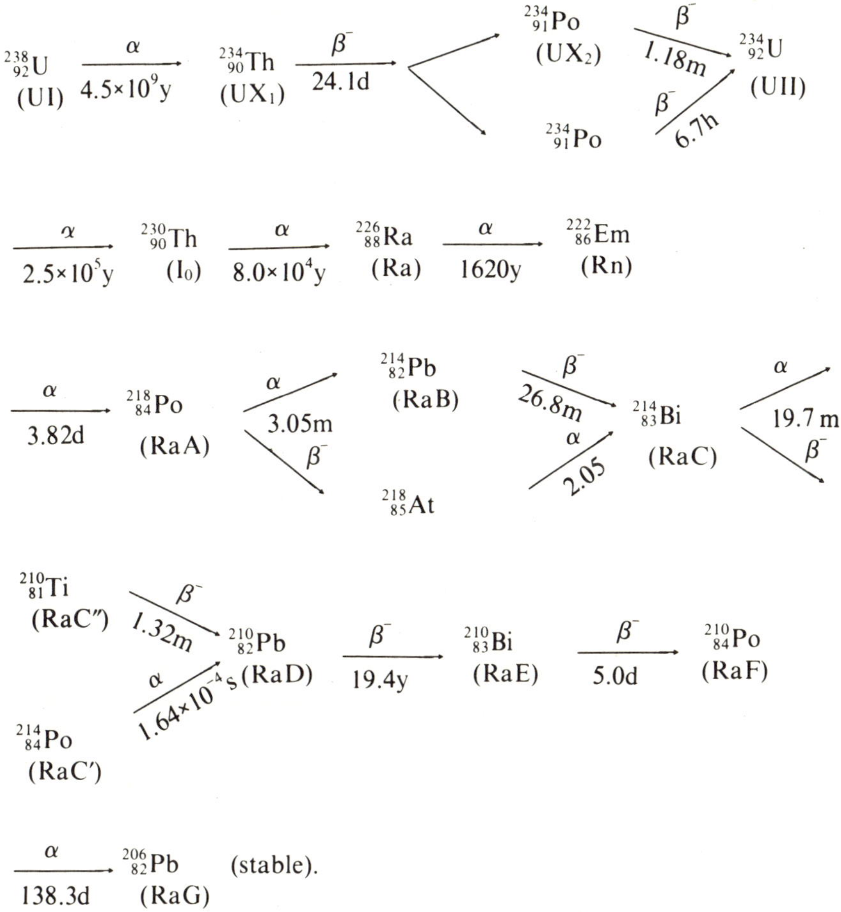

Fig. 18.1: Uranium series.

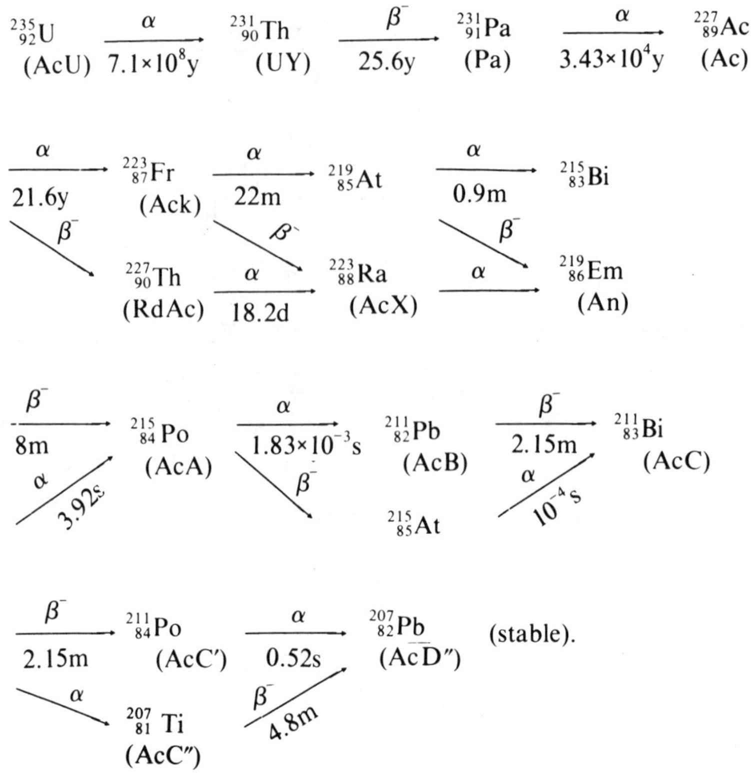

Fig. 18.2: Actinium series.

The first artificial production of radioisotopes coincided approximately with the completion of the first particle accelerator. The quantities of radioisotopes produced by the accelerators were so small that initially they were used only for research. It was only after the appearance of nuclear reactors that their large scale production became possible. Radioisotopes produced by nuclear reactors were used for the first time in 1946. At that time a few millicuries of the radioisotope C-14 were made and used for cancer research. However, from 1946 to 1964 at least 2.4 million curies from the Oak Ridge National Laboratory alone were used in industry, medicine, agriculture and other fields. The growing applications of radioisotopes in various fields stimulated the production of a large variety of radioisotopes in huge quantities in numerous countries.

The various facilities which are available in nuclear reactors for the production of radioisotopes are now described.

a) Wet Tubes

The tube can be placed inside or outside the core depending upon the

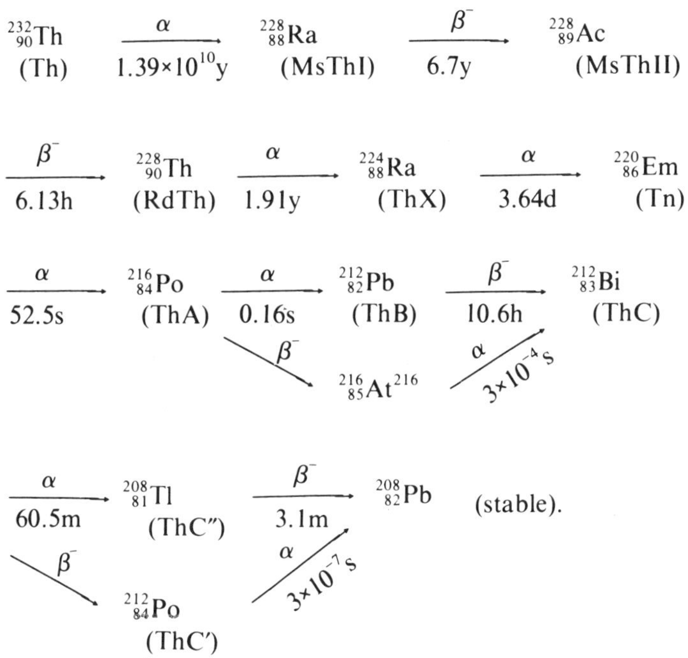

Fig. 18.3: Thorium series.

flux requirements. Water in the tube acts as a shielding material. The sample is placed in a water tight container before it is inserted in the tube. In-core and out-core wet tubes are shown in Figure 18.4.

b) Dry Tubes

Wet tubes are good for water tight samples. It is sometimes desired to have a dry tube to facilitate the handling of samples and also when the fast flux is desired for irradiation. Such tubes have radiation streaming problems and hence curved or bent tubes are used to minimize streaming effects. The samples which are properly sealed in containers are put in an aluminum ball which can be rolled down into the heart of the reactor. The position of the sample is controlled by an outside mechanism. After the desired irradiation, the ball containing the sample falls into a shielding case as shown in Figure 18.5.

c) Rotating container facility

Sometimes when large quantities of radioisotopes are to be produced a large number of containers are arranged around the core. In order to expose

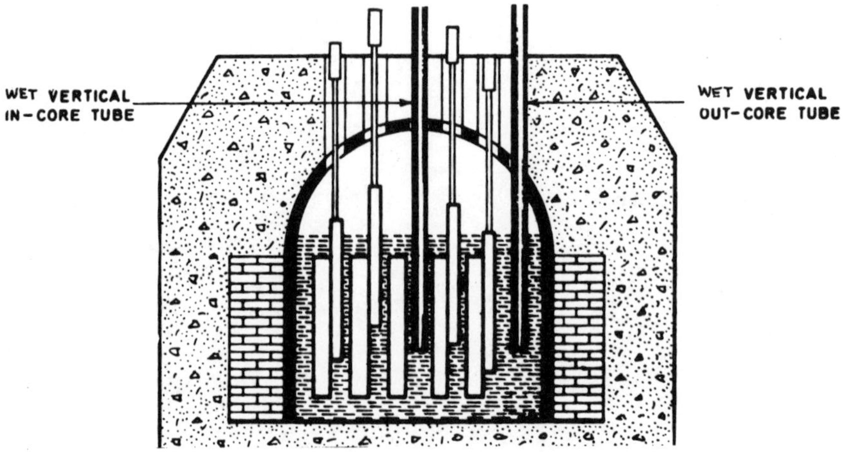

Fig. 18.4: Wet tube activation facility.

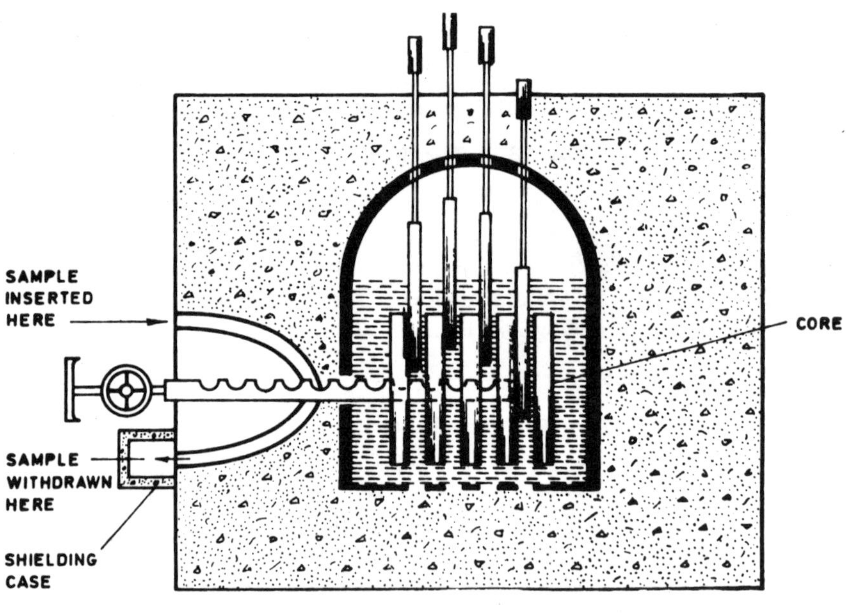

Fig. 18.5: Dry tube activation facility.

all the containers uniformly they are allowed to rotate with uniform speed. This complicated arrangement results in the large scaled production of different types of radioisotopes in a single irradiation. Extra precautions are taken to make the containers water tight in order to avoid any interaction with the surrounding water. Figure 18.6 illustrates the rotating container facility for radioisotope production.

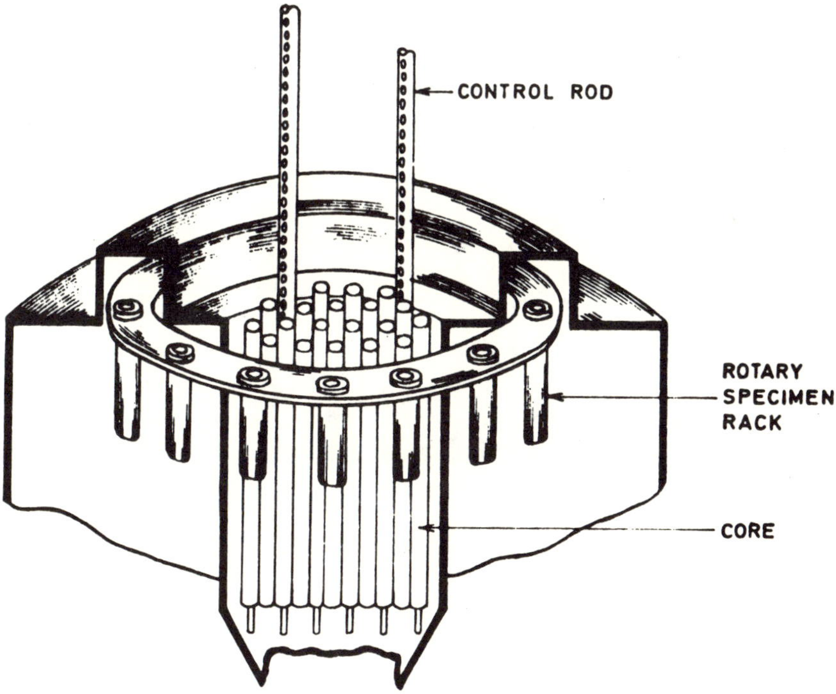

Fig. 18.6: Rotating container facility.

d) Moderator Island

When relatively high thermal neutron fluxes are needed, the principle of thermal flux peaking is used. The value of the thermal flux is higher in a moderator island in the central region of a reactor. Nuclear reactors having a moderator island in the center are generally known as 'flux-trap reactors'. The critical mass and the size of such a reactor is generally larger than for other equivalent reactors. Samples to be irradiated are placed in numerous tubes located in a high flux region in the moderator island. This is illustrated in Figure 18.7.

e) Bulk Irradiation Facility

For irradiation of large samples which cannot normally be

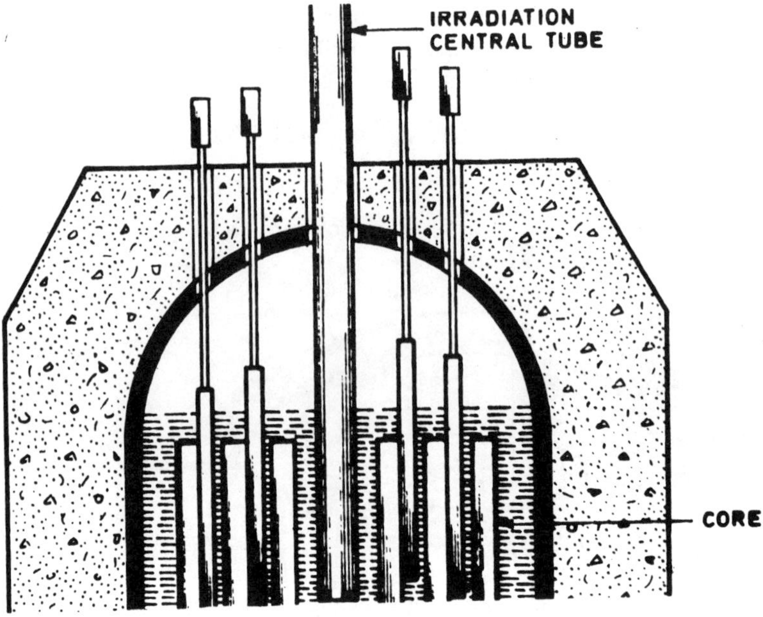

Fig. 18.7: Moderator island.

accommodated into small vertical or horizontal tubes, special arrangements are made for irradiation in water tight compartments. These are lowered and placed at a predetermined position close to the core when irradiation is required. Figure 18.8 illustrates the bulk irradiation facility.

f) Hollow Fuel Element Facility

When it is desired to irradiate samples at very high fast neutron fluxes, irradiation facilities are incorporated into hollow or partial fuel elements located in the center of the core. One or several irradiation tubes can be arranged in a hollow fuel element for irradiating numerous samples simultaneously.

g) Fissile Material Plate

At times it is desired to obtain a fast neutron flux away from the core. This is generally done by placing an enriched plate of fissionable material at the end of a beam tube having predominantly thermal neutrons. Fast neutrons are produced as a result of the interaction of thermal neutrons with fissionable material. The neutrons produced at the outer boundary of the reactor shield can be utilized for fast neutron activation work. The principle of the fissile fuel plate is illustrated in Figure 18.9.

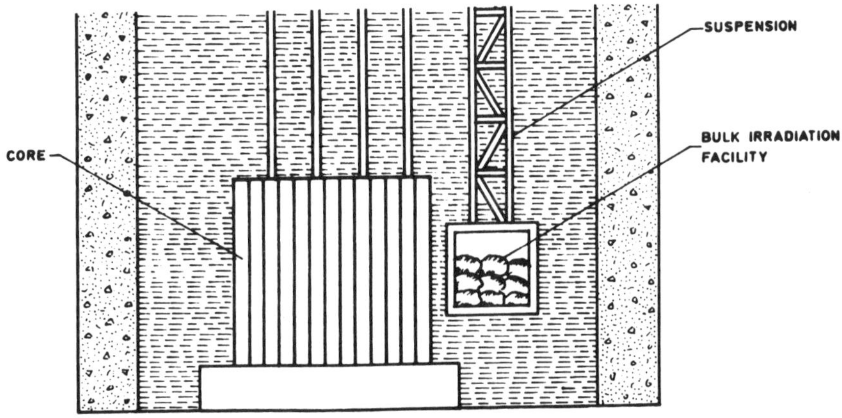

Fig. 18.8: Bulk irradiation facility.

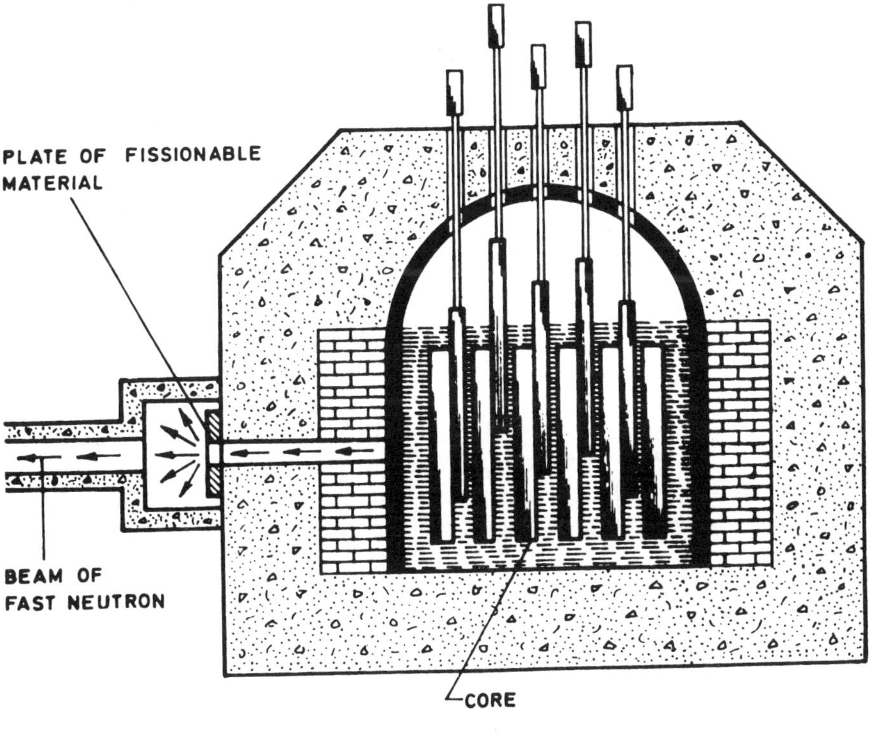

Fig. 18.9: Fast neutron facility.

In order to take the sample to and bring it back from the predetermined position near or inside the core, both manual and remote control techniques are employed. For the production of short lived isotopes, it is desired to insert and remove the samples in the core while the reactor is operating. The following remote and fast handling devices are used:

i) Hydraulic tube system

The system is known as the hydraulic tube system because water is used to push samples in and out ot the reactor. Water supplies the necessary flow and pressure for inserting, removing, and cooling the samples. Valves at the loading stations allow the water flow to be directed towards the reactor thus carrying the sample into the core. The water is then discharged through a cooling tube to the decay tank. For removing the sample, the water flow is directed back by the hydraulic tube from which the sample is to be removed, carrying the sample back to the loading station.

ii) Pneumatic tube system

The pneumatic tube system uses compressed air for the very rapid transportation of samples to be irradiated in the region of very high flux in the core. A small container called a rabbit carries the material to be irradiated in both directions, i.e. to the core of the reactor and back to the sending point. Along with a rabbit tube another tube is installed for the air supply.

The production of radioisotopes involves various stages. Firstly, the raw material needed for the preparation of radioisotopes is obtained in the required form. If it is not available as such, this is accomplished either by separation of the desired isotopes of an element or by chemical techniques. The nuclear, chemical, and physical properties of this material are investigated under the same conditions of temperature and radiation as are expected in nuclear reactors. The material, sealed in suitable containers, is placed by remote handling techniques in the spaces provided in a nuclear reactor for irradiation. Another technique used for the large scale production of radioisotopes involves the isolation of the required isotopes from the fission product solutions. This method is generally employed for those radioisotopes which cannot be produced by irradiation in a reactor to have the required specific activity. The irradiated material or the fission products are transferred to storage facilities until the short lived components have decayed. Perhaps the most tedious task is the extraction of useful irradiated material and the preparation of labelled compounds and radiation sources from the extracted material. Measurements are then made and various radioisotopes are dispensed in standard sizes with proper labels indicating the amount of radioactivity, the date on which it was labelled, and the half-life of the radioisotope. The shipment of the radioisotopes to various centers of utilization also involves proper attention in order to avoid any hazards involved. All these stages are summarized in Figure 18.10.

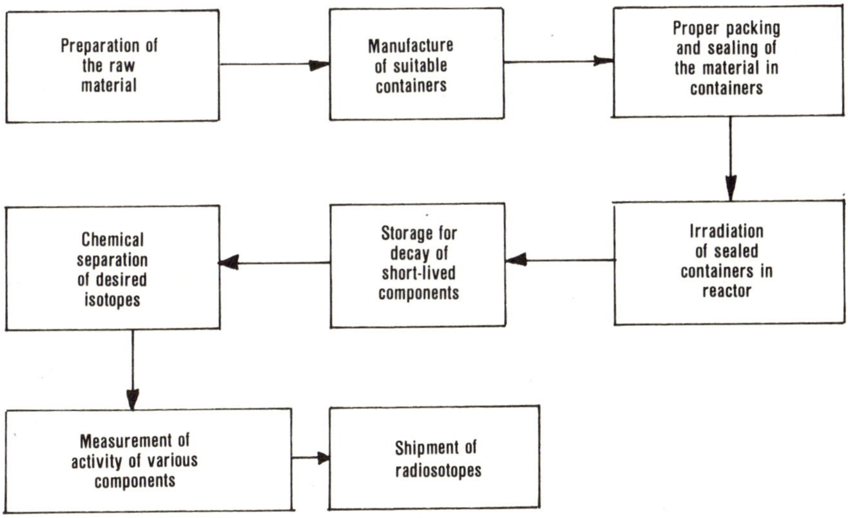

Fig. 18.10 : Various stages involved in the preparation and shipment of radioisotopes.

Generally, radioisotopes are used in two ways, i.e. as tracers or as sources of radiation. The use of radioisotopes as tracers is possible because their chemical behavior is identical with that of stable isotopes of the same element and they decay by the emission of radiation, which can be detected so as to indicate their position and quantity. Hence, the addition of a very minute quantity of radioisotope of the same element into a system acts as a marker which, without changing the properties of the material under test, enables the following of the path of the tracer atoms. Tracer technique is very sensitive and the amount of the radioactive substance which can be detected by radiation is generally orders of magnitude less than that which can be found by weighing, chemical means, and even spectroscopic techniques. Besides, the process under investigation need not be stopped if tracer technique is used. This technique has widespread applications because practically all elements and compounds can now be 'tagged' by suitable radioisotopes.

As sources of radiation, radioisotopes are utilized because they are detectable after having been absorbed or scattered by matter, or because they initiate chemical or biological reactions. These reactions are produced without significantly raising the temperature of the absorbing medium and without making this medium radioactive, if due attention is paid to the field intensity, period of exposure, radiation energy, and the target's thermal conductivity, thickness, and composition.

18.2 Radiation Engineering

RADIATION ENGINEERING IS THE BRANCH OF NUCLEAR ENGINEERING CONCERNED WITH THE PRODUCTION AND APPLICATION OF RADIOISOTOPES AND THEIR RADIATION. Usually it is divided into two categories: radioisotope engineering and radioisotope applications. The former field deals with medium to high level radiation while the latter field is mainly concerned with low level radioisotopes. One special feature of radioisotope engineering is its emphasis on the engineering design and evaluation of high level irradiation processes. The increasing trend of utilizing radiation or penetrating isotopes in industry, agriculture, medicine, and other fields demands an increasing number of nuclear engineers specialized in this field. Figure 19.11 presents some of the subjects in the field of radiation engineering.

In the following chapters various techniques of utilizing radioisotopes are discussed. Many of these methods are so versatile that they may be used in many different fields. For example, neutron activation analysis has been used in the petroleum industry, agriculture, electronics, space applications, metallurgy, criminology, numismatics, geology, medicine, and many other fields.

Radioisotope engineering is a recently formalized field of study in nuclear engineering, derived from the progressively wider use of irradiation processes in many fields. Except for some radioisotopes such as I-131, Kr-85, Sr-90, radiation sources in the range of microcuries are usually considered as being within the safety limit so that their handling does not require a detailed and elaborate engineering design. In irradiation applications, however, the radiation sources that are used are usually in the Megacurie range. Therefore, the irradiation facility design must be carefully executed to overcome problems of safety and for ease of operation.

Geographical engineering pertains to the peaceful applications of nuclear explosives. One common characteristic of these applications is the change of the geography. While large scale earth-moving is the frequently mentioned usage, geographical engineering applications are certainly not

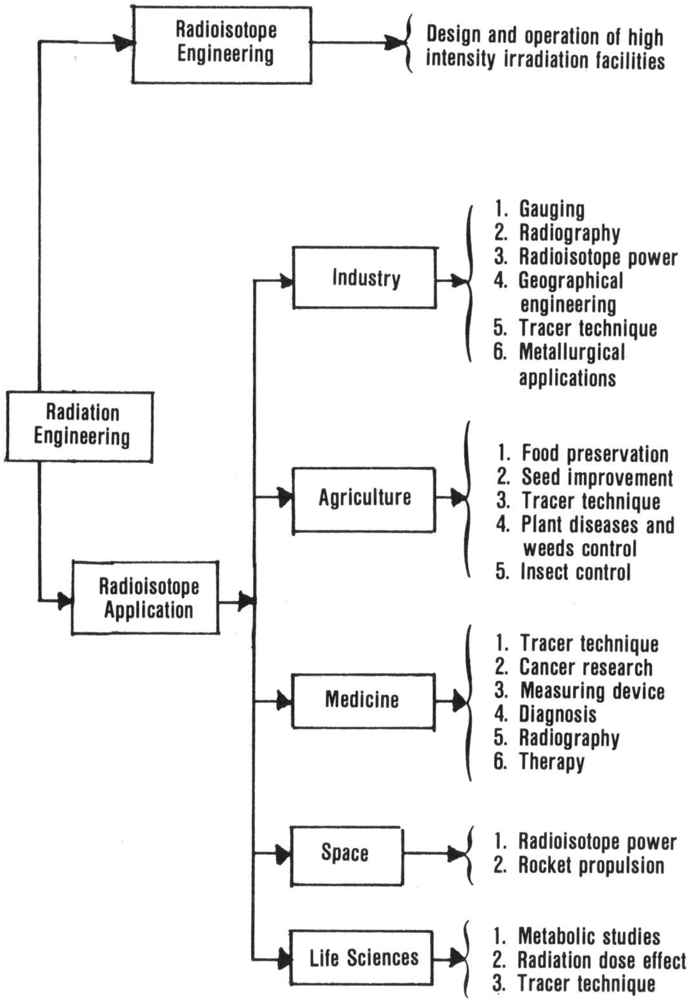

Figure 18.11 : Radiation Engineering

limited to this. These and other uses of radioisotopes will be discussed in the chapter on geographical engineering.

Radioactive tracers are used in many fields such as agriculture, medicine, and industry. They have been very widely used for research work in laboratories. Special radiation equipment is designed for measuring minute quantities of tracers. Carbon dating is a very commonly used method for the determination of the age of specimens in archeology and other applications. Wider use of this technique can be achieved if more systematic studies are carried out and improvements in procedures and equipment can be made.

Radioisotope power is a source of energy that needs some special attention. Due to the very nature of the radiation and energy emitted from radioisotopes, these units are suitable for providing small to medium amounts of electric power for instrumentation. A series of systems for nuclear auxiliary power (SNAP) has been designed and used in space applications. The use of such sources may be possible for transportation some time in the future. If this becomes a reality many of the economical and social problems caused by the energy shortage might be alleviated.

Nondestructive testing, radiography using either gamma rays or neutrons, and many other uses of radioisotopes are also discussed in this Part. Although efforts have been made to introduce the subject matter, the discussions are far from exhaustive. With increasing knowledge of radiation and its interaction with matter, many new and improved uses of radioisotopes in our everyday life can be foreseen.

CHAPTER 19

RADIOISOTOPE POWER

Every radioisotope emits energetic radiation. The energy associated with the radiation can be transformed into electric energy. Such a conversion can be achieved in principle by any of the following methods: The dynamic method is the most frequently used conversion technique. Nuclear energy is used to boil a working fluid which turns a turbine; the turbine then drives an electric generator. The second method is thermionic conversion and is used when a charged subatomic particle is radiated from the chosen radioisotope. Electrons emitted from a heated source are collected to form the output electric current. Another method is thermoelectric conversion, based on the thermoelectric effect. When two dissimilar metals are joined at two ends at different temperatures, a potential difference between these two metals is established. Usually the voltages are in the microvolt to millivolt range. However, if semiconductor materials are used, the efficiency of this type of conversion increases.

Radioisotopes that can be used to supply electric power must have the following characteristics: a) relatively long half-lives of several years, b) high energy radiation so that it would yield higher power densities, c) high mechanical resistance to heat and radiation, and d) radiation emission that is easy to shield. Economic availability is, of course, a main factor in choosing certain radioisotopes for the purpose of producing electricity. Among the presently known radioisotopes, about 1400 or more in number, only a few satisfy the above mentioned conditions. Some of these are listed in table 19.1.

The major sources of these radioisotopes are fission products (Cs-137, Pm-147, Sr-90) and reactor irradiation products (Cm-244 and Pu-238). The beta sources from fission products are abundant but require chemical separation and purification. Alpha sources, although not from fission products, also require chemical processing to separate them from other chemical species contained in the irradiated sample.

Table 19.1: Radioisotopes suitable for power production.

Radioisotope	Half-life (years)	Power density (W/g)	Type of radiation emission and energy (MeV)	Fuel form
Cesium-137	33	0.26	beta (0.66)	CsCl
Curium-244	18	2.8	alpha (5.80)	Cm_2O_3
Promethium-147	2.6	0.36	beta (0.23)	Pm_2O_3
Plutonium-238	89	0.55	alpha (5.50)	PuC
Strontium-90	28	0.93	beta (0.54)	$SrTiO_3$

Figure 19.1 illustrates the fuel preparation from fission products using Sr-90 as an example. This fuel was first used in an electric power generator for weather station instruments at Axel Heiberg Island in Canada. Strontium-90

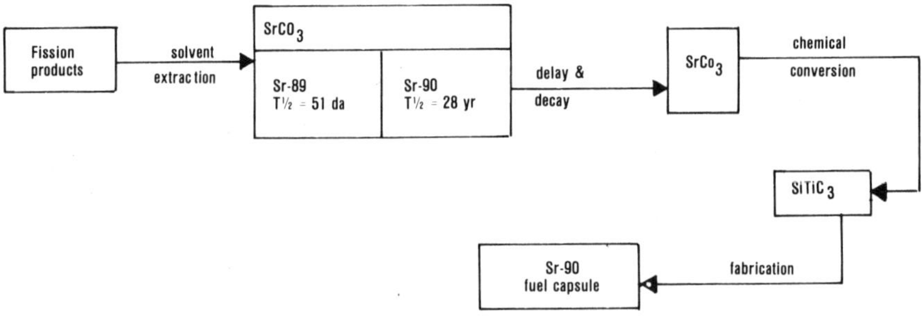

Fig. 19.1: Sr-90 fuel production.

is available in Megacurie quantities in fission products. After solvent extraction $SrCO_3$ is obtained which contains both Sr-89 and Sr-90. Since the half-life of Sr-89 is only 51 days as compared to 28 years for Sr-90, a one year delay and decay process would effectively eliminate all the Sr-89 components in $SrCO_3$. In order to meet the mechanical specifications, strontium carbonate ($SrCO_3$) is further converted to strontium titanate ($SrTiO_3$). This chemical compound of strontium has an excellent mechanical strength for enduring heat, pressure, and radiation. This is next fabricated into small cylindrical fuel pellets, which are then enclosed in the fuel capsules. Solid cylindrical or spherical fuel capsules have the added advantage of maintaining higher fuel temperatures which would yield a higher thermal to electric conversion efficiency. However, due to other problems such as gas release, discussed in the later part of this chapter, hollow fuel elements are fabricated frequently.

The methods used for converting radioisotope heat to electric energy are now discussed. As heat carried by the radiation enters the converter it encounters insulation. Most of this heat usually escapes from the system and is radiated out as waste heat. The basic principle of thermoelectric converters is illustrated in Figure 19.2.

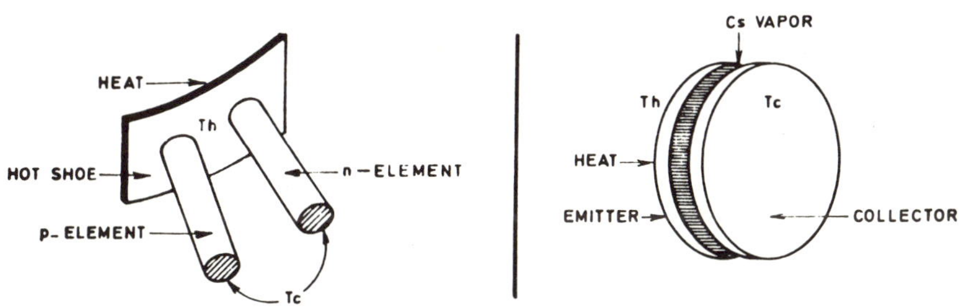

Fig. 19.2 : Thermoelectric converters.

The principle underlying the operation of a thermoelectric converter is the generation of a potential difference due to temperature differences at the joints of dissimilar conductors. The ideal efficiency of this system is given by

$$\eta = \frac{T_h - T_c}{T_h}, \qquad (19.1)$$

where T_h and T_c are the temperatures of the hot and cold joints respectively.

When radiation heat enters the hot shoe of a thermoelectric converter element, it serves to maintain or raise the temperature T_h at this end. With the proper choice of materials and geometric arrangement of these elements, the cold end temperature can be reduced, resulting in increased converter efficiency. The performance of a thermoelectric converter can be judged by the figure of merit Z defined by

$$Z \equiv \frac{\alpha^2_{pn}}{KR}, \qquad (19.2)$$

where α_{pn} is the Seebeck coefficient of the material and is equal to the potential difference produced in the system per unit temperature difference between the hot and cold ends. R is the sum of the resistances of both elements and K is the sum of the thermal conductances.

For a practical thermoelectric converter based on the Peltier effect and the Seebeck effect, the efficiency cannot reach the ideal Carnot cycle efficiency given by equation (19.1). To investigate the energy balance, consider the system in Figure 19.3. Heat from the fuel establishes the hot shoe

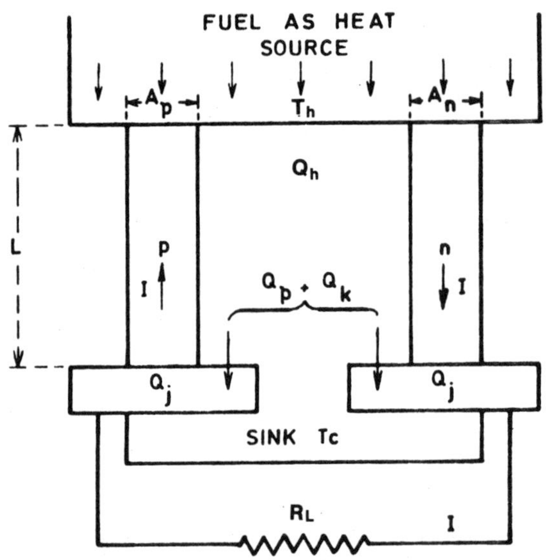

Fig. 19.3: A simplified network diagram of a thermoelectric converter.

temperature T_h. The converter elements are a semiconductor pair of p-type and n-type of materials. These elements are connected to a heat sink and serve as the output voltage terminals. The sink temperature is given by T_c. The current I flowing through the load resistor R_L is generated due to the Seebeck voltage $\alpha_{pn}(T_h - T_c)$. When this current is flowing through the system the energy balance equation is

$$Q_h + Q_j = Q_p + Q_k, \tag{19.3}$$

where Q_h is the heat input rate, Q_j is the Joule heating rate due to the current flow I, Q_k is the heat transfer rate through the p and n elements, and Q_p is the Peltier heat rate. Evaluating these heat rates,

$$Q_h = Q_p + Q_k - Q_j$$

$$= \alpha_{pn} IT_h + (C_p + C_n)(T_h - T_c) - \tfrac{1}{2}I^2(R_p + R_n), \tag{19.4}$$

where C_p, C_n and R_p, R_n are the conductances and resistances of the p-element and n-element respectively.

The output power is given by

$$W = I^2 R_L = \frac{V_L}{R_L}, \tag{19.5}$$

where the terminal voltage V_L is

$$V_L = \alpha_{pn}(T_h - T_c) - I(R_p + R_n). \tag{19.6}$$

Therefore the current is

$$I = \frac{\alpha_{pn}(T_h - T_c)}{R_L + R_p + R_n}. \tag{19.7}$$

From equations (19.4), (19.5), (19.6), and (19.7) the efficiency of the generator can be evaluated. Since $\eta = W/Q_h$, this can be written as

$$\eta = \frac{T_h - T_c}{T_h} \times M, \tag{19.8}$$

where the correction factor M is given by

$$M = \frac{R_L}{(R_L + R_p + R_n) + (\frac{C_p + C_n}{T_h \alpha^2_{pn}})(R_L + R_p + R_n)^2 - (\frac{R_p + R_n}{2T_h})(T_h - T_c)}. \tag{19.9}$$

This factor is always less than one. To increase the value of M, recall the definitions of C and R for a p-element

$$C_p = k_p \frac{A_p}{L_p} \text{ and } R_p = \rho_p \frac{L_p}{A_p}, \tag{19.10}$$

where k_p is the thermal conductivity, ρ_p is the resistivity while L_p and A_p are the length and cross sectional area of the element, respectively. By varying L_p and A_p or by selecting materials according to their thermal and electric properties as given by k and ρ, a generator designer can optimize the value of M to attain a higher efficiency. If the geometric arrangement is to be altered to optimize M, one seeks

$$\frac{\partial M}{\partial x} = 0, \tag{19.11}$$

where $x = A_n / A_p$ is the ratio of the cross sectional areas of the n-element to the p-element. When this is done one obtains

$$\left(\frac{A_n L_p}{A_p L_n}\right) = \left(\frac{\rho_n C_p}{\rho_p C_n}\right)^{1/2}. \tag{19.12}$$

From these the optimum efficiency can be obtained

$$\eta_{opt} = \frac{T_h - T_c}{T_h}(M_{max}), \tag{19.13}$$

where M_{max} is the maximum value of M.

The major problems encountered in thermoelectric conversion are the heat loss and the Joule loss. In the first case, heat enters the hot end and propagates through the entire length of the element without being converted. To avoid this tendency one needs to choose materials with low thermal conductivity. On the other hand, some materials with high resistivity would use up large amounts of electric energy in the form of the I^2R loss. This factor must be kept at a minimum by choosing materials with small values of resistivity. From equation (19.2) notice that reducing the values of R and K increases the figure of merit of the converter. Apparently these two requirements are contradictory to each other because a good electric conductor is almost always a good thermal conductor. Fortunately, the recent developments in semiconductor technology have at least partially solved the material problems for converters. Typical semiconductors that can be used for converter elements are lead telluride, bismuth telluride, and germanium silicide. Some other possibilities are cadmium sulfide, cobalt silicide, and gadolinium selenide. The choice of a converter is therefore mainly a problem of material research to improve the converter element performance.

Thermionic conversion is quite different in principle and operation from thermoelectric conversion. It has the advantages of smaller weight and greater efficiency. The second factor leads to reduced waste heat rejection problems. Its efficiency can be obtained from the following considerations of the energy balance. The current density in a thermionic converter can be expressed as

$$J = A_0 T^2 e^{-\phi/KT}, \tag{19.14}$$

where A_0 is a constant, T is the absolute temperature, K is the Boltzmann's constant and ϕ is the work function of the emitter. In a thermionic converter the net current density is the algebraic sum of the current densities passing through the cathode and the anode.

$$J = J_c - J_a = A_{0c} T_c^2 e^{-eV_c/KT} - A_{0a} T_a^2 e^{-eV_a/KT}, \tag{19.15}$$

where V_c and V_a are the potentials of the cathode and the anode, respectively. Therefore the net power output per unit area of the conductor can be written as

$$\frac{W}{A} = J(V_c - V_a). \tag{19.16}$$

The heat input rate to the converter must be equal to the power needed to release the electrons from the cathode or the heat required for compensating cathode electron emission cooling. The average energy carried by electrons to overcome the cathode potential may be expressed as $eV_c + 2KT_c$. The electrons emitted from the anode would give up an energy $eV_c + 2KT_a$ at the cathode. If the radiation and heat losses through the gap between the electrode and the electric conducting materials such as switches and wirings are denoted by Q_L, the total power input to the thermionic converter can then be summed up as

$$\frac{Q}{A} = J_c\left(V_c + \frac{2KT_c}{e}\right) - J_a\left(V_c + \frac{2KT_a}{e}\right) + Q_L . \tag{19.17}$$

From equations (19.16) and (19.17), the efficiency can be calculated,

$$\eta = \frac{W}{Q} = \frac{(J_c - J_a)(V_c - V_a)}{(J_c - J_a)V_c + \frac{2K}{e}(J_c T_c - J_a T_a) + Q_L} . \tag{19.18}$$

This equation can be rewritten as

$$\eta = \frac{V_c - V_a}{V_c} M, \tag{19.19}$$

where the correction factor M is given by

$$M = \frac{1}{1 + \frac{2K}{e(J_c - J_a)V_c}\left(J_cT_c - J_aT_a + \frac{Q_L}{V_c(J_c - J_a)}\right)}. \tag{19.20}$$

Note that in idealized cases, $Q_L = 0$ and $J_cT_c = J_aT_a$ so that $M = 1$. In this case, since the potentials at the electrodes are proportional to their temperatures, the thermionic converter efficiency reduces to the ideal efficiency $(T_c - T_a)/T_c$.

The two flat plates serving as hot (cathode) and cold (anode) terminals are separated by a metallic vapor. Electrons are emitted from the hot plate through the outside load back to the collector which is the cold plate of the converter. In this converter, there are still many problems regarding the energy losses. The most serious problem is the direct bombardment of the collector (anode) by the emitter electrons through the gap between them.

The next discussion deals with problems involved in the radiation shielding and the waste heat rejection of a radioisotope power plant. In such a plant, up to 90% of the heat from the fuel may have to be ejected as waste heat from the plant. The converter material used is a critical factor in waste heat ejection considerations. There is usually a narrow range of temperatures in which the converter has the highest efficiency. For example, if it is lead telluride, the hot-junction temperature must be within the range of 900°C to 1000°C. Usually the amount of heat extracted from the fuel is more than what is needed to maintain this temperature. As a result excess energy must be ejected from the system. On the outer surface of the generator plant ordinary heat radiation and convection methods can be used to achieve waste heat ejection. A more serious problem is, however, radiation shielding and safety. Because of the nature of the radiation emission and decay of the fuel, gases will accumulate in the fuel capsule. For example, if Ce-144 is used, a series of decays of the following forms will occur due to changes of valence:

$$CeO_2 \xrightarrow{\beta^-} Pr_2O_3 \xrightarrow{\beta^-} Nd_2O_3. \tag{19.21}$$

When CeO_2 decays to Pr_2O_3, oxygen is released. One way to solve this problem is to fabricate the fuel in the form of hollow cylinders, thereby providing space for gas retention within the fuel.

Radiation hazards are also a very real problem in such plants. The problem is even more pronounced if plutonium is used as a fuel because of its very hazardous nature if it is ingested. When Sr-90 is used, a special biological problem exists because of its bone seeking characteristics. To alleviate this problem, the fuel capsule must be designed carefully. The capsule must be able to withstand fire, collision and radiation damage.

The major radiation emission from a radioisotope generator is Bremsstrahlung. It is the secondary radiation produced inside the generator due to the absorption of alpha or beta particles by the converter and shielding materials. Lead up to four inches thick must be used as the casing to house the converter and the fuel capsule to prevent this radiation from escaping.

A radioisotope thermoelectric generator (RTG) consists of four major parts as illustrated in Figure 19.4. For convenience and economy, a RTG is

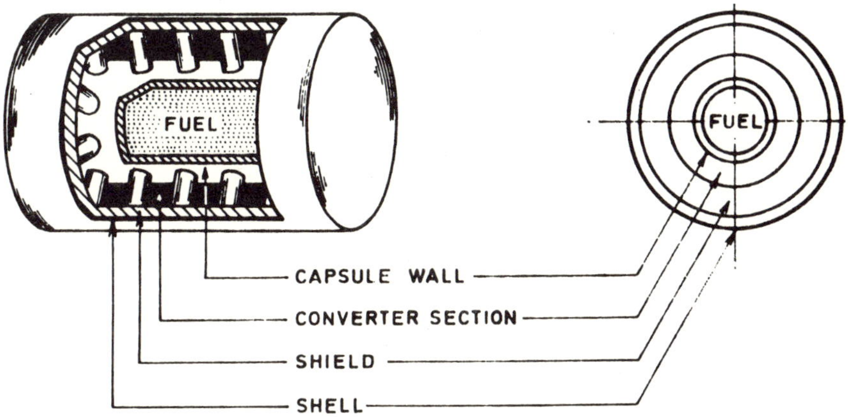

Fig. 19.4: A radioisotope generator.

designed to have a cylindrical geometry. The innermost part of the generator is the encapsulated fuel. The heat generated in the fuel is carried off by radiation (alpha or beta particles) to the energy conversion section. A number of converter elements absorb the heat energy and convert it into electric energy by thermionic emission. The converter is surrounded by a massive radiation shield for safety purposes. Several inches of lead or cast iron are used to shield the secondary radiation. Finally, in the outermost region, a metal shell serves as a radiator for waste heat ejection and as a container for the entire assembly. A cutaway view of an actual operating radioisotope generator (SNAP-1A) is shown in Figure 19.5.

Various kinds of radioisotope generators have been used in space applications. The shielding and safety problems are most carefully examined

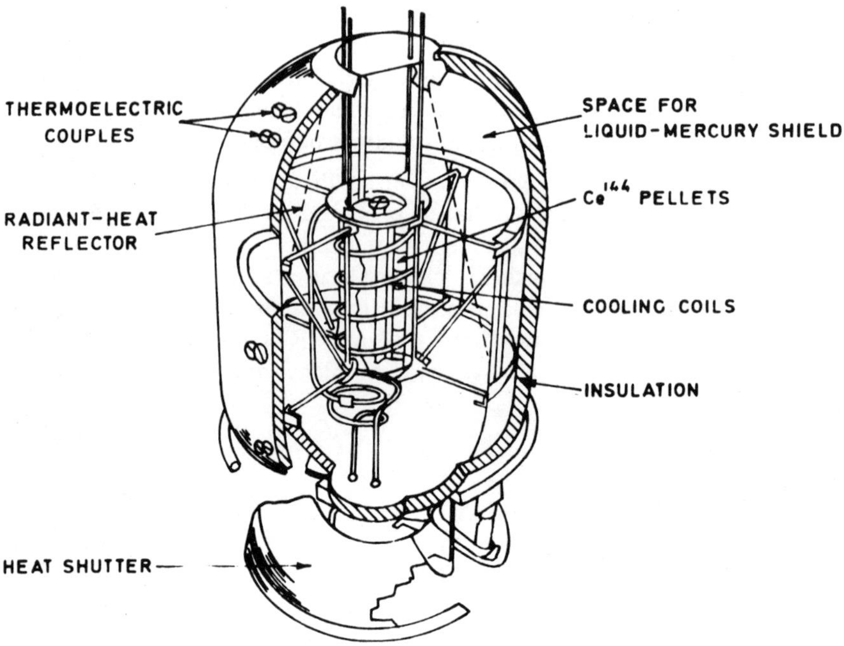

Fig. 19.5: Cutaway view of SNAP-1A.

in many NASA laboratories. Some fuel capsules are tested under severe temperatures (~ 2500° C or higher) and pressure (~ 1000 psi) conditions. In a specific experiment, a capsule was subjected to a pressure of 10,000 psi upon impact with a block of granite. Even under such extreme conditions, there was no damage to the capsule. This illustrates the ruggedness of radioisotope power sources as well as their safety under various circumstances.

The basic engineering design of a radioisotope generator involves i) uniformity in the power supply, ii) structural integrity, and iii) load matching. These tasks must be performed or accomplished from trade offs of the following factors:

a. safety — radiation and heat,
b. reliability,
c. weight, and
d. cost.

The choice of a relatively long lived radioisotope as the fuel automatically solves the problem of uniformity in the power supply. The generator power is not excessive in the initial stages of its operation; therefore large waste heat ejection is not required. After long periods of operation, however, the generator power will not be reduced too much. The problems of structural integrity and load matching must be solved by compromising weight and cost. Structural integrity involves the safety and reliability of operation. Fuel capsules must be designed and tested in such a way that they prove to be practically rupture proof. This was achieved in many practical systems as mentioned before. The load matching is usually achieved by DC to DC coupling. The advantages of this method are many; above all, it is static. The entire system of load matching includes the converter, switching network, transformer, rectifier, and filter. The wear and tear problems encountered in a conventional large scale power plant do not exist in this case.

The most famous radioisotope generator usage is in space applications. The series of the systems for nuclear auxiliary power (SNAP) with odd-numbered designations are radioisotope generators. Even-numbered SNAPS are small nuclear reactors. Table 19.2 lists this series of generators.

Table 19.2: Characteristics of various SNAP generators.

Generator designation	Power (watts)	Weight (pounds)	Fuel Isotope	Fuel half-life	Generator life
SNAP-3	2.5	4	Polonium-210	138 da	90 da
SNAP-3A	2.7	4.6	Plutonium-238	89.6 yr	5 yr
Undesignated	5	1680	Strontium-90	28 yr	2 yr
SNAP-7A	10	1870	Strontium-90	28 yr	2 yr min.
SNAP-7B	60	4600	Strontium-90	28 yr	2 yr min.
SNAP-7C	10	1870	Strontium-90	28 yr	2 yr min.
SNAP-7D	60	4600	Strontium-90	28 yr	2 yr min.
SNAP-7E	7.5	600	Strontium-90	28 yr	2 yr min.
SNAP-7F	60	4600	Strontium-90	28 yr	2 yr min.
SNAP-9A	25	27	Plutonium-238	89.6 yr	5 yr
SNAP-11	21-25	30	Curium-242	162 da	90 da
SNAp-13	12	4	Curium-242	162 da	90 da
SNAP-15A	0.001	1	Plutonium-238	89.6 yr	5 yr
SNAP-17	25	30	Strontium-90	28 yr	5 yr
SNAP-19B	30	30	Plutonium-238	89.6 yr	5 yr
SNAP-21	10	500	Strontium-90	28 yr	5 yr
SNAP-23	60	900	Strontium-90	28 yr	5 yr
SNAP-27	60	30	Plutonium-238	89.6 yr	5 yr
SNAP-29	500	500	Polonium-210	138 da	90 da

With the increasing energy shortage, radioisotope power utilization furnishes a very attractive alternative. Its small size and relatively minor radiation hazard problems make its application that much easier. Its fuel can be obtained from radioactive wastes from a nuclear fuel reprocessing plant. It is quite conceivable that solidified radioactive wastes can be utilized as the fuel for small radioisotope power palnts. Some studies have been carried out in this direction. At present there are numerous technical difficulties in the design of radioisotope generators in the kW or MW range. The largest SNAP unit designed so far is capable of supplying only around 500 watts of electric power. Besides supplying small rugged power sources for numerous applications, the radioisotope power generators show some promise of solving the radiation hazard problem by utilizing the otherwise useless radioactive wastes. The development of radioisotope generators is hence a very beneficial endeavor.

Example 19.1

A thermal electric converter is operating between temperatures of 300°C and 25°C. The Seebeck coefficient of the semiconductors is $\alpha_{pn} = 400\mu V/°C$. $k_p = 0.011\ W/cm°C$, $k_n = 0.012\ W/cm°C$, $R_p = 0.005\Omega$, $R_n = 0.005\Omega$. Calculate the efficiency of this converter if the load resistance is 0.01Ω. The dimensions of the semiconductors are the same i.e. $L = 0.2$ cm and $A = 0.1\ cm^2$.

Solution

From equation (19.9), the efficiency correction factor is

$$M = \frac{R_L}{(R_L + R_p + R_n) + \frac{(C_p + C_n)(R_L + R_p + R_n)^2}{T_h\, \alpha^2_{pn}} - \frac{(R_p + R_n)(T_h - T_c)}{2T_h}}$$

$$= \frac{0.01\Omega}{0.02\Omega + \frac{(0.02\Omega)^2(C_p + C_n)}{(573°K)(4 \times 10^{-4}V/°C)^2} - \frac{(0.01\Omega)(275°C)}{2(573°K)}}$$

Since $C_p = A_p k_p / L_p$ and $C_n = A_n k_n / L_n$,

$$\therefore C_p + C_n = \frac{A}{L}(k_p + k_n) = \frac{0.1\ cm^2}{0.2\ cm}(0.011 + 0.012)\ W/cm°C$$

$$\cong 0.016\ W/°C.$$

$$\therefore M = 0.01\Omega/\ [(0.02\Omega + \frac{(0.02\Omega)^2(0.016W/°C)}{(573°K)(4 \times 10^{-4}V/°C)^2} - \frac{(0.01\Omega)(275°C)}{2(573°K)}]$$

$$\cong 0.45.$$

$$\therefore \eta = \frac{T_h - T_c}{T_h} M = \frac{275°K}{573°K}(0.45) \cong 22\%.$$

Example 19.2

The open circuit voltage of a thermionic converter is 3 volts. The work functions of the cathode and the anode are 4.2 eV and 4.1 eV, respectively. If the cathode temperature is 2500° K and is at a potential of 2.7 volts, calculate the current density. $A_0 = 50A/cm^{2\circ}K^2$.

Solution

The cathode potential energy with respect to that of the anode is given by

$$\phi_c = eV + \phi_a = 2.7 + (4.2 - 3.0) = 3.9 \text{ eV}.$$

Therefore,

$$J = A_0T^2e^{-\phi_c/KT}$$

$$= (50A/cm^{2\circ}K^2)(2500°K)^2e^{-3.9eV/(8.62 \times 10^{-5}eV/°K)(2500°K)}$$

$$= 4.3 \text{ A/cm}^2.$$

Problems

19.1 Is the efficiency of 22% calculated in example 19.1 an optimum value? Explain your answer.

19.2 What is the mass of Co-60 needed to produce an initial power of 200 W in a generator if the efficiency is 20%?

19.3 Construct a list of radioisotopes the power densities of which are larger than 5 watts/g. Discuss the possibilities of using any one of the listed radioisotopes as a power source.

19.4 Explain why ordinary thermocouples are not suitable as converters for a thermoelectric generator.

19.5 A Co-60 thermoelectric generator is operating between a high temperature of 290° C and a sink temperature of 30° C. It is desired to have a total power output of 5 watts. If the generator has an efficiency of 20%, what is the required activity of Co-60?

19.6 Three kilocuries of Po-210 are used as the heat source of a RTG. Calculate the total power output if the efficiency is 8%.

19.7 If the thermionic work function of tungsten is assumed to be 4.5 eV, calculate the ideal current density from a tungsten surface when it is heated to 2000° K?

19.8 A thermionic converter uses tungsten as the cathode and ZrO_2 as the anode. The cathode temperature is 2500° K and that of the anode is 1000° K. If the cathode open circuit voltage is 3 volts and the work function is 4.5 eV, calculate a) the anode work function, b) the current density, and c) the ideal efficiency of this converter.

19.9 An important factor in the selection of materials for the cathode and anode of a thermionic converter is the work function. If the work functions of the cathode and anode are equal, what would be the open circuit voltage across the terminals? Also calculate the power output.

19.10 Calculate the figure of merit Z of a material in a thermoelectric cooler if the maximum temperature drop is 100° C at a cold temperature of 200° K.

Ans: 5×10^{-3}.

19.11 If the Peltier coefficient of a certain material is given by $\nu = T(100 - 2T) \times 10^{-7}$, what is the Seebeck coefficient at 900° K?

19.12 The cold temperature of a thermoelectric generator is kept at 300° K. The hot temperature is 500° K. What sould be the hot temperature if the generator efficiency is to be increased by 50%?

CHAPTER 20

NEUTRON ACTIVATION ANALYSIS

20.1 Introduction

The exposure of a sample to some subatomic particles such as protons, deuterons, or neutrons, generally results in a new product material. As the most frequently used bombarding particles are neutrons, the discussion in this chapter will be concerned with neutron activation. The new product from neutron activation is almost always radioactive with a specific half-life, emitting gamma rays with characteristic energies. From the measurements of these characteristic radiations the identification of the original sample material and the product material can be made.

Subatomic particles were first used to produce new isotopes as early as 1933. In 1936, George Hevesy and Hilde Levi activated a sample of yttrium to detect the dysprosium impurity contained in it. Since that time, numerous discoveries, research and development work have shown the usefulness of neutron activation. The availability of large neutron sources of all kinds, including nuclear reactors, combined with great improvements in radiation detection systems have now established the effectiveness of neutron activation analysis as a tool for applied and pure research. Due to the ability to detect tracer quantities of many kinds of isotopes, neutron activation analysis is now used widely in many fields. Table 20.1 gives some applications of neutron activation analysis (NAA).

Table 20.1: Applications of neutron activation analysis.

Field	Specific Applications
Space	1. Moon rock examination 2. Remote controlled moon surface exploration
Criminology	1. Hair identification 2. Wiping test 3. Gun powder analysis
Medicine and Biology	1. Trace element measurement 2. Identifying dangerous chemicals 3. Identifying opium sources 4. Drug and pharmaceutical material analysis 5. Toxicant analysis 6. Manganese content of insects 7. Kidney stone studies 8. Tracing metals in metabolism 9. Therapeutic applications
Agriculture	1. Pesticides control analysis 2. Fertilizers and feeding stuff analysis 3. Food and food additive examinations 4. Seed improvement 5. Crop examination
Geology	1. Rock analysis 2. Relative abundance of elements 3. Meteorites analysis 4. Petroleum geology 5. Mineral and ore examination 6. Soil, dust, and water analysis 7. Sea sand examination
Industry	1. Fuel analysis 2. Glass and ceramics examination 3. Lubricant examination 4. Metal and alloy analysis 5. Paint and pigments impurity 6. Petroleum products test 7. Plastic and resinous materials testing 8. Oil refinery feeds analysis 9. Semiconductor impurity studies 10. Oxygen impurity in steel

The basic operations in neutron activation analysis are

a) activation of a sample by neutron bombardment and
b) measurement and analysis of radioactivity of the activated sample.

Before carrying out a specific type of neutron activation analysis, some preliminary studies must be made. The choice of the best nuclear reaction and the suitable neutron sources must be determined. At present there are three main categories of neutron sources available for neutron activation applications:

1) Nuclear reactor — the energies of reactor neutrons vary over a wide range, from thermal (0.025 eV) to fast (up to 10 MeV or higher). The neutron flux density of the order of $10^{15} n/cm^2 s$ can be made available.

2) Accelerator neutron generators — the neutron energy is usually single-valued. However, the emitted neutron density from the target of such a generator is only in the range of 10^5-$10^8 n/cm^3$.

3) Encapsulated or sealed tube neutron sources — the neutron energy is also single-valued. Depending on the type of nuclear reactions chosen for the neutron production and the source size, the neutron emission rates range from 10^5 to $10^{10} n/cm^3$. This high yield portable neutron source development is a very promising factor in NAA. Sometimes, due to the neutron source characteristics and irradiation facility set up, samples might have to be prepared to suit the particular conditions of irradiation.

The various steps involved in neutron activation analysis are summarized in Figure 20.1.

20.2 Neutron Activation

There are several neutron reactions of interest in NAA. The major interaction for thermal neutrons is (n, γ). For fast neutrons, besides (n, γ) reactions, (n, p), (n, α), and (n, 2n) are also encountered, depending on the type of sample material. For low atomic number materials both (n, α) and (n, p) might occur. In either thermal or fast neutron activation, however, the (n, γ) reaction is still of major importance. All of our discussion on NAA is therefore based on neutron activation followed by gamma ray emission of the activated sample. The amount of radioactive nuclide produced in an irradiated sample can be calculated from the balancing of the production and the decay rates. The production rate is proportional to the neutron flux density ϕ. It is also proportional to the macroscopic activation cross section, Σ_{ac}, of the sample if the sample volume is so small that the inner part of the sample is exposed to the same amount of neutron flux density as the outer portion. The decay rate of the activated radionuclides is λN, where λ is the decay constant

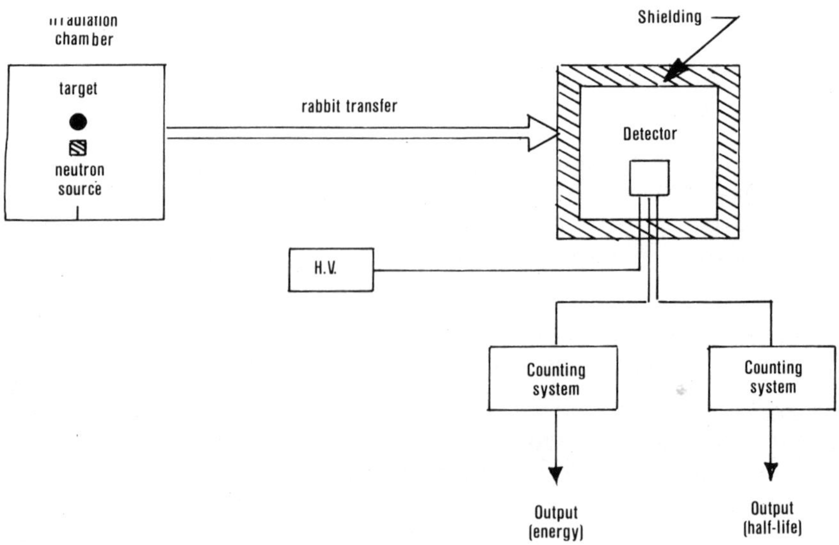

Fig. 20.1 Neutron activation analysis system.

and N is the number of radionuclides per unit volume. On the basis of the above discussion, the following equation is obtained,

$$\frac{dN}{dt} = \Sigma_{ac}\phi - \lambda N, \tag{20.1}$$

where dN/dt is the rate of change of radionuclide density. If the number of nuclei of the sample is very large as compared with the number of activated radionuclides, Σ_{ac} and ϕ may be considered as constants. Then the solution of equation (20.1) is

$$N = \frac{\Sigma_{ac}\phi}{\lambda}(1 - e^{-\lambda t}). \tag{20.2}$$

The induced activity per unit volume of the sample is thus given by

$$A = |\lambda N| = \Sigma_{ac}\phi(1 - e^{-\lambda t}), \tag{20.3}$$

where t is the time of irradiation. The special case of $t \rightarrow \infty$ yields the so-called saturation activity (per unit volume),

$$A_{sat} = \Sigma_{ac}\phi. \tag{20.4}$$

This is the maximum activity per unit volume obtainable from such an irradiation arrangement. From this, one can calculate the maximum number density of the activated radionuclides.

Since activities are usually expressed in terms of curies, we can use equation (20.3) to calculate the specific activity of the irradiated sample and write the result in the following form,

$$A_s = \frac{A_0 \sigma_{ac} \phi}{M(3.7 \times 10^{10})} (1 - e^{-\lambda t}), \tag{20.5}$$

where A_s in curies/g is the specific activity induced in the sample, A_0 is Avogadro's number, σ_{ac} and M are the microscopic activation cross section in cm^2 and the atomic mass of the sample in g/mole, respectively, and ϕ is in terms of n/cm^2s.

In a practical irradiation exercise, the actual induced activity is always less than what is calculated from equation (20.5). The ratio of these values is called the irradiation efficiency. The factors that influence the irradiation efficiency are the 1) self-shielding of the sample, 2) target impurities, and 3) the number of the nuclide of the sample. The third source of error is a very important factor when high neutron flux density and activation cross section are involved. The irradiation efficiency ϵ ranges from 20% to 80% depending on the activation process. Therefore, in a practical system, equation (20.5) must be corrected to read as follows,

$$A_s = \frac{A_0 \sigma_{ac} \phi \, \epsilon}{M(3.7 \times 10^{10})} (1 - e^{-\lambda t}). \tag{20.6}$$

Even with this correction this equation is still an approximation. Accurate results can be calculated from equation (20.1) only if the time variation of the macroscopic cross section and the neutron flux density, the self shielding and other factors are taken into consideration. This usually requires computer aided calculations.

20.3 Measurement and Analysis of Activated Sample

The basic radiation measurements in NAA are the gamma ray differential spectrum and the induced activity measurements. In the first case, the energy of the radiation (gamma rays) is determined. The induced activity measurement yields information for the determination of the half-life of the radioisotope produced from neutron activation. The block diagram of a typical laboratory set-up for carrying out these measurements is given by Figure 20.2.

A multichannel analyzer (MCA) can be used to measure the gamma ray differential spectrum of the irradiated sample. If the sample contains

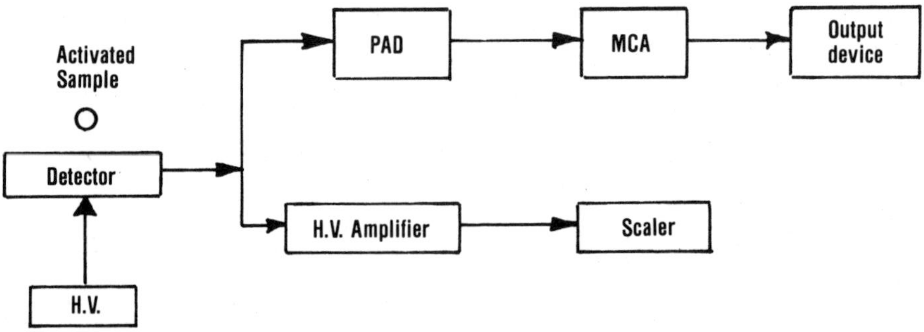

Fig. 20.2: The block diagram of a NAA counting system.

several different isotopes, this spectrum might be very complicated. In this case, one has to use many standard spectra from numerous radioisotopes. Successively the standard spectra can be subtracted from the differential spectrum. This process is called the peeling process. Such a process is usually quite involved. It requires elaborate and careful analysis. A very simple example of a mixture of Cs-137 and Co-60 is illustrated in Figure 20.3. If a

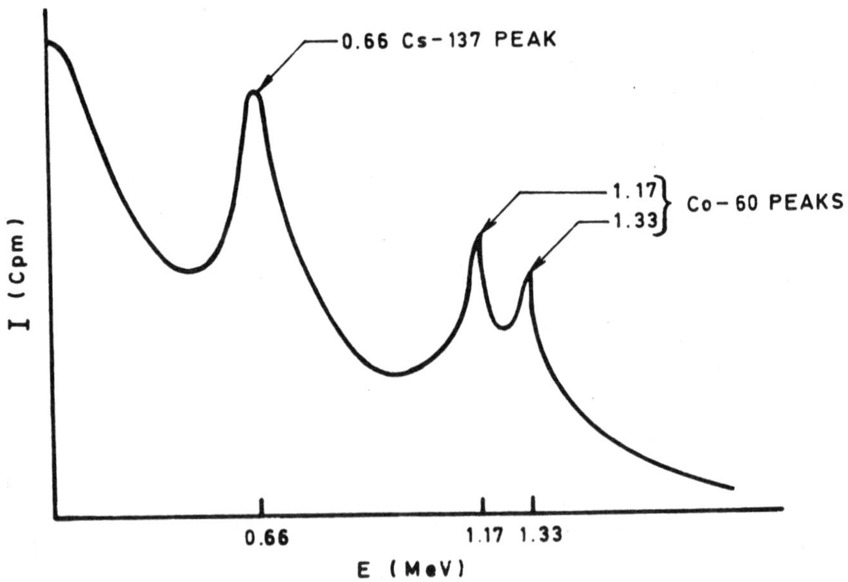

Fig. 20.3: Cs-137 and Co-60 gamma ray differential spectrum.

standard differential spectrum of Co-60 is subtracted from this curve, the only remaining peak of the resulting spectrum would be that due to Cs-137 alone. Caution must be exercised here because the height of a peak depends on various irradiating and couting conditions.

The quantity of a sample that can be measured to yield a specified activity is called the sensitivity. It varies with the neutron flux density and the time of irradiation. It also depends on the characterisitcs of the counting system. In some cases the NAA sensitivity has been arbitrarity defined as the amount of material in micrograms that yields a total of 100 disintegrations per minute initially. If this definition is used, based on equation (20.6), the sensitivity can be calculated by

$$m(\mu g) = \frac{10^7 M}{6 A_0 \sigma_{ac} \phi (1 - e^{-\lambda t})} . \tag{20.7}$$

It is obvious that m decreases with increasing neutron flux density. If the neutron source strength is 10^{10} n/s or more, it is quite possible that the sensitivity of a certain irradiated material might be as low as a few billionths of a gram.

The measurement of the gamma ray differential spectrum using a MCA is straight forward. The result can be printed out, shown on an oscilloscope, or plotted on an x-y plotter.

The half-life or the activity measurement of the irradiated sample can be done simultaneously with the spectrum measurement. It requires a detailed recording of all NAA operations. After the sample is irradiated for a time t, and is removed from the irradiation chamber, the induced activity in the sample at any time t after an irradiation time t_1 is given by

$$A = A_s m = \Sigma_{ac} \phi \epsilon V (1 - e^{-\lambda t_1}) e^{-\lambda (t - t_1)}$$

$$= \Sigma_{ac} \phi \epsilon V (e^{\lambda t_1} - 1) e^{-\lambda t} , \tag{20.8}$$

where V is the volume of the sample and A is the activity in disintegrations per second. If the sample is being counted from t_2 to t_3, the average value of the counting rate can be obtained from

$$\bar{A} = \frac{1}{t_3 - t_2} \int_0^{t_3 - t_2} A(t_2) e^{-\lambda t} \, dt$$

$$= \frac{\Sigma_{ac} \phi \epsilon V}{(t_3 - t_2)\lambda} (e^{\lambda t_1} - 1) e^{-\lambda t_2} [1 - e^{-\lambda (t_3 - t_2)}].$$

$$\therefore \bar{A} = \frac{\Sigma_{ac} \phi \epsilon V}{(t_3 - t_2)\lambda} (e^{\lambda t_1} - 1)(e^{-\lambda t_2} - e^{-\lambda t_3}). \tag{20.9}$$

This average counting rate can be regarded as the counting rate of the sample at a time $t = (t_3 - t_2)/2$. From this point onward, successive counting measurements can be carried out. The plotting of the counting rate versus time can then be used to estimate the half-life of the activated sample. In case two or more radioisotopes with markedly different half-lives are contained in the sample, the activity curve is no longer a simple exponential decay curve. In such a case, the data peeling method must be used to identify the half-lives of both of the components. Taking silver, for example, naturally occuring silver has two isotopes Ag-107 and Ag-109, and the abundance of each isotope is approximately 50%. After irradiation and counting, the activity curve obtained may be similar to the one shown in Figure 20.4. In this figure, the

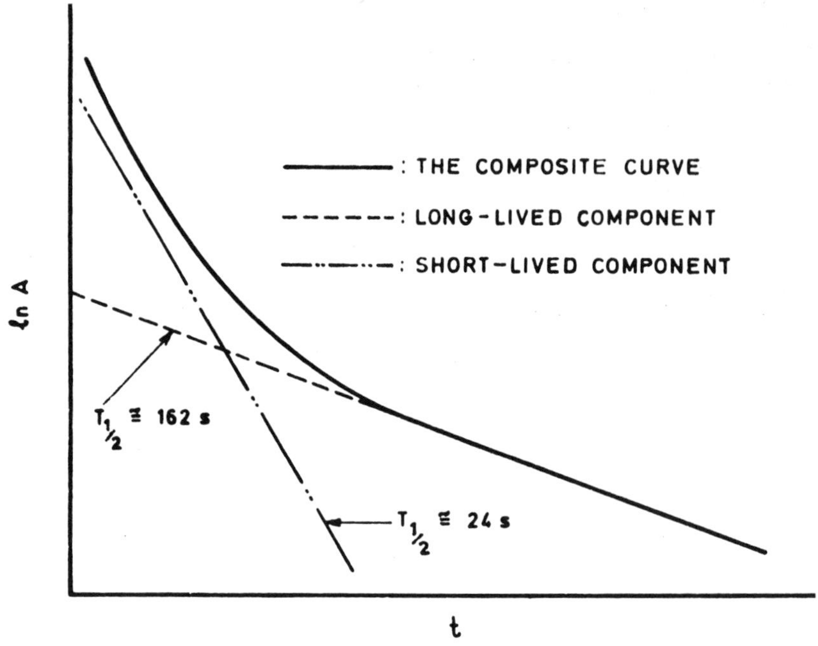

Fig. 20.4: Activated silver decay curve.

composite curve is based on the data from the activity measurement. It includes both the long lived and the short lived radioisotope activities. The peeling procedure is done by 1) identifying the straight line portion of the composite curve and extrapolating it to the y-axis which represents the initial time (the time at which the first counting was made), 2) subtracting the long lived component from the composite curve to obtain the short lived component, and 3) calculating the half-lives of both components.

In order to make a precise determination of the tracer quantity of an element in an unknown sample, a standard sample with a known mass of the element being determined is used. Both the unknown and the standard samples are irradiated under identical conditions. The activity of the element is isolated and then the standard and the unknown are counted under identical counting conditions. A comparison of the radioactivities of the standard and the unknown can then be made. This comparison involves both the radiation energy measurement and the half-life determination as mentioned before. From this analysis the impurity in the unknown can be quantitatively determined.

Note that all foil neutron detector operations are NAA with the only difference that, in neutron detection, neutron flux density is calculated when the foil mass and volume are given. Also, in reactor operations, both the fuel burnup and the absorber calculations are based on the same principle as that of NAA.

Example 20.1

When boron is used as burnable control poison in a reactor in the form of borated water, tritium is produced. The major interactions are the following:

1) $^{10}B + n \rightarrow 2(^{4}He) + T.$

This is a fast neutron reaction with a threshold energy of ~ 1 MeV and an average cross section of ~ 0.05 b in the neutron energy range from 1 MeV to 10 MeV.

2) $^{10}B + n \rightarrow {}^{4}He + {}^{7}Li.$

$^{7}Li + n \rightarrow {}^{4}He + T + n.$

The first reaction is a thermal neutron reaction with a cross section of ~ 670 b. The second is a fast neutron reaction with a threshold of 3 MeV. Its cross section varies linearly with the neutron energy from zero at 3 MeV to 400 mb at 6 MeV and is constant from 6 MeV up to 14 MeV. Over the reactor fast neutron spectrum, the average value of the cross section may be taken to be 0.05 b.

A reactor is operating at a constant neutron flux density of 10^{15} n/cm^{2}s. Assuming that this flux density is for both thermal and fast neutrons, derive relations for calculating the tritium release from the reactor. The initial boron concentration is 2000 ppm.

Solution

Tritium is produced both from the fission products and the

neutron activation of borated water in a reactor. The tritium release of the reactor is mainly due to the latter because practically all the tritium produced by fission is confined within the capsule.

1) Let N = boron concentration per unit volume,
N_1 = tritium concentration per unit volume, and
σ_{aB} = boron activation cross section in barns.

$$\frac{dN}{dt} = -N\sigma_{aB}\phi. \quad \therefore N = N_0 e^{-\sigma_{aB}\phi t} \approx N_0.$$

$$\frac{dN_1}{dt} = N\sigma_{aB}\phi - \lambda N_1.$$

$$\therefore N_1 = \frac{N_0\sigma_{aB}\phi}{\lambda_1 - \sigma_{aB}\phi}(e^{-\sigma_{aB}\phi t} - e^{-\lambda_1 t}).$$

Here N_0 is the initial boron concentration.

$$N_0 = 0.034 \times 10^{24} \times 2000 \times 10^{-6}\text{cm}^{-3} = 6.8 \times 10^{19} \text{ per cm}^3.$$

$$\sigma_{aB}\phi = (670 \times 10^{-24} \text{ cm}^2)(10^{15} \text{ n/cm}^2\text{s}) = 6.7 \times 10^{-7}/\text{s}.$$

$$\lambda = \frac{0.693}{T_{1/2}} = \frac{0.693}{12.3 \text{ yr}} = 2.7 \times 10^{-7}/\text{s}.$$

$$\therefore N_1 \cong 1.1 \times 10^{20} (e^{-2.7\times10^{-7}t} - e^{-6.7\times10^{-7}t}).$$

2) Let N_2 and N_3 be the concentrations of Li-7 and tritium, respectively. σ_{aLi} is the Li-7 microscopic activation cross section. The balance equations are

$$\frac{dN}{dt} = -N\sigma_{aB}\phi,$$

$$\frac{dN_2}{dt} = N\sigma_{aB}\phi,$$

and

$$\frac{dN_3}{dt} = N_2\sigma_{aLi}\phi - \lambda_3 N_3,$$

where λ_3 is the tritium disintegration constant. Solving these equations with the initial conditions

$N = N_0$ and $N_2 = N_3 = 0$ when $t = 0$,

$N = N_0 e^{-\sigma_{aB}\phi t}$,

$N_2 = N_0(1 - e^{-\sigma_{aB}\phi t})$,

and

$$N_3 = N_0\sigma_{aLi}\phi\left[\frac{1}{\lambda_3}(1 - e^{-\lambda_3 t}) - \frac{1}{\lambda_3 - \sigma_{aB}\phi}(e^{-\sigma_{aB}\phi t} - e^{-\lambda_3 t})\right].$$

$N_0 = 6.8 \times 10^{19}/cm^3$, $\sigma_{aB}\phi = 6.7 \times 10^{-7}/s$, $\lambda_3 = 2.7 \times 10^{-7}/s$, and $\sigma_{aLi} = 0.05$ b.

$$\therefore N_3 = 3.4 \times 10^{15}\,[2.7(1 - e^{-2.7\times10^{-7}t}) + 2.5(e^{-6.7\times10^{-7}t} - e^{-2.7\times10^{-7}t})].$$

Example 20.2

Calculate the specific activity of a gold foil irradiated by a thermal neutron flux density of 10^{12} n/cm^{2}s for 5.4 days.

Solution

Gold has a 100% natural abundance of Au-197. It has an average thermal neutron activation cross section of 85 b. The activated radioisotope Au-198 has a half-life of 2.7 days. The density of gold is 19.3 g/cm^3.

The disintegration constant of Au-198 is

$$\lambda = \frac{0.693}{T_{1/2}} = \frac{0.693}{2.7\text{ d}} = 0.26\text{ d}^{-1}.$$

From equation (21.5),

$$A_s = \frac{(6.023 \times 10^{23})(85 \times 10^{-24}\text{cm}^2)(10^{12}\text{n/cm}^2\text{s})}{197\text{ g }(3.7 \times 10^{10}\text{ dis/s/curie})}[1 - e^{-(0.26\text{d}^{-1})(5.4\text{d})}]$$

$$\cong 5.3\text{ curies/g}.$$

In practice the actual specific activity would be less than this value. The reasons are many. Firstly, irradiation efficiency due to self shielding and other factors must be considered as well as the flux density variation. Target impurity affects the specific activity. Finally, target and product material burnup must be evaluated. Not until all these factors are considered can a true specific activity be estimated. In general such a specific activity lies in the range of 20% to 80% of the value calculated from equation (21.5). Therefore, when gold is irradiated under the conditions stated in the example, the specific activity would be anywhere from ~ 1 curie/g to ~ 4 curies/g.

Problems

20.1 Prepare a list of various foils for thermal neutron flux measurements. Repeat this for the resonance and fast neutron fluxes. Note that, in the latter cases, the energy range must be clearly specified.

20.2 A segment of a silver wire is exposed in a uniform thermal neutron flux density of $\phi = 10^{10}$ n/cm^{2}s. The dimensions of the wire are L = 10 cm and R = 0.1 mm. What is the total number of silver atoms lost due to neutron activation for an irradiation time of two minutes? What is the total induced activity? Consider Ag-107 and Ag-109 separately and then sum up the results.

20.3 What fraction of the saturated activity does a sample reach if its activation time is two half-lives?

20.4 On examination of the gamma ray differential spectrum, three distinctive photopeaks are observed. Their energies are 0.662MeV, 1.17MeV, and 1.33MeV. What is the composition of the unknown?

20.5 Boron is used as a burnable control poison in a natural uranium heavy water moderated CANDU-type of reactor. The initial boron concentration is 1500 ppm. The reactor is operating at an average flux density of 10^{14} n/cm^{2}s. Calculate the tritium concentration from the neutron activation of boron after 100 days of steady state reactor operation.

20.6 A thin foil of Pu-239 0.56 cm in diameter and 0.013 cm thick is irradiated in the graphite column of a nuclear reactor. The thermal neutron flux density at the location of the sample is 10^8 n/cm^{2}s. What is the neutron absorption rate in the foil? What is the fission rate in the foil?

20.7 A multichannel analyzer is used for data analysis in NAA. Develop a formula for calibrating the scales in pulse height versus channel number using Cs-137 and Co-60 as standards.

20.8 A gold wire with length l = 10 cm and radius r = 0.5 mm has been exposed for 20 hours to a thermal neutron flux density of 5×10^{14} n/cm^{2}s. What is the induced activity of the wire two hours after the irradiation? The density of gold = 19.3 g/cm^3, the half-life of gold = 2.7 d, the thermal neutron capture cross section of gold = 9 b.

20.9 Liquid sodium is one of the coolant materials considered for a fast reactor. Discuss the problem of neutron activation of sodium in such a reactor.

20.10 Describe a BF_3 neutron detector.

20.11 Explain why indium foils are frequently used as thermal neutron detectors.

20.12 What does saturation activity mean? The saturation activity of a foil is 0.1 mCi. If the macroscopic activation cross section is 7 cm^{-1}, what is the neutron flux density? The volume of the foil is 0.001 cm^3.

Ans: 5.3×10^8 n/cm^{2}s.

CHAPTER 21

RADIOGAUGING AND IONIZATION APPLICATIONS

21.1 Introduction

Physical properties and characteristics can be determined from the transmission and scattering behavior of radiation from various materials and systems. **Radiogauges** are based on this concept and consist, essentially, of a radiation source and a detector along with the measuring equipment to indicate the detector output. The selection of a particular type of gauge and the radiation source depends on the specific application of the gauge. Backscatter beta gauging is generally used for the measurement of small thicknesses while transmission type gamma gauging is used for large thickness and level measurements. Neutron gauging is used when it is desired to determine the content of the lighter material mixed with the heavier ones. Radiogauging applications in industry and other fields are so many that it is not possible to go through all of them here. An effort is made, however, to outline the basic principle underlying some of these applications.

The ionization properties of nuclear radiations contribute to the numerous applications of small sealed and unsealed radiation sources in various areas. Most of these are based on the interaction of radiation with gases. The commonly used applications of radiation sources using ionization in gases are static charge elimination, smoke detection, ignition of gas discharges, lightning conductors, vacuum measurement, measurement of gas flow rates, etc. Some of the other applications are also discussed in the later part of this chapter.

21.2 Radiogauging Applications

a) Thickness and Density Measurements

Specially designed alpha, beta, and gamma gauges are used for thickness and density measurements in increasingly large numbers in numerous industries. One of the main advantages of radioisotope instruments is that no contact is required with the material whose thickness or density is to be determined. Besides, measurements can be made on a continuous basis on various materials moving at high speeds and at high temperatures. Such measurements save millions of dollars by cutting down the waste of raw materials and improving the quality of the products.

In a transmission type of gauge, a suitable beta or gamma source is placed below the sheet or object under investigation and a detector is located above the source and the material. The paper, plastic or any other material whose thickness has to be continuously measured is allowed to pass between the source and the detector. Any variation or non-uniformity in the thickness of the material results in a variation in the transmitted radiation intensity. This triggers a radiation meter for controlling the roller settings so that uniformity in thickness or density is regained. Figure 21.1 illustrates the working of such a gauge.

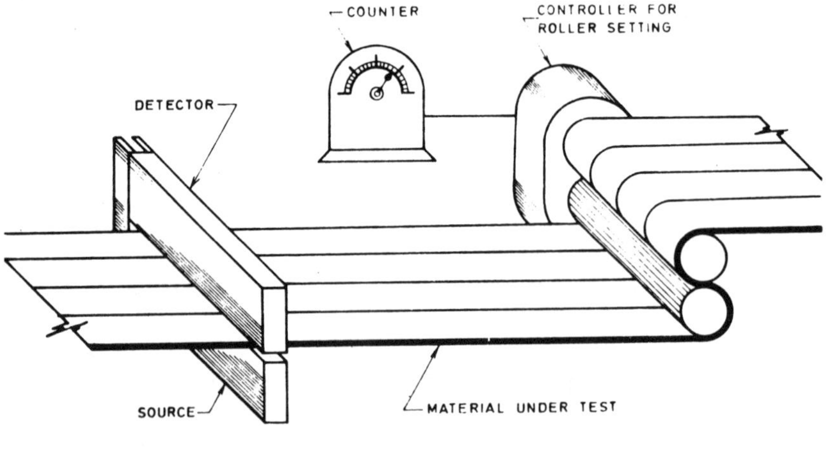

Fig. 21.1: Transmission type thickness gauge.

Multiple beta gauges are used for successively controlling the thicknesses of more complex operations required in the manufacture of sandpaper. The material passes through numerous rollers, under some hoppers and finally into ovens for the application of adhesive coatings, feeding of abrasives, and heat treatment operations, respectively. During each operation a thickness gauge applies necessary control on the thickness and makes the required corrective adjustments. Figure 21.2 illustrates the use of numerous thickness gauges required for the manufacture of sandpaper.

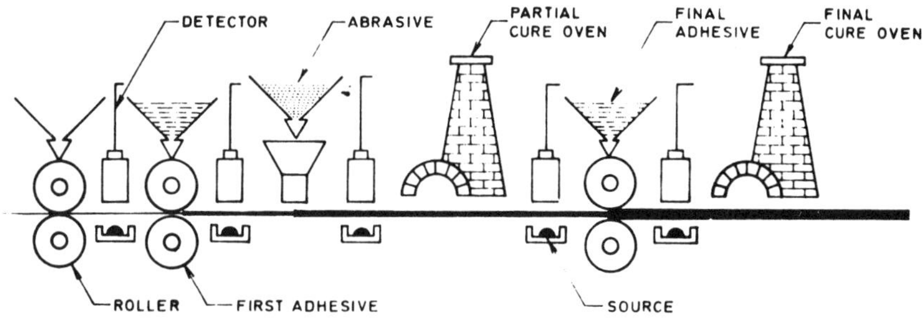

Fig. 21.2: Use of multiple radiogauges for thickness adjustments.

Backscattering beta particle gauges are used extensively for thickness measurements. The principle is similar to that of the transmission type thickness gauges. Radiation is scattered when it strikes any material. The intensity of the backscattered radiation depends on the radiation energy as well as the nature and thickness of the material. As the source and detector are located on the same side, the detector is shielded from all but the scattered radiation. The backscatter gauge is illustrated in Figure 21.3.

The intensity of the backscattered radiation does not remain constant with increasing thickness and attains a saturation value for a thickness of material much smaller than the range of the radiation. For this reason the method is suitable for the measurement of smaller thicknesses. The chief advantage of a backscatter gauge is that the thickness of an object can be measured if access is possible from one side only like the thickness of a pipe. Figure 21.4 illustrates the continuous measurement of thickness of industrial products like metallic sheets, paper, etc.

The principle underlying the density and thickness measurements is identical since the absorption of radiation is a function of the density of the substance. Radiogauges for the measurement of density have become an important tool in numerous industries for on the spot and continuous measurements of various products. One other application is to evaluate the density of fluids circulating on pipes. The method can even be used for inflammable and corrosive fluids flowing under pressure.

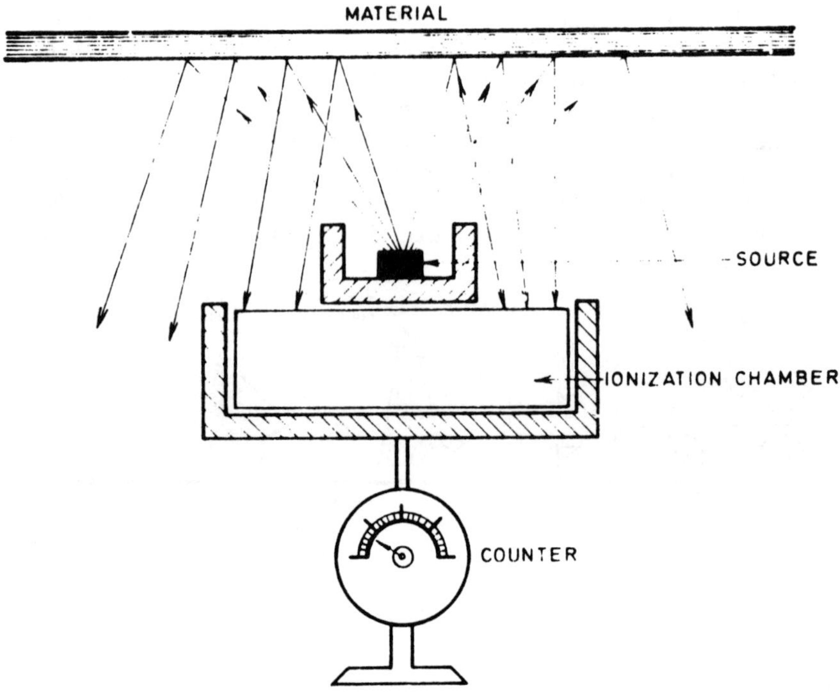

Fig. 21.3: Backscatter thickness gauge.

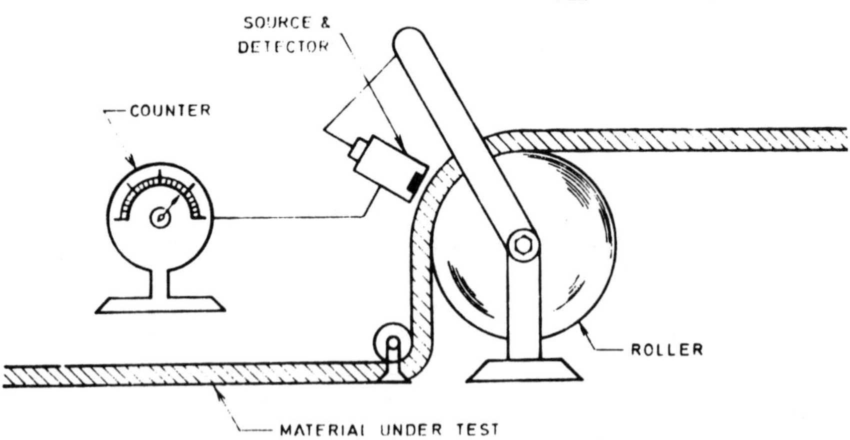

Fig. 21.4: Backscatter gauge for continuous thickness monitoring.

Density gauges are in common use in the tobacco industry. Such gauges not only eliminate manual control but also result in the saving of tobacco and more uniform firmness in the cigarettes. The density of the cigarettes is compared with a standard sample using two independent density gauges. The difference of these gauges is fed into an electical control unit which regulates the tobacco feed for uniformity in the firmness of cigarettes. The working of such gauges is shown in Figure 21.5.

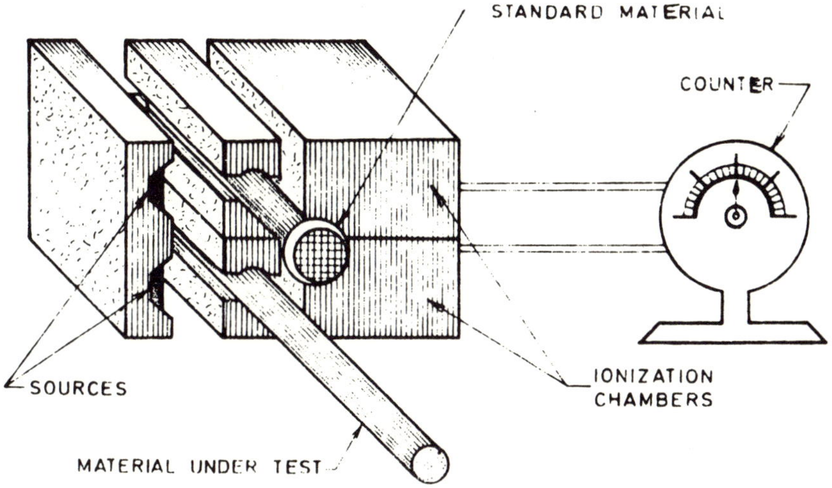

Fig. 21.5: Radiogauging in the tobacco industry.

Radiogauging techniques using gamma or neutron sources are also used for the measurement of density or moisture content in soils. For the determination of density of the soil close to the surface, a gamma source is placed in a hole and the intensity of radiation at the surface is measured using a radiation detector. The attenuation of radiation by the soil between the source and the detector provides the data for the density determination. The backscattering principle is used for the density measurement of the soil surface. For the measurement of density of the soil at greater depths, a special type of source-detector arrangement is lowered into a vertical hole in the soil at various depths. A shielding material is inserted between the source and the detector so that the detector detects only that part of the radiation which has been scattered by the soil. As the scattering process is a function of the soil density, the detector response indicates the soil density at various depths. Figure 21.6 illustrates the determination of soil density at the surface and also at various depths underground.

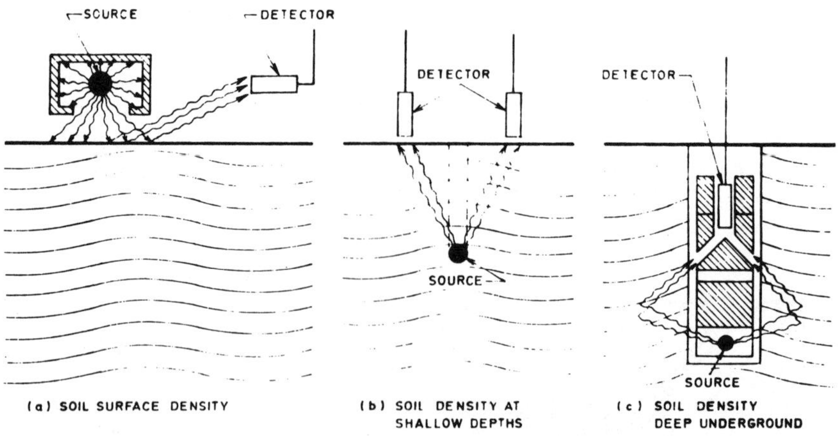

Fig. 21.6: Soil density measurements.

b) Fluid Level Measurement

One of the very versatile radiogauging applications is the measurement of fluid levels in containers as they come off the assembly line. This technique is also used for the continuous monitoring of fluid level in storage vessels. The radiogauging technique is well suited for such measurements as it can be used outside of thick-walled vessels.

Level gauges can be arranged in a number of ways. A source-detector system may be mounted, as shown in Figure 21.7, to determine the fluid level by means of a servomotor controlled by the instrument signal. For large containers, the source may be introduced as a float within the container and the detector is then placed outside to find the position of the source.

Nuclear level gauges are being successfully used for determining levels in airplane and spaceship fuel tanks which are not normally held in constant positions. A suitably located radiation detector and numerous properly located radiation sources are adjusted in such a way that the detector response is always a function of the total amount of fuel available in the tank and is not affected by its position. This is illustrated in Figure 21.8.

In addition to the measurements of thicknesses, density, and fluid levels, radiogauges are used for numerous other applications. One example is the measurement of the wear of the firebrick lining in high temperature furnaces. Gamma sources are incorporated at various depths in the firebrick lining. A suitably located radiation detector indicates the wear of the thermal lining during normal furnace operation. This is depicted in Figure 21.9.

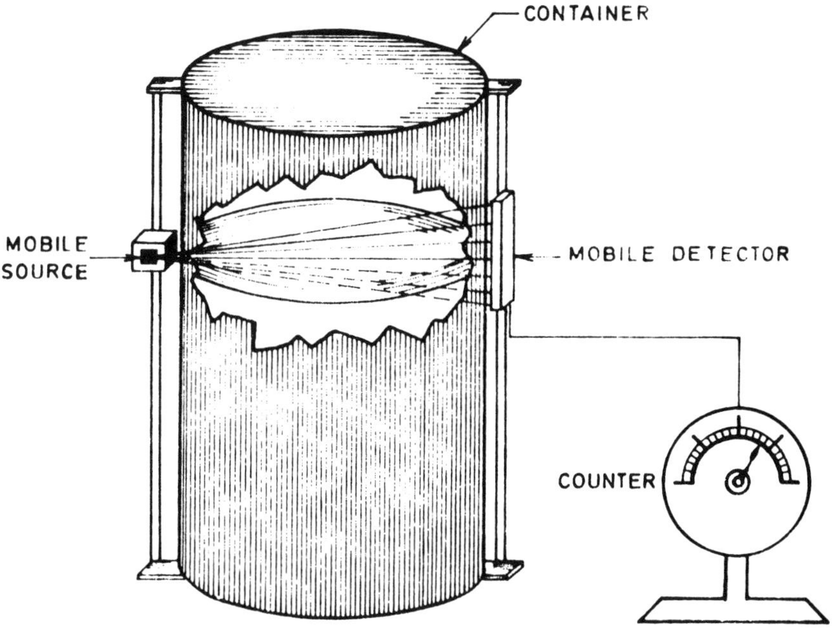

Fig. 21.7: Liquid level measurement by radiogauging.

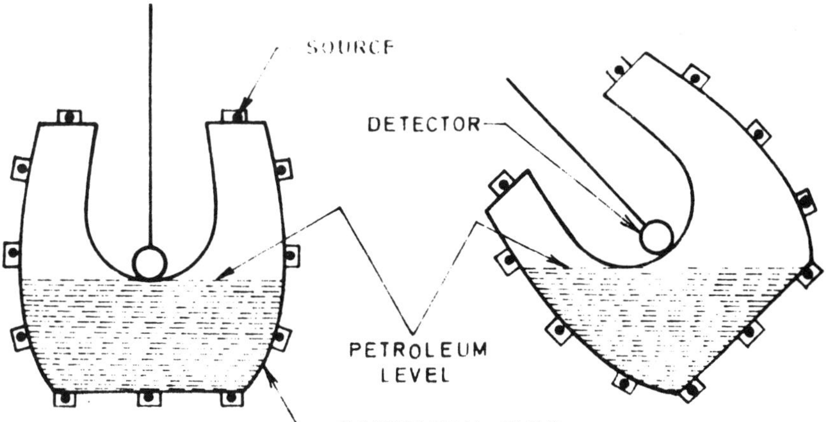

Fig. 21.8: Determination of fuel volume in an aircraft fuel tank.

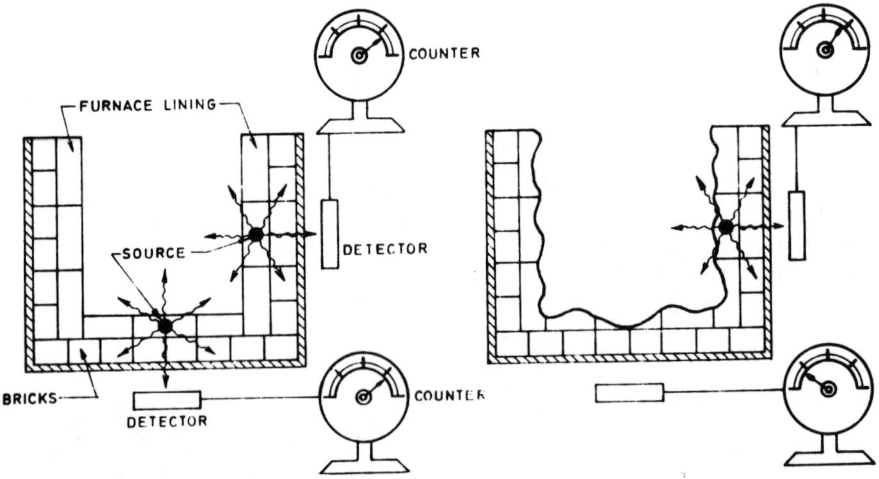

Fig. 21.9: Measurement of wear of the internal lining in furnaces.

Table 21.1 summarizes the application of various types of radioisotope gauges in different industries.

Table 21.1: Applications of radioisotope gauges in different industries.

Nature of industry	Type of gauge	Applications
1. Food and beverage	Thickness and mass per unit area	Slabs of chocolate, bread, pastry, etc.
	Density	Fish, meat, condensed milk, syrup, juice, tomato paste, etc.
	Level	Level of soup, meat vegetables, coffee, liquids in cans, bottles, packets, and containers.
2. Tobacco	Density	Control of cigarette weight, monitoring of empty and underfilled packages, etc.

Nature of industry	Type of gauge	Applications
3. Textiles, clothing and footwear	Thickness and mass per unit area	Fabric impregnation, coated surgical dressing, footwear materials, fabric wear, etc.
	Density	Sulphuric acid, polymer and synthetic yarn solutions, etc.
	Level	Contents of reaction vessels, etc.
4. Wood, paper and rubber	Thickness and mass per unit area	Paper of various thicknesses and quality, coatings on paper, rubber for floor coverings, tires, rubber coated fabrics, etc.
	Density	Timber, compounds in wood and rubber industries, etc.
5. Chemicals, plastics, glass, minerals and cement	Thickness and mass per unit area	Plastic sheets, tapes, coated metals, fabrics, adhesives, glass and asbestos fibers, cement panels, walls of pipes and tanks, etc.
	Density	Process solutions, plastic and organic solutions, cement, mineral ore slurries, etc.
	Level	Liquids, liquified gases and slurries in containers and reaction vessels, molten glass in furnaces, sand and cement slurries, plaster, etc.
	Analysis	Contents of chlorine and sulphur in plastics, components in glass, etc.

Nature of industry	Type of gauge	Applications
6. Petroleum, coal and basic metals	Thickness and mass per unit area	Coated and laminated metal products, hot and cold rolled-sheet metal, galvanized iron, etc.
	Density	Petroleum and petroleum fraction, coal, coke, catalysts, sulphuric acid, powdered and slurried ores, etc.
	Level	Coal, coke, metal and ores in bunkers, catalysts and liquids in reaction vessels, hydrocarbons in cracking units and in distillation columns, sodium in tubes, ores and leaching solutions in pressure vessels, etc.
	Analysis and others	Carbon-hydrogen ratio of hydrocarbons; sulphur cobalt and lead content of hydrocarbons, hydrogen-sulphide content of gas, ash content of coal, various contents of metallic ores, logging, etc.

21.3 Ionization Applications

a) Elimination of Static Electricity

Static electricity is generated in industries where material with high electrical resistivity is produced, such as paper, textile, plastic, rubber, and printing. In addition to being a fire hazard due to the generation of spark discharges, the static electricity results in the attraction of atmospheric dust and the attraction or repulsion of sheets of materials being produced. Mild electric shocks can also result from touching various charged surfaces.

If the air in the vicinity of the charged surfaces is ionized, the ions having opposite charges will move to the charged spot and neutralize it. The dust elimination action of a radioactive source is hence dependent on the ion-pairs produced by the source per unit time. Theoretically alpha emitters are the most suitable sources for use as electrostatic eliminators since they have a very intense specific ionization. However, due to a very short range and radio-toxicity, alpha emitting nuclides are not used very often. Beta sources are generally employed for this purpose.

b) Smoke Detection

Some detectors using radioisotopes are widely used for the early detection of fire. The smoke particles can be ionized by the charge exchange with the ionized air molecules. These heavier ionized smoke particles have a very small mobility as compared with the mobility of air molecules. Besides, the probability of recombination increases due to the presence of low mobility ionized smoke particles. For these reasons, when smoke particles are introduced into a saturated ionization chamber containing a suitable radioactive source, the current decreases. A small amount of smoke particles introduced into the chamber greatly changes the current in the detector.

A smoke detector essentially consists of two similar ionization chambers. One of the chambers is open to the smoke particles while the other is completely sealed and is operated in a saturated state. The mid-point of the two chambers is fed to the trigger voltage of a cold-cathode discharge tube. This is shown in Figure 21.10. When there is no fire, and hence no smoke, the

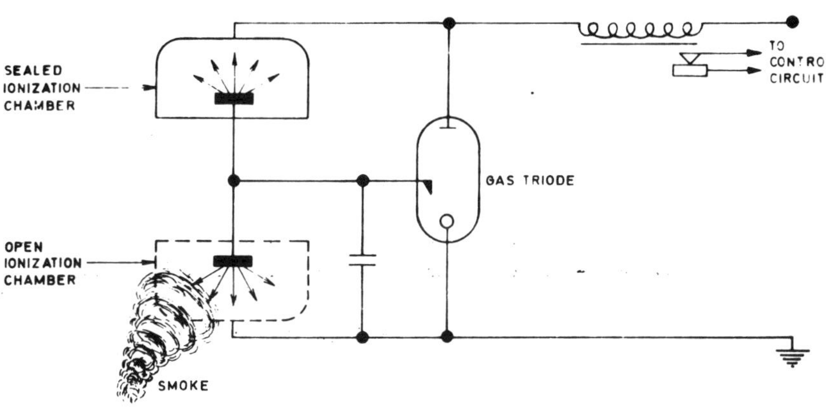

Fig. 21.10: Ionization smoke detector.

two chambers are in equilibrium and the trigger electrode potential is too low to cause the tube to ignite. In case of fire, the current in one ionization chamber increases. As a result, the trigger electrode potential increases, which in turn fires the tube. The discharge current is used to actuate a relay warning system at the desired locations in or outside the building.

c) Ignition of Gas Discharges

Ignition of fluorescent tubes and certain cold cathode tubes is based on gas discharges. The ignition of the discharge depends on the presence of sufficient primary ionization in the gas volume of the tube. Normally this background is provided by light, cosmic and radioactive background radiation. Sometimes the ionization density is too small for an immediate ignition and a certain delay results. This delay can be completely eliminated if a radioactive substance in the form of a gas is contained in the tube or a small radiation source is placed at some suitable location in the tube. Generally, a small quantity of tritium-3 or krypton-85 is added to the gas filling. Pure gamma ray sources like cobalt-60 have also been used. The radiation source provides dependable firing characteristics by ionizing the gas as indicated in Figure 21.11.

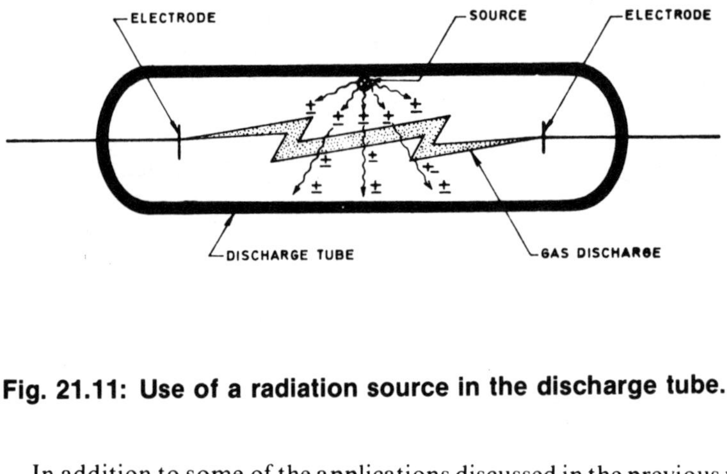

Fig. 21.11: Use of a radiation source in the discharge tube.

In addition to some of the applications discussed in the previous pages there are numerous other applications of ionization produced by radiations. Radioisotopes are now used in a large number of applications such as vacuum measurement, lightning conductors, gas flow measurements, gas chromatography, etc. Intense ionization and localized damaging effects are being investigated for neutron capture therapy for brain and some other types of cancer. The capacity of nuclear reactors to provide large numbers of thermal neutrons, combined with the fact that certain elements, when injected into the human body and irradiated by thermal neutrons produce short lived radioisotopes has made neutron capture therapy a promising field. The main advantage lies in the fact that the radioactive material is not injected in the tumor tissue, but elements like boron-10, lithium-6, and uranium-235, with high thermal neutron capture cross sections, are injected into the cancerous tissues. The elements become radioactive only when bombarded by thermal neutrons. Heavy particles and fission fragments so emitted in the diseased tissue have a very strong and localized damaging influence, hopefully without affecting the normal tissues.

Problems

21.1 Calculate the maximum thickness of a piece of paper of density 1.2 g/cm^3 if P-32 is used. The detection system is capable of detecting 50 disintegrations per second.

21.2 A gamma ray beam is incident normally upon a 25 cm thick slab of aluminum. The reduction in beam intensity is 5%. The gamma ray energy is one MeV. What is the half value thickness?

21.3 Charged particles can be used to determine the thickness of some scattering materials. In a scattering experiment, a beam of protons is directed at a gold foil. It was found that one particle out of 3×10^6 is scattered into an area of 0.5 cm^2 at 20 cm from the foil along a 45° scattering angle. Use the basic formula of the Rutherford scattering experiment to determine the thickness of the foil.

21.4 Prescribe a method of using fast neutrons to perform a transmission type of experiment in order to measure the thickness of a slab.

21.5 The thickness of rolled steel sheets is to be controlled by a thickness guage and a process controller. The guage uses Ir-199 which emits several gammas. If the density of the steel sheets is 8 g/cm^3 and their mass absorption coefficient is 0.1 cm^2/g, what is the difference in thickness of the sheets to cause a 10% difference between the transmitted intenstiy and the intensity without a sheet of steel between the source and the detector?

Ans: 0.13 cm.

21.6 Draw a schematic diagram to indicate the use of a radioisotope guage system to control the cigarette density.

21.7 What are the major advantages in guaging to use a gamma source compared with an alpha source? a neutron source? a beta source?

21.8 The linear absorption coefficient of sugar for gamma rays from a Co-60 source is 0.1 cm^{-1}. The mass flow rate of sugar on a conveyor belt, 15 cm wide, moving with a speed of 3 cm/s is to be controlled by a transmission gauge. The density of sugar is 0.3 g/cm^3. If the mass flow rate is to be set at an intensity ratio of 0.5, what is the mass flow rate?

Ans: 94 g/s.

CHAPTER 22

RADIOACTIVE TRACER TECHNIQUES

22.1 Introduction

Tracer techniques have been in use even before radioisotopes were discovered or artificially produced in laboratories. Salts and dyes were in common use for tracing fluid movements and many other transport and transformation phenomena. However, the widespread applications of tracer techniques were developed after the discovery of radioactivity when radioisotopes began to be produced on a commercial scale. The radiotracer technique consists of labelling the object or the process under study with a specific agent having a similar behavior throughout the investigated period as the object itself. The two major requirements of a tracer are that

a) the tracer and the material under investigation (traced material) must be identical and that
b) the tracer should have one distinguishable property from the traced material.

These two requirements are apparently contradictory. In actual practice, the first requirement is not as stringent and can be taken to mean that the tracer behavior should be like that of the traced material with respect to the characteristics under study in the system. As far as the second requirement is concerned, radioisotopes have now practically eliminated the previously used tracers based on properties such as color, refractive index, conductivity and density of the additive substances. Some additional advantages of radio-tracers over the conventional tracers are outlined here:

a) As an atom is the smallest element of a radioisotope that retains the properties of the traced material, the radiotracers can be used at the atomic level for investigation.

b) Radioisotopes of almost every known element are now available. This means that radiotracers can now be used to trace almost all elements.

c) Nuclear radiation can be measured with high sensitivities and hence the amount of tracer can be kept so small that no interference is expected with the process under study. Radioisotopes with half-lives up to a few months can be detected in quantities as small as 10^{-16} to 10^{-17} grams!

d) The radioisotope tracer measurement is absolutely specific and no other substances, except other radioactive elements, can disturb the qualitative measurements. Conventional tracer techniques such as color and refractive index are affected by the interference of other materials in the system.

e) Radioisotopes and the radiation detection equipment required for tracing applications are quite inexpensive and much simpler to use than other types of tracers.

f) If proper detection methods are used, the counting rate of a sample of radioactivity is directly proportional to the amount of radiotracer present. This is not so for the conventional tracing techniques where color, for example, may exhibit a maximum or a minimum in the measurement calibration curve. The radiotracer technique therefore allows unambiguous measurements unlike other conventional techniques.

g) As the tracer is required to be identical with the traced material only for the particular process, the isotopic identity between the two is not required. This means that a large number of nuclei are available to suit the experimental conditions.

h) The penetrating capacity of gamma rays emitted by radioactive tracers makes the measurements possible through pipes or vessel walls and the system need not be disturbed during measurement.

Due to all of these features, radiotracers are used in numerous applications in practically all spheres of life. Some of the applications in industry, medicine, agriculture, etc. are outlined in the following pages.

22.2 Applications in Industry

The applications of radiotracer techniques in industry are now so numerous and diversified that it is not possible to give a full account of all of them here. Only a brief survey of some of the important radiotracers employed in industry is therefore being presented.

a) Leakage Detection

Various methods using radioisotope tracer technique are in common use for leak detection and locating underground pipelines carrying water or other fluids. In the most commonly used methods, some suitable radiotracer is injected into the pipeline under pressure. The radiotracer, along with the fluid in the pipe, leaks out and accumulates in the neighbourhood of the leak as shown in Figure 22.1.

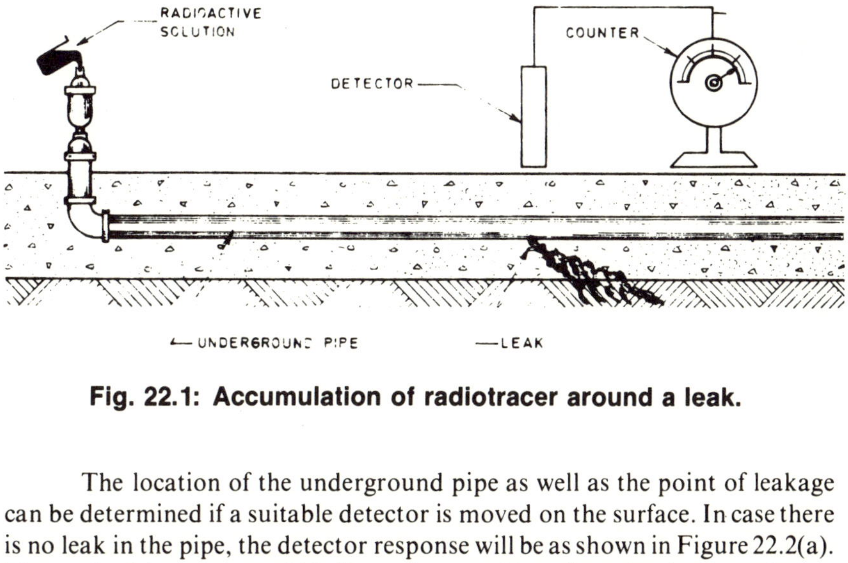

Fig. 22.1: Accumulation of radiotracer around a leak.

The location of the underground pipe as well as the point of leakage can be determined if a suitable detector is moved on the surface. In case there is no leak in the pipe, the detector response will be as shown in Figure 22.2(a). The peak of the curve would indicate the actual location of underground pipe. In case of a leak, some of the radiotracer will leak out along with the liquid. The detector response in this case would be similar to that shown in Figure 22.2(b).

Another method consists of injecting a radiotracer in the pipeline and applying pressure so that an appreciable quantity of the tracer flows out of the leaks. The pipeline is next flushed with water so that no tracer atoms remain in the line. A radiation detector is then moved along the pipeline to locate the

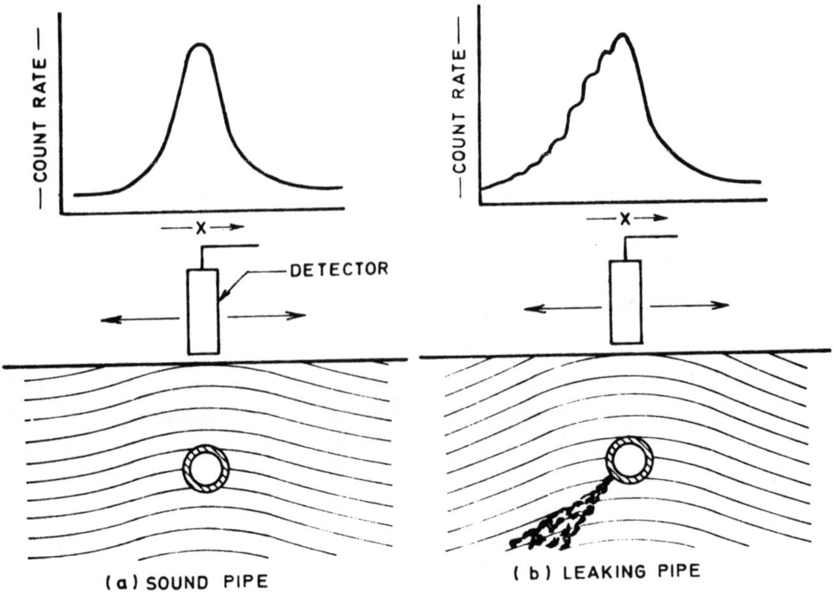

Fig. 22.2: Location and leak detection in an underground pipe.

places where the radioactive tracer has leaked out. For obtaining the exact positions of the leaks, gamma sources with long half-lives are placed along the pipeline at fixed distances. A graph of the count rate versus position indicates the exact location of the leak. This is illustrated in Figure 22.3.

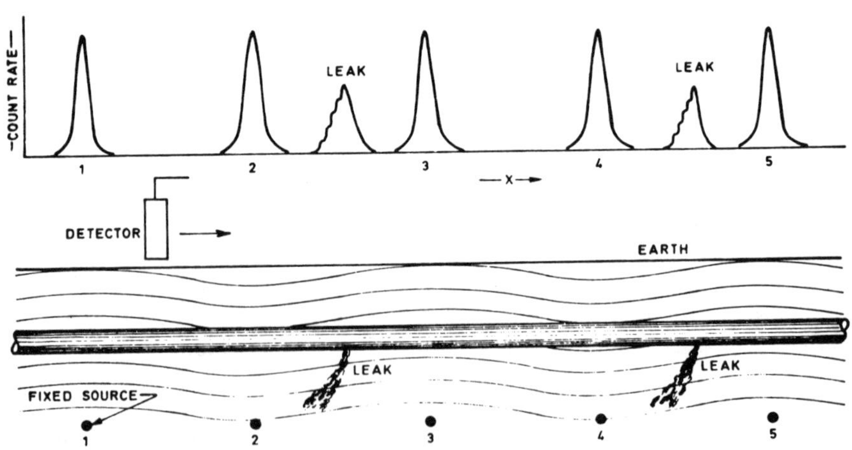

Fig. 22.3: Location of leaks using permanent gamma sources.

More recently, techniques have been developed to locate leaks using a special type of detector fitted with a minute battery-driven wire magnetophone. The location system is allowed to move in the pipe and record pulses corresponding to all the fixed gamma sources and radioactivity that might have leaked out when the test run was made with radiotracer in the pipe. The recordings of the magnetophone are transferred to a recorder. The location of the leaks can be pinpointed by measuring the time intervals from the position of fixed sources.

In addition to leak detection in pipelines, radiotracer techniques can be used for the detection of leaks in the sheathing of telephone cables, diffusion rates of natural gas from underground reservoirs, position and magnitude of leaks in underground coal gasification systems and locations of leaks through valves, heat exchangers, etc.

b) Flow Studies

Radiotracing techniques are very useful for finding out volume/mass flow rates of different fluids over long distances. This is very important as the total quantity of fluid shifted from one place to another through pipelines can be easily evaluated. Flow rates of liquids in pipes are measured by introducing a slug of a radioactive solution and subsequently using the signals given by two radiation detectors placed at known distances. Oil distributors have been using radioactive markers to indicate the boundary between two different types of oil being transported through the same pipeline. If at the end of the journey, one type of oil is to be directed in one direction and the other type in the other direction, then a Geiger counter, held over the pipe, signals the arrival of the second batch of oil as a small amount of radiotracer soluble in oil is introduced between the two types of oil. Figure 22.4 illustrates this scheme.

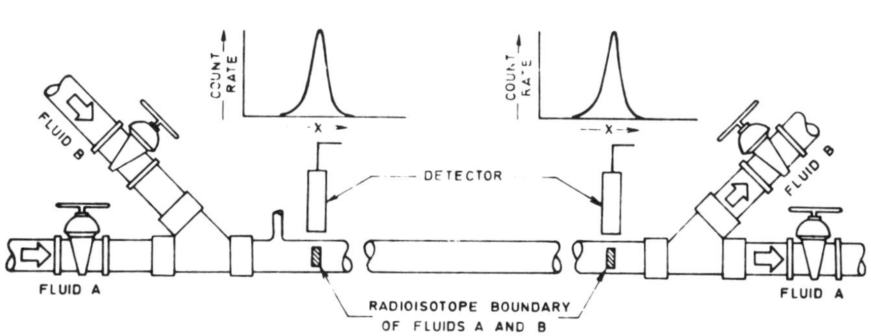

Fig. 22.4: Utilization of a single pipeline for the transportation of different grades of oil.

Sometimes a metallic ball is used in place of a ring of radioactive material to separate two fluids of different types. As the diameter of the ball is exactly equal to that of the pipe, the internal surface of the pipe can be cleaned using a sharp edged cleaning device. This device is pushed through the pipe under great hydraulic pressure to scrape and clean all irregularities, impurities and other extra materials that might hamper movement. If this device gets stuck somewhere in the pipe, its location can be found if it was previously tagged with some suitable radioactive marker. Figure 22.5 illustrates the working of the separating ball mechanism and the tagged pipe cleaning device.

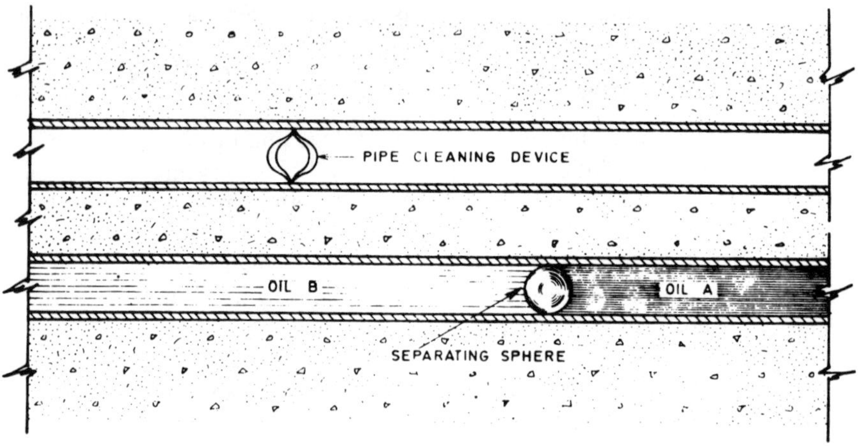

Fig. 22.5: Separating ball and tagged pipe cleaning device.

c) Friction and Wear Studies

Radiotracers are used in friction and wear measurements of various materials like piston rings, etc. The popularity of the tracer wear method results from the fact that it eliminates the long running times, engine dismantling, and careful cleaning and weighing involved in the old methods based on the measurement of the decrease in weight. In the tracer wear method, the component whose wear rate is to be determined is made radioactive in a neutron flux. When the radioactive component undergoes wear in an engine, the worn radioactive material is carried through the engine lubricating oil. A continuous measurement of radioactivity of the oil yields precise information as to the quantity of radioactive element in the oil. The radioactivity measurements can be continuously made using a G-M counter or a scintillation detector around which radioactive oil circulates. Alternatively, samples of oil can be taken from the engine and measured using some suitable radiation detector. Equipment for wear studies is shown in Figure 22.6. The wear of tires has also been studied with tracer methods. The wear is measured by studying the activity on the road. The investigation has yielded very useful information about the influence of the quality of roads, the tires and other parameters on wear rates.

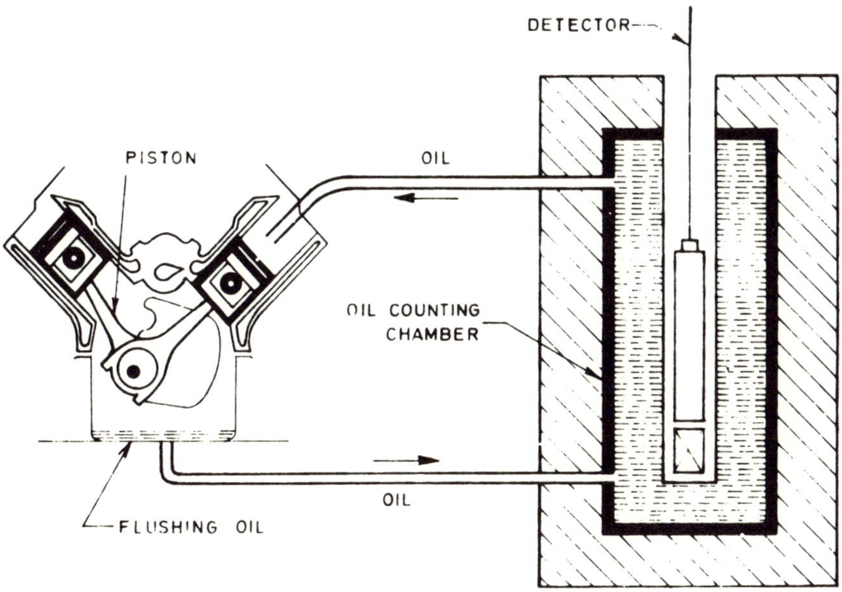

Fig. 22.6: Friction and wear studies.

Measurement of tool wear rate and tool life is very efficiently done with the help of tracer techniques. The technique is faster and more reproducible than other methods. The tool under investigation is made radioactive by placing it in a neutron flux. When it is used in machining work, the amount of radioactivity in the scrap indicates the wear rate. This is illustrated in Figure 22.7.

Radioisotopes are also finding applications in studies relating to lubricants. The thickness and location of the tagged lubricant additives on bearings is determined by measuring the activity using suitable radiation detectors. A specific example would be compounding a mineral oil with diphenzyl disulfide containing radioactive S-35. The gears are lubricated with this lubricant while operating under controlled conditions. The location and thickness of the S-35 containing film on the load bearing surfaces can then be studied.

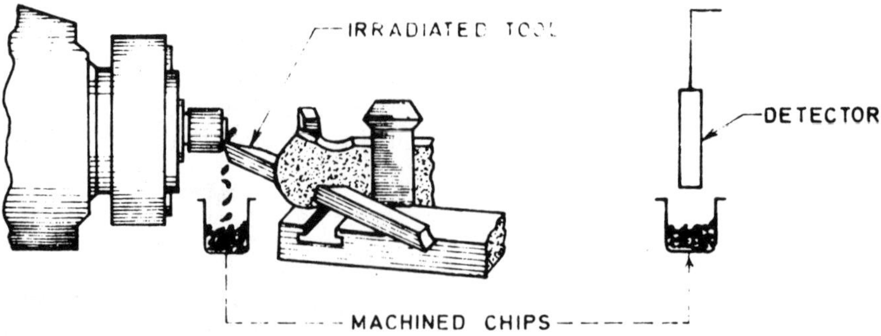

Fig. 22.7: Measurement of tool wear rate.

d) Diffusion Studies

Diffusion and mass transfer studies using radiotracers allow the possibility of performing analyses without the need for sampling, which could result in the disruption of the normal industrial process. The technique is extremely sensitive and is useful in analyzing the very slow diffusion processes. The diffusion of various fluids can be determined using the apparatus shown in Figure 22.8.

Two containers are connected by a small diameter tube and two fluids under investigation are continuously fed to both the containers. The overflow rates in the two containers are kept constant in order to maintain constant concentrations in both of them. A selected radiotracer is mixed into one of the fluids. After the steady state condition is reached, the concentration of radio-

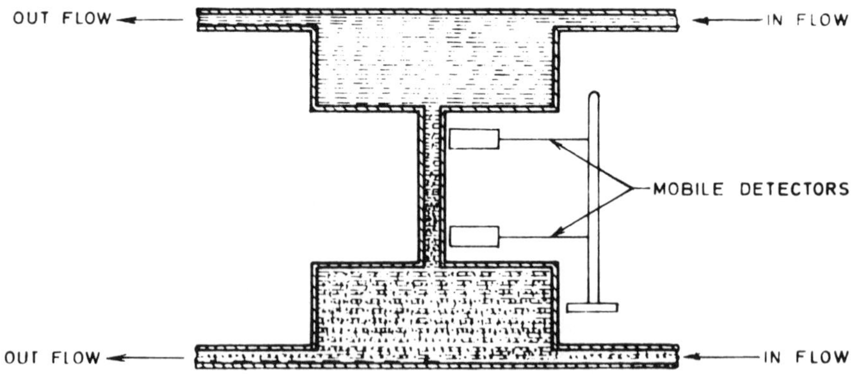

Fig. 22.8: Diffusion studies with radiotracers.

tracer at all points in the narrow connecting tube becomes independent of time. The relative concentrations at the points are measured by observing the counting rates at these positions. Proper collimation and shielding should be used so that the response is only due to the concentration at the desired points. The fluids from the top and bottom tanks are taken and counted in a similar manner. The experiment yields information on the diffusion and mass transfer rates. The experiment described here measures the diffusion parameter for steady state diffusion.

e) Mixing Studies

In numerous industrial processes it is desired to find out the degree of mixing of various solutions and powders. Mixing operations are often time consuming and generally mixers have to be operated for time intervals longer than necessary if the degree of mixing is not determined during operation. Chemical analysis of the material during the mixing operations is one technique but it is not only time consuming but also requires a large amount of labor for analyzing large series of samples. Radioactive tracers can be used very effectively for optimizing mixing operations and hence yielding a better process economy.

Radiotracing using gamma emitting radioisotopes is ideally suited for studying the mixing of liquids, colloids, slurries, and gases. The technique eliminates the necessity of analyzing the samples and allows the checking of mixing during operation. The basic principle underlying such studies involves the addition of a small amount of tracer to a batch and the attainment of a uniform concentration throughout the material under consideration. An internal or external detector provides a continuous measurement of mixing. The time required for complete mixing in a given container can be determined by recording the count rate versus time from one or several points of a mixer. Complete mixing is attained when the activity becomes constant with time. This principle of the radiotracer is outlined in Figure 22.9.

22.3 Applications in Medicine

Radioisotope techniques in medicine may be divided into two broad categories, i.e. diagnostic studies and therapeutic applications.

a) Diagnostic Applications

In diagnostic applications small and harmless quantities of radioisotopes are injected into the body. These isotopes serve as tools in gathering information about the normal and abnormal life processes. A large number of different radioisotopes are being used in metabolic studies and in the investigation of disease conditions in the human body, as well as in the measurement

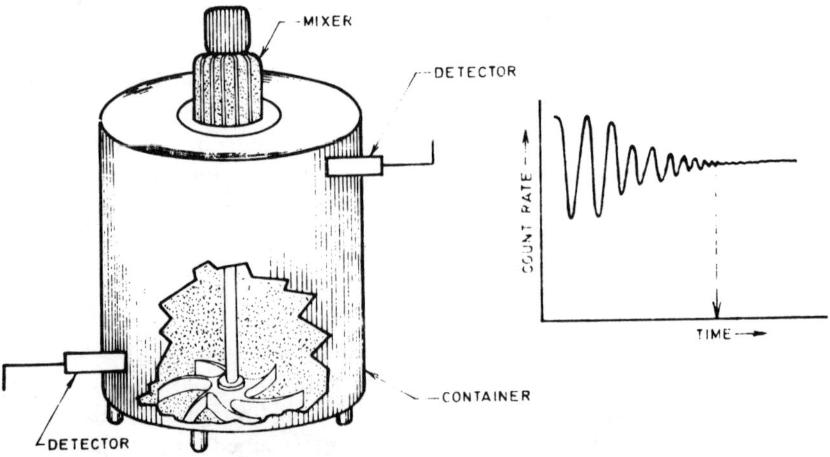

Fig. 22.9: Radiotracers for studying mixing operations.

of liquid volume and the rates of flow through various organs in the body. In addition, these studies determine the behavior of internal organs and differentiate between normal and malignant tissues. The movement of various components like blood, fat, protein, and hormones inside the body are studied by labelling an appropriate molecule with a radioactive isotope and injecting it into the body. A detailed analysis of radioactivity and the known chemical composition of blood, urine, and other body fluids yields information about the amount of a particular substance in the different parts of the body as well as the rates at which the substance enters and leaves these areas. Such studies have helped in diagnostic tests and applications.

Diseases and disorders of the thyroid gland are diagnosed with the help of iodine-131, due to the ability of the thyroid gland to concentrate and store any iodine containing material and gradually release it to the blood stream. If the compound thyroxin is not produced in proper quantities by the thyroid gland, various ill effects, such as the expansion of the gland, increased body weight, etc., start appearing. For diagnostic purposes, the patient is given a small dose of some compound, say, sodium iodide containing I-131. After a certain interval of time, the amount of iodine in the gland is determined on the basis of the amount of radiation emitted from the gland. This provides information on the functional state of the thyroid gland as well as its physiology.

Blood cell and blood volume studies have also been carried out using various isotopes like iron-59, phosphorous-32, and chromium-51. Most of the information regarding the iron absorption in various clinical conditions has been obtained using iron-59, while the measurement of the rate of formation of red cells, their lifetime and volume can be determined by using iron-59 or chromium-51. Consider a case in which it is desired to investigate the reasons for the short lifetime of red blood cells in a patient's blood. Normally red blood cells have a lifetime of about 120 days. There can be two reasons for the shorter lifetime: either the cells may be abnormal or the cells might be normal but some substance present in the patient's blood destroys them. For diagnosis, a sample of the patient's blood is withdrawn, stabilized to prevent clotting, and a small quantity of sodium chromate is added to it. Some of the sodium chromate attaches itself to the red blood cells while the excess chromate that is not taken up by the cells is washed away. These tagged blood cells are now injected back into the patient and if the survival time is abnormal in the healthy recipient as well as in the patient, it may be concluded that the red blood cells in the patient are normal. On the other hand, if the cell survival time is normal in the healthy individual and short in the patient, there must be something in the patient's blood that destroys the red cells.

The volume of the red blood cells in a patient can also be determined by preparing a chromium-51 tagged blood sample as described earlier. The radioactivity of this sample is measured very accurately just before injection, the sample is injected into the blood stream, and a suitable time is allowed for the complete mixing of this material. A new sample of the blood is taken and its radioactivity is measured. The volume of the red blood cells can be calculated using the radioactivity of these two samples; the total blood volume in a patient can also be determined. Such information is very important in deciding whether a blood transfusion is needed or not. Many other disorders can be diagnosed by studying the blood circulation by tagging it. For example, the effectiveness of the heart can be found by measuring the amount of blood pumped per unit time. The functions of many other parts like the kidneys and liver can also be investigated from blood flow studies.

Tracer techniques have also been quite successful in the localization of brain tumors. It appears that the barrier which prevents the migration of certain ions from the plasma into the brain tissues is disrupted at the surface of a growing tumor. An increase in radioactivity at the tumor surface is therefore expected if a suitable radioisotope is injected intravenously. Radioisotopes like iodine-131 and phosphorous-32 have been used successfully for investigation of brain tumors. For diagnostic measurement of trace elements in any part of the human body, various sensitive measuring devices such as scanners, scintillation cameras, and whole body counters are now in operation.

b) Therapeutic Applications

Although they are not as widespread as diagnostic and biological research applications, therapeutic applications of radioisotopes are important in treating certain diseases. The general principle is that of selective localization of radioactive elements in the tissue in quantities great enough to destroy it without damaging the surrounding tissue.

Such therapeutic applications may be divided into two categories. The first one makes use of the penetrating radiations to alleviate the effects or to destroy the cancerous tissues. The techniques most commonly used are teletherapy and neutron capture therapy. The second type involves the deposition of suitable radioisotopes in the diseased organs. This may either be accomplished by implanting radioisotope sources inside the organs or by making use of a preferred concentration of certain radiotracers in specific organs. Some of the uses of these techniques are now briefly mentioned.

'Polycythemia vera' was the first disease cured successfully with the help of radioisotopes. The blood disease is characterized by the enlargement of the spleen and a tendency for bleeding to occur. The number of red cells in the blood increases abnormally and there is a markedly increased activity of the blood-producing elements of the bone marrow. Previously, various drugs were used to destroy the excess red cells, resulting in temporary relief. The radioisotope phosphorous-32, in the form of a sodium-radiophosphate solution, works at the root of the disease by slowing down the formation of red blood cells. This compound mixes with the blood and is quickly concentrated in the blood-forming tissues where the rapidly dividing blood forming cells are extremely sensitive to radiation. A considerable lengthening of human life has been achieved by such treatment. Precautions are necessary, however, to determine the required dose of phosphorous-32, since excessive nuclear radiation may cause leukemia. Leukemia is characterized by too few red cells. Abnormal thyroid glands have also been successfully cured by injecting a compound of iodine-131 into the patient. The thyroid selectively removes iodine from the blood stream and deposits the radioactive iodine in the gland. The radiation emitted by the radioisotope inhibits the abnormal activity of the thyroid glands.

22.4 Miscellaneous Applications

a) Agriculture and Food

In agriculture, experiments with radiotracers have proved very successful in the utilization of fertilizers to produce high yield crops. For example, for the study of the uptake of phosphate fertilizer by plants, the original method involved cutting the plants, burning them and analyzing chemically for the phosphorous content. This method would only determine how much phosphorous had been taken up by the plant as a whole and not how much of it had been taken up from the soil and from the fertilizer separately. Using radiotracer techniques, phosphorous-32 is incorporated into a fertilizer and the tagged fertilizer is added to the soil. With the help of a radiation detector it is possible to indicate the utilization of the fertilizer very accurately. In addition to determining its uptake, tracer studies also show the fixation of a fertilizer by soil, proper type and placement of the fertilizer and the efficiency of various fertilizers. Tests using radiotracers have indicated that certain crops, such as cucumbers and strawberries, are able to absorb fertilizers through their leaves as well as their roots.

Tracer techniques have been used in studying photosynthesis. In the process, the plant takes carbon dioxide from the air, water from the soil and, under the action of sunlight, converts these into carbohydrates and other substances. Photosynthesis is a complicated but very important process because it is the source of much of the food that is eaten. A study of photosynthesis by plants can be carried out with the help of a carbon-14 tracer. Radioactive carbon dioxide added to the air is continuously fed into a jar in which plants are grown. The radio chemical analysis of the plant products at various intervals of growth provide the material for understanding the photosynthesis process.

Radiotracer techniques are finding widespread applications in studying trace deficiencies in diets, the utilization of food by chickens and other farm animals, large scale movement of birds in the air and insects inside the soil. All of these techniques result in better farm and agricultural products.

b) Hydrology

In hydrology the tracer techniques are very effective in providing information regarding the origin of springs, the origin of infiltration of water, the age of water, the influence of ocean currents and the wind on the movement of sand. Many of these applications have now become routine and are being widely used in many countries.

c) Radiocarbon Dating

By the radioactive carbon-14 technique, it is possible to determine the age of very old objects. This cannot be done by other methods. Living things acquire carbon-14 in the same way as they absorb ordinary carbon from the air. Carbon-14 is continuously produced in the atmosphere due to the interaciton of cosmic ray neutrons with nitrogen. The radiocarbon thus produced interacts with oxygen in the air and is converted into carbon dioxide. The carbon dioxide is taken up by the plants and living matter. Since this carbon comes from the atmosphere, all the living organisms, i.e. trees, vegetables, fish, animals, etc., must have exactly the same ratio of carbon-14 to carbon-12 as that present in the air. Whenever an organism dies, the exchange of carbon atoms with the surroundings ceases and hence the ratio of carbon-14 to carbon-12 diminishes. The life of an ancient object can hence be predicted from the measured carbon-14 to carbon-12 ratio.

Example 22.1

Organic objects absorb carbon from the atmosphere directly or indirectly. A certain amount of the absorbed carbon is C-14. During the lifetime of the object a certain equilibrium amount of C-14 is accumulated in the body. After death this C-14 decays into N-14. The initial specific activity due to the equilibrium value of C-14 is determined to be 15.3 disintegrations per minute per gram of carbon. If the half-life of C-14 is taken to be 5600 years, what is the age of an archeological specimen which contains only 80% of the C-14 found in living material, i.e. the specific activity is about 12.2 disintegrations per minute per gram?

Solution

The specific activity of the specimen is given by

$$\frac{A}{m} = 15.3\ e^{-\lambda t} = 15.3\ e^{-0.693t/5600yr}.$$

Since A/m is 12.2 dis/g min

$$0.8 = e^{-0.693t/5600yr}.$$

$$\therefore t = -\frac{(\ln 0.8)(5600\ yr)}{0.693} = 1800\ yr.$$

Example 22.2

A solution of CsCl labelled with Cs-134 having a half-life of 2.2 yr is dissolved in a one percent NaCl solution for the measurement of flow of water. The amount of the labelled compound yields a calibration factor of (1200 cpm/μCi)(liters/min). The activity is 4.3 mCi. At down stream the detecting system registers a counting rate of 1200 cpm. What is the water flow rate?

Solution

The measurement of liquid flow can be represented by a simple formula given by

$$Q = F\frac{A}{N},$$

where Q is the flow rate, A is the activity of the tracer, and N is the detector counting rate. The quantity F is a calibration factor in (cpm per μCi)(liter/min).

$$\therefore Q = (1200\ cp/\mu Ci)\ \frac{liter}{min} \cdot \frac{4300\ \mu Ci}{1200\ cpm}$$

$$= 4300\ liters/min.$$

Problems

22.1 An organic specimen recovered from an excavation was found to have an activity of 10^{-5} μCi. If the mass of the sample is 100 grams, what is its age?

Ans: 34000 yr.

22.2 Discuss the possibility of measuring high electric resistance by using an ionization chamber.

22.3 Oil used in a diesel engine is mixed with 100 μCi/liter of Zn-65. What is the activity of the exhaust gas if the oil consumption rate is 0.5 liters per day?

22.4 Discuss the advantages and disadvantages of an alpha gauge.

22.5 Lay out a detailed testing scheme for measuring the wear of a steel part. Start with a discussion of the different elements in steel. Describe the neutron activation properties of these elements including iron. Then discuss the radiation measurements to be made for testing the amount of wear.

22.6 A radiation detection system is capable of detecting 50 disintegrations per second.

(a) What are the specific activities of Co-60, P-32, Al-28, and Cs-137?

(b) Determine the minimum amount of each of these radioisotopes necessary if it is to be used as a tracer when the above mentioned detection system is used.

22.7 Phosphorus-32 is a widely used tracer material. It can be produced from thermal neutron activation of P-31, i.e.

$$^{31}P + n \rightarrow {}^{32}P \xrightarrow{\beta^-} {}^{32}S.$$

Use numerical examples to show that the specific activity of a reactor produced sample of P-31 and P-32 is usually very small.

22.8 Natural uranium contains 99.3% of U-238 and 0.7% of U-235. In uranium ore all of the daughter nuclides are in secular equilibrium with the parent nuclides. This means that $\lambda_1 N_1 = \lambda_2 N_2 = \cdots = \lambda_n N_n$. Find the atomic concentrations of all uranium daughters in parts per billion.

22.9 Radioisotopes can be used to determine the age of some materials. For example, the determination of the amount of Pb-206 (RaG) in a uranium-containing ore can lead to the age of the mineral. Prove that the age of a mineral can be calculated from

$$\text{Age} = (\text{Pb}/\text{U} + 0.6\ \text{Pb}) \times 7.5 \times 10^9 \text{yr}.$$

In deriving this equation the minute presence of U-235 (0.7%) may be neglected.

22.10 Ionizing radiation can be used to produce heat for many applications. One specific example is the tritium heat generation. A sample of hydrogen contains 1.3 cm^3 of tritium. It was found that this sample produces 0.1 cal/hr of heat. What is the tritium beta ray energy?

Ans: 5.8 KeV.

22.11 Sodium-24 is used to measure the thickness of a machine part. The material has a density of 8.5 g/cm^3. Na-24 emits two gamma rays of 1.38 MeV and 2.75 MeV energies. If the linear absorption coefficients of the material are 0.52 cm^{-1} and 0.31 cm^{-1} for the softer and the harder gammas from Na-24, respectively, and the ratio of the emergent beams of softer gammas to the harder gammas is 1.5, what is the thickness of the part?

Ans: 1.93 cm.

22.12 Uranium-238 has a half-life of 4.5×10^9 yr. What is the specific activity of U-238?

Ans: 12300 dis/g s.

22.13 Sometimes minute quantities of radioactivity in liquids are measured by evaporating the solution to dryness. A sample of $BaSO_4$ has a specific activity of 1.7×10^7 dis/mg s. Ten milliliters of this sample in the form of a saturated aqueous solution is dried. What is the counting rate when a G-M detector system with an efficiency of 1% is used to measure the activity?

CHAPTER 23

GEOGRAPHICAL ENGINEERING

23.1 Introduction

The history of explosives dates back to about 1000 A.D. when gunpowder was first discovered and used by the Chinese. Ever since, chemical explosives have been in use not only in warfare but also for numerous peaceful applications. These range from digging ditches and starting reluctant oil wells to digging huge canals like the Panama, Erie and Corinth. Since the nuclear explosives were initially developed for destructive purposes their use for peaceful applications did not draw too much attention until the early fifties. Uses of nuclear explosives appear to be as varied as those of the high yield chemical explosives. While the chemical explosive yield is measured in kilograms and tons, the traditional unit for nuclear explosive is kilotons or even megatons of TNT (one ton of TNT is equivalent to 10^9 calories!).

Nuclear explosives are not only capable of doing what was considered to be impossible with conventional explosives, but at the same time they are extremely attractive economically for large scale excavations or earth moving operations. Keeping these factors in view, the USAEC initiated 'Project Plowshare' in 1957 at an initial cost of 45 million dollars with the following objectives:

(a) Development and understanding of the basic processes involved in nuclear detonations in various environments.

(b) Development of explosives specially designed for specific programs.

(c) Conducting various field experiments using different types and sizes of nuclear explosives.

(d) Assessment of safety and cost of nuclear explosives.

(e) Considerations of legal and diplomatic procedures to allow the use of perfected nuclear explosives.

The use of nuclear explosives for large scale applications pose additional problems besides those of a scientific and engineering nature. The first major problem is, perhaps, an emotional one. Since the first use of nuclear explosives displayed only their destructive potential, this characteristic still captures the human imagination. Besides, as the effects of nuclear explosives are neither familiar nor visualized, there is a natural resistance to their use for peaceful applications.

Another major problem of using nuclear explosives for peaceful applications is a political one. For example, a major nuclear explosive project of geographical engineering may draw the attention of the entire world. Furthermore, international agreements or treaties, which are designed to limit the development of nuclear weapons, may apply to the peaceful uses of nuclear explosives, as there is not too much difference between a nuclear weapon and a peaceful nuclear explosive. Added to all of these is the problem of environmental contamination.

An effort has been made in the following pages to summarize the nature, type and safety aspects of nuclear explosives as well as their peaceful applications concerning geographical engineering.

23.2 The Nuclear Explosives

In a nuclear explosive, it is necessary to handle a supercritical mass of some fissionable material in such a way so as to result in the release of a large amount of energy in an extremely short time. A nuclear explosive must therefore be based on the fast fission chain reaction. The fissionable material in excess of the critical mass as well as the containment time of the material determines the yield of a nuclear explosive. This is entirely different from a chemical explosion, where large amounts of explosives can be accumulated without any event because the reaction is not self-sustaining at room temperature and the concept of critical mass is not involved. Nuclear explosives may hence be termed inherently inefficient as far as the utilization of fissionable material is concerned. In any nuclear explosive device, an amount of fissionable material at least equal to the critical mass is always wasted.

In order to explode a nuclear explosive it is necessary to assemble two or several subcritical pieces of some fissionable material into a supercritical configuration. For an efficient nuclear explosive the assembling must be done extremely rapidly. As neutrons are always present in the atmosphere due to nuclear reactions induced by cosmic rays and the spontaneous fission of some elements, the time to assemble all the subcritical pieces of the fissionable material must be less than the average time of the capture of these unavoidable neutrons. For a plutonium device this is around 10^{-8}s. In case the subcritical pieces of some fissionable material are assembled slowly, a neutron would be able to initiate the chain reaction before all the pieces join to form a solid supercritical mass. This would result in the heating up of individual pieces which would blow themselves apart before reaching a highly supercritical state. To bring substantial amounts of fissionable materials together quickly and keep them thus for some time is a technological problem presenting difficulties in the engineering aspects. An increase in the containment time by a fraction of a microsecond results in a difference in the yield of the nuclear explosion by orders of magnitude. Two basic methods are described in the literature for detonating a fission explosive.

a) Gun Method or Ballistic Technique

Here two or more pieces of fissionable material that will become supercritical when assembled together are used. The subcritical pieces are fired into each other by using some high chemical explosive, as projectiles are fired from a gun. The fissile material is surrounded by a heavy tamper of very dense material (generally U-238) to delay the expansion of the supercritical assembly. In addition, the U-238 tamper results in a decrease of the critical mass due to the reflection of some outgoing neutrons and the transformation of some of the U-238 into fissionable Pu-239. To ensure the instantaneous occurrence of a chain reaction, it is necessary to provide a neutron source at some suitable place within the assembly. The neutron source may consist of a radioactive alpha emitter like polonium. The alpha particles so emitted react with the beryllium nuclei in order to produce neutrons. During the normal subcritical phase of the fissionable assembly the interaction of alpha particles with beryllium is prevented with the help of a thin gold foil or some other form of containment. At the moment when the subcritical pieces attain the most favorable shape, the foil or the containment is pierced and the neutrons are produced in the newly formed supercritical assembly. Figure 23.1 illustrates the gun type of nuclear device.

The device is generally recommended when highly enriched (about 95 percent) U-235 is used as the fissionable material. The only source of spontaneous fission in U-235 is cosmic ray neutrons. As most of these neutrons can be shielded, the chances of predetonation during the assembling can be minimized. Besides, the generation time for fast fission neutrons is much larger in U-235 than in Pu-239, which means that the time in which the subcritical

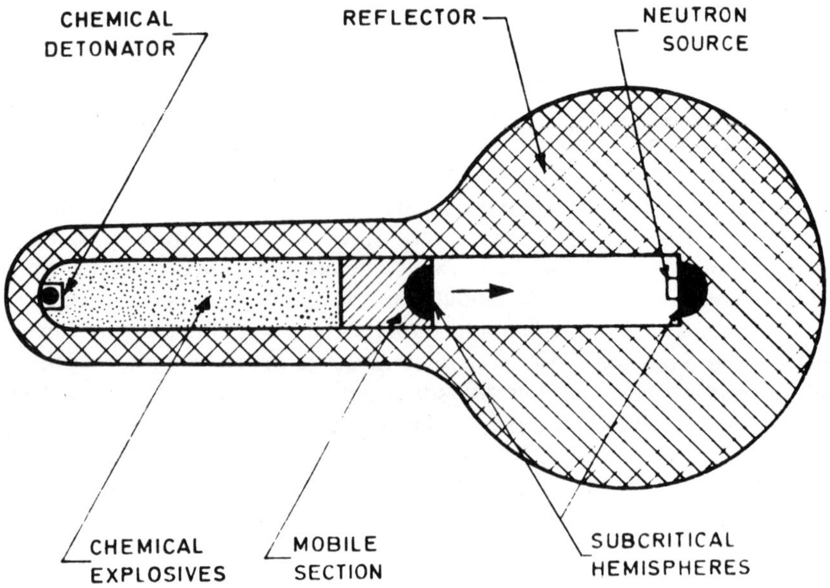

Fig. 23.1: Ballistic or gun type device.

pieces of U-235 are to be brought together is within the limits achieved by present day firing devices. The amount of highly enriched U-235 needed for a 20 kT device is estimated to be around 20 to 30 kilograms.

b) Implosion Technique

When the fuel for a nuclear explosive is Pu-239, the supercriticality must be obtained in time intervals much shorter than those allowed by the gun or the ballistic method. The Pu-240 present in plutonium extracted from nuclear reactors provides sufficient neutron background. Besides, since plutonium is an alpha emitter, any light element impurities could result in neutron production. Also, the generation time for fast fission in Pu-239 is much smaller than that for U-235. A gun device could be used for plutonium only if very high purity Pu-239 could be obtained and only if the projectiles could be fired at the extreme upper limits of practical artillery velocities. Both of these difficulties have resulted in the development of an entirely new explosion device for plutonium commonly known as the implosion technique.

A subcritical hollow sphere of fissionable material is designed in such a way that it can be compressed into a solid symmetrical supercritical mass. The technique consists of the use of a number of chemical explosives, shaped and located in such a way as to produce a symmetric implosion. The proper design of the shaped charges for implosion is a challenging problem. In peacetime, shaped charges are used for perforators, oil well casing and tappers for open hearth furnaces. The speed of intense compression could be made high enough to avoid the predetonation caused by neutrons within the assembly. This technique is more efficient, particularly for obtaining larger yields from a given quantity of excess critical mass. The implosion type of nuclear weapons are, however, bulkier because of the incompatible materials like chemical explosives, metals, etc. Figure 23.2 illustrates the salient features of an implosion type of nuclear explosive.

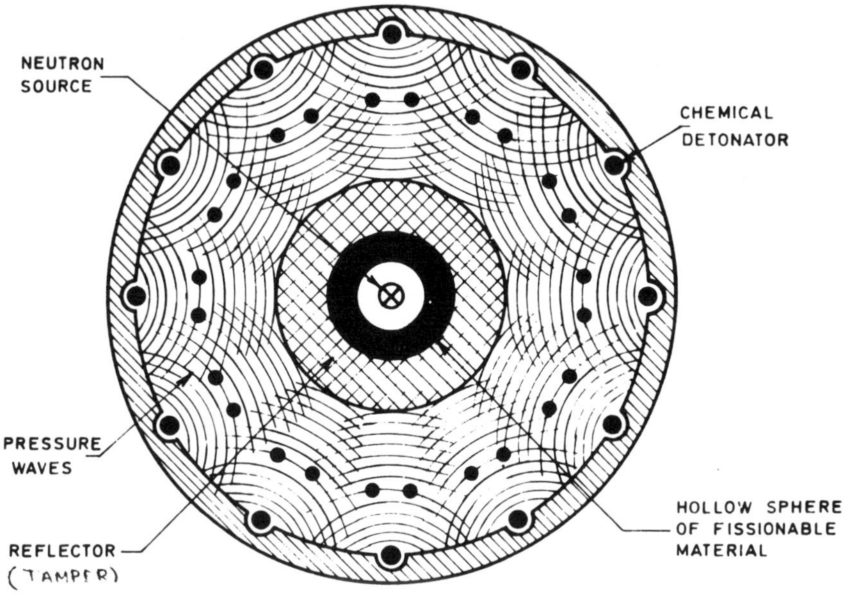

Fig. 23.2: Implosion Technique.

The detonation of chemical explosives is arranged in such a manner as to send high pressure spherical waves inward. The converging shock waves initiate further chemical reactions and in the process, gain energy and momentum. The pressure waves reach the nuclear tamper and compress it along with the hollow core of fissionable material. This results in a super-critical condition and the release of large quantities of energy. For increased yield efficiency and a decrease in the chances of predetonation, extremely high implosion velocities are required. The main difficulty in this technique is the achievement of symmetric implosion. By performing numerous experiments on metallic dummies the difficulties in achieving symmetric implosion can be removed.

A 20 kiloton plutonium explosive requires 7 to 10 kilograms of plutonium of sufficient purity. The technique offers a marked economy in the use of fissionable material since any increase in the plutonium density would greatly reduce the critical mass requirements. Besides, the implosion device allows the use of plutonium containing significant impurities of higher isotopes, easily available from nuclear reactors. The implosion technique can be used for all fissionable materials, i.e. U-233, U-235, and Pu-239.

When nuclear explosives are to be used for peaceful purposes it is desirable to reduce the radioactivity content in the debris. This concept has given rise to the so-called 'clean devices'. In such a device the proportion of the energy produced by fission reactions as compared to the fusion reactions is kept minimum. A fission trigger is used in such devices to raise the temprature of light elements by several million degrees. For better efficiencies of the fission trigger, the extra materials which could absorb heat and neutrons should be minimized. For the infinitesimal time and extremely high temperatures, conduction plays an insignificant role and only radiation and the motion of particles is of consequence. The use of ballistic detonators reduces the amount of extra materials and hence produces a higher concentration of energy leading to a higher temperature to initiate the fusion reaction. All these considerations indicate that a fission trigger using a ballistic detonator is preferable compared to an implosion type of trigger.

Since the most common materials for thermonuclear reactions are in gaseous form at ordinary temperatures, these cannot be used directly as the primary material for the device. In the liquid form these materials are not suitable because of the requirements of low temperatures and high pressures for keeping them in the liquid form. In addition tritium does not exist naturally. Moreover, it has a half-life of 12.5 years and is extremely expensive to produce. The storage of tritium for any period of time will hence be very uneconomical. The recommended technique is the production of tritium by the bombardment of neutrons produced by the fission trigger on solid lithium deuteride. This triggers the thermonuclear phase of the explosion. The high temperature thus produced vaporizes the lithium deuteride into lithium nuclei and deuterons. The lithium nuclei produce tritons according to the following reaction

$$^{6}Li + n \rightarrow {}^{4}He + T + 4.8\ MeV.$$

The deuterons and triton so produced result in an uncontrolled thermonuclear reaction. It has been estimated that even at 2 percent efficiency a mass of 1000 kg of lithium deuteride would yield energies equivalent to several million tons of TNT in a time of the order of few microseconds. As the thermonuclear reactions are not limited by the consideration of critical size, they are capable of producing extremely large yields without producing too much radioactivity. Fission devices, on the other hand, are not very clean and their yield is limited to below 1 million tons of TNT, mainly due to the criticality considerations.

Nuclear explosives have been made in a variety of sizes and shapes. The USAEC provides canisters of four different diameters for containing nuclear explosives. Explosive yields of 10 kilotons can be accommodated in 12 inch diameter canisters, 25 kilotons in 16 inch canisters, 100 kilotons in 18 inch canisters and 1 Megaton in 24 inch canisters. The height to diameter ratio of the canisters containing nuclear devices is quoted to be 5 to 1. The approximate weight of small and medium sized nuclear explosives including the canister, etc. is around a few tons.

23.3 Explosion Phenomena

A nuclear device may be exploded in the atmosphere, near the earth's surface or at various depths underground. A surface device creates a shallow crater by compacting the ground beneath the zero point and blowing away the overburden due to the extremely high pressure. If the explosion is too close to the surface, much of the energy escapes into the air and only a shallow crater is formed. On the other hand, if the explosion takes place deep underground, most of the broken and shattered rocks settle back in the crater itself and a very shallow crater results also. The best location for obtaining the largest crater size lies between these two limits. This is illustrated by Figure 23.3.

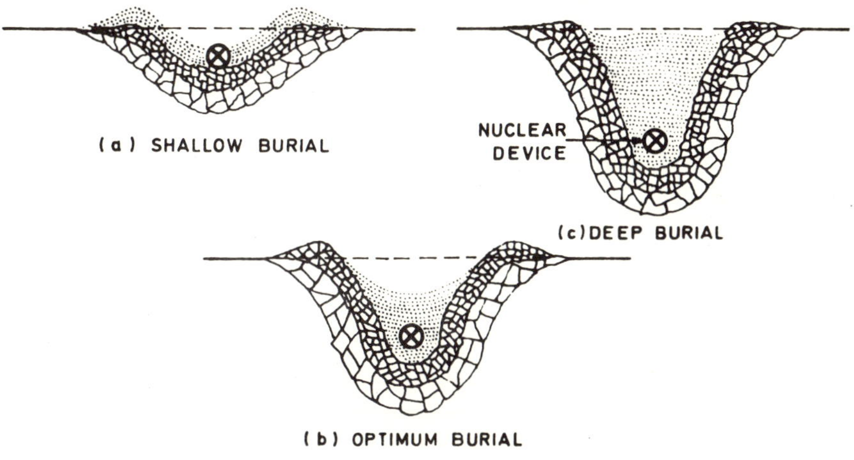

Fig. 23.3: Excavation with nuclear explosives.

For safe and completely contained nuclear explosives, the depth of the burial should be such that the rubble core ceases its growth before it reaches the surface. If the cavity does not reach the surface, any initial uplifting of the surface remains there. A chimney like structure is formed for deep underground explosions. The chimney is filled with broken material, having generally a small cavity at the top. As both shallow and very deep burial nuclear explosives are used for geographical engineering and other peaceful applications, a brief description of the sequence of events of such explosions is discussed here.

As soon as a nuclear device is detonated, a large amount of energy is produced in a period less than a microsecond. The material in the device is vaporized and raised to a temperature of several million degrees. Within a few microseconds, the extremely high pressure and temperature thus produced generate a strong shock wave which propagates spherically from the source through the rock medium and vaporizes some of the surrounding earth. In common silicate rocks, the vaporization continues up to a radius of about 2 meters for a 1 kiloton yield. This radius increases roughly as the cube-root of the yield. Figure 23.4 illustrates the initial events.

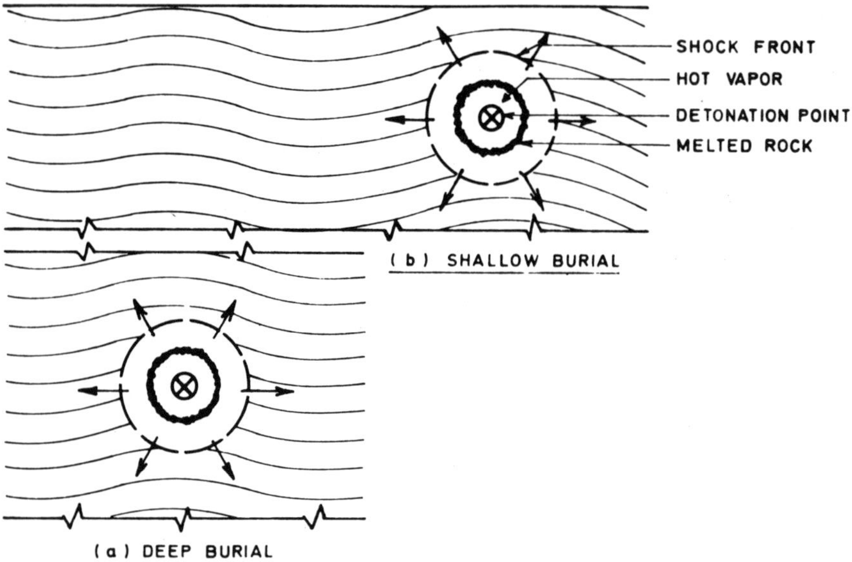

Fig. 23.4: Initial phase of a nuclear explosion.

The cavity so produced continues to expand since the pressure of the vaporized rock in the cavity is in the neighbourhood of one million atmospheres. This expansion continues until its pressure drops and finally equals the pressure from the weight of the rock overburden. This pressure is about 1 pound per square inch for each foot of burial depth. It has been estimated that a cavity of about 50 feet in radius results from a 1 kiloton explosion. This increases as the cube-root of the explosion yield. For shallow burials the cavity begins to grow preferentially towards the surface. The energy contained in the explosive lifts up most of the rock and dirt in it, as indicated in Figure 23.5.

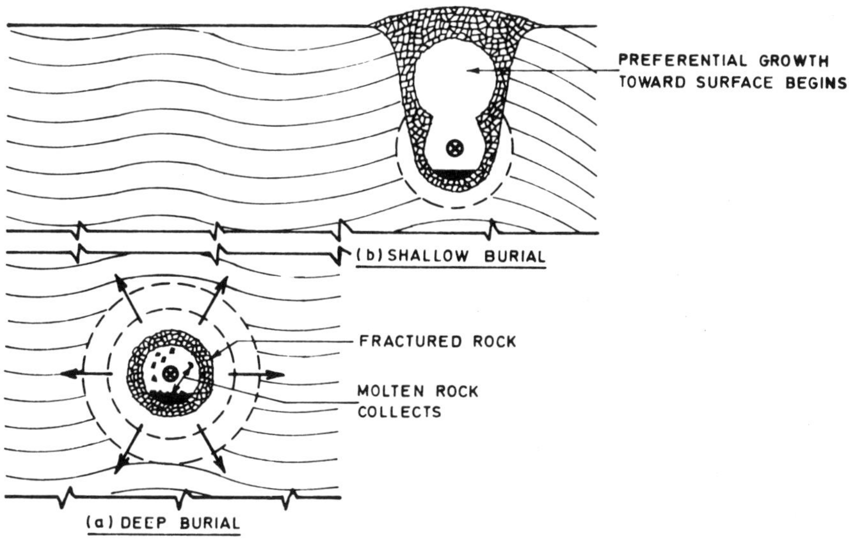

Fig. 23.5: Second phase of a nuclear explosion.

Soon after the stabilization of the cavity volume, rock vapors quickly condense, leaving water vapor as the main component of the cavity gas. The cavity cools down by heat transfer to the surrounding medium. The fractured rock begins to collapse within the cavity. For shallow burials, the rock and dirt is further lifted up. As a result of this, vapors as well as radioactive gases and particles escape to the outer atmosphere through the broken rock and other materials. This phenemenon occurs within a few seconds after the nuclear explosion. This is shown in Figure 23.6.

After a time ranging from a few minutes to a few days, the cavity pressure falls to such a level that, combined with the natural strength of the rock in the cavity roof, it is no longer sufficient to support the overburden. This results in the collapse of the cavity roof which reduces the cavity pressure still further by the condensation of water vapor present in the cavity. Once started, the collapse proceeds upwards by the mechanism of successive collapses of the roof into the cavity. If the chimney does not encounter a rock strong enough to support a roof or if it does not get filled completely with more bulky broken rocks, the collapse continues to propagate upwards. On the average, it has been observed that the cavity height ranges from 4 to 5 times the cavity radius in common rock formations. If the burial depth of the

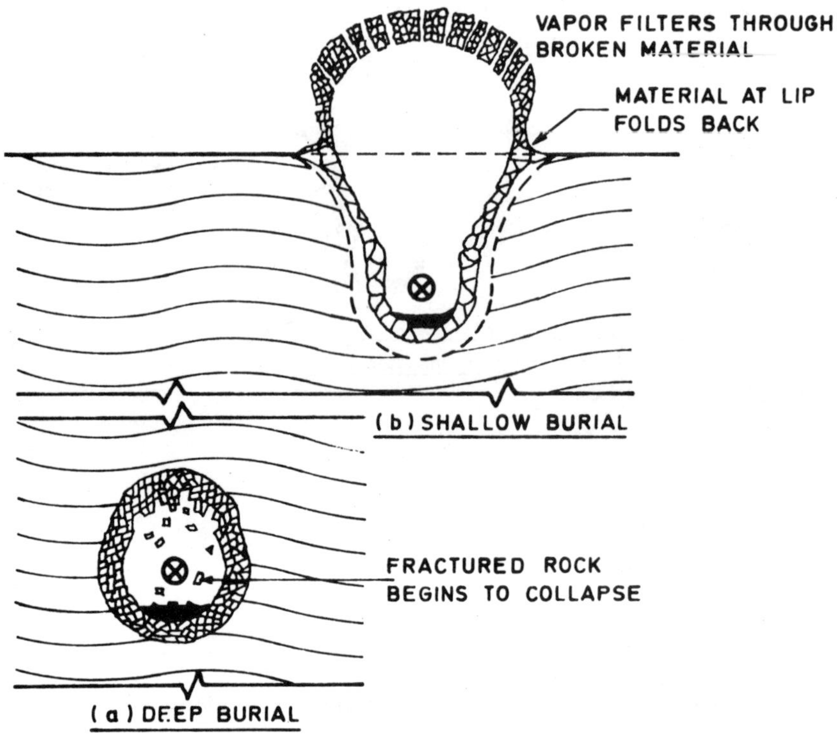

Fig. 23.6: Third phase of a nuclear explosion.

nuclear explosive is small, most of the lifted broken rock and dirt settles back in the crater produced by the explosion. The final shapes of the chimney and the crater are shown in Figure 23.7.

As the nuclear reactions involved in peaceful nuclear explosives and nuclear weapons are basically the same, it is essential to make a safety report before using a nuclear explosive for any specific application. The report should be based on the maximum credible upper limit of the yield rather than the most probable yield. Even if no release of radioactivity is predicted the tests are executed only under favorable weather conditions.

23.4 Geographical Engineering

GEOGRAPHICAL ENGINEERING INCLUDES THE APPLICATIONS OF NUCLEAR EXPLOSIVES TO CHANGE THE GEOGRAPHY OF THE EARTH, such as the construction of waterways, earth moving from deep mineral deposits, creating harbors, lakes and dams, the construction of highways and railways in difficult terrains, the distribution and

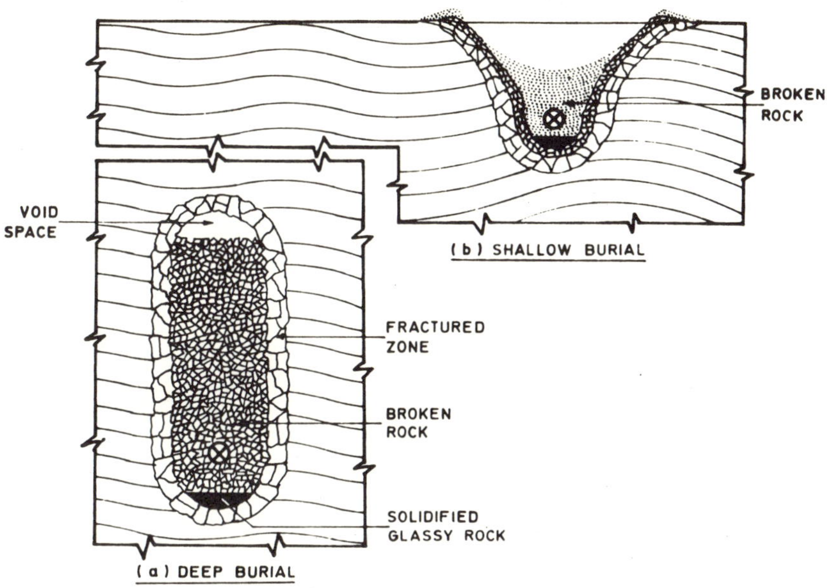

Fig. 23.7: Final shapes of chimney and crater.

controlling of water movement, the stimulation of gas and oil wells, the production of underground reservoirs, and the mining of low concentration minerals, etc.

a) Waterways and Sea Level Canals

The most publicized application of nuclear explosives is the construction of huge waterways and sea level canals. Experiments with both chemical and nuclear explosives show that a row of properly spaced charges can be used to produce a huge waterway or a sea level canal. The use of conventional explosives for such a gigantic task is extremely expensive, and at the same time very time consuming. The same canal built by nuclear excavation methods would cost only a fraction of the expense of conventional excavation. A marked characteristic of such cratering is very high lips on the canal sides as shown in Figure 23.8.

b) Harbors, Dams and Groundwater Recharge

The ease with which huge craters can be produced by nuclear explosions under the sea led to the speculation of constructing useful harbors with the help of nuclear explosives. The main advantage of such harbors is their location where neeed exists. Building such harbors near the locations where hard rocks are predominant increases their life. Besides, the entrance channel could be lengthened to reach some inland point. It is expected that

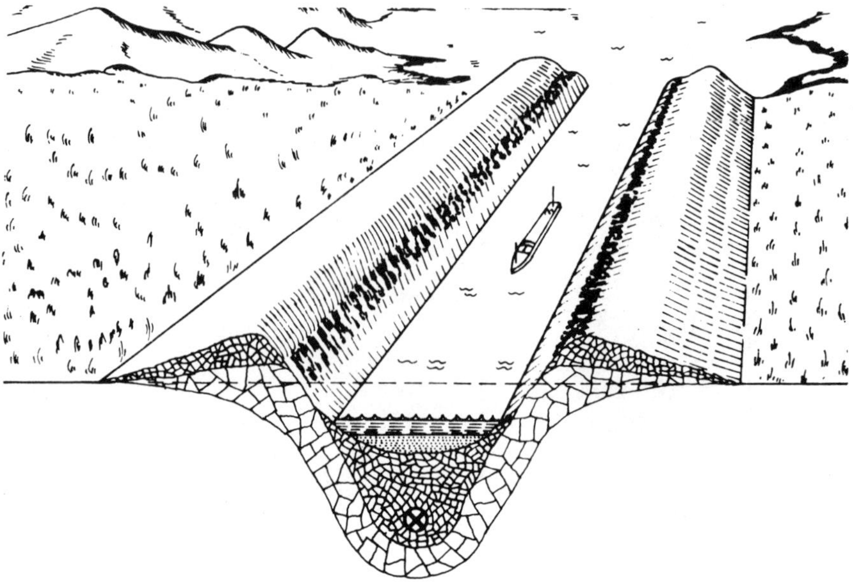

Fig. 23.8: Construction of waterways by nuclear explosives.

the ocean-floor cratering nuclear explosives would result in shallower and wider craters than those on the earth. The cost of drilling emplacement holes for exploding nuclear devices under the sea is expected to be quite high. The construction of harbors at remote but suitable sites is expected to result in an accelerated economic return and the development of the surrounding area. Such projects need detailed study for the selection of locations. Numerous areas such as the western coasts of Africa, Australia and South America could gain great economic advantages with adequate harbor facilities. These coasts are very close to areas which are rich in mineral resources and to some of the most fertile fishing areas.

Numerous studies have been made for creating water storage facilities based on nuclear cratering experiments. For very large capacity water reservoirs, nuclear explosives are very attractive economically. In addition, the nuclear explosives can be used to break through a barrier to allow the unused water to recharge the underground aquifiers as shown in Figure 23.9.

c) Highways and Railways

Nuclear explosives offer a very attractive economic alternative to conventional explosives for cutting passes through difficult and mountainous

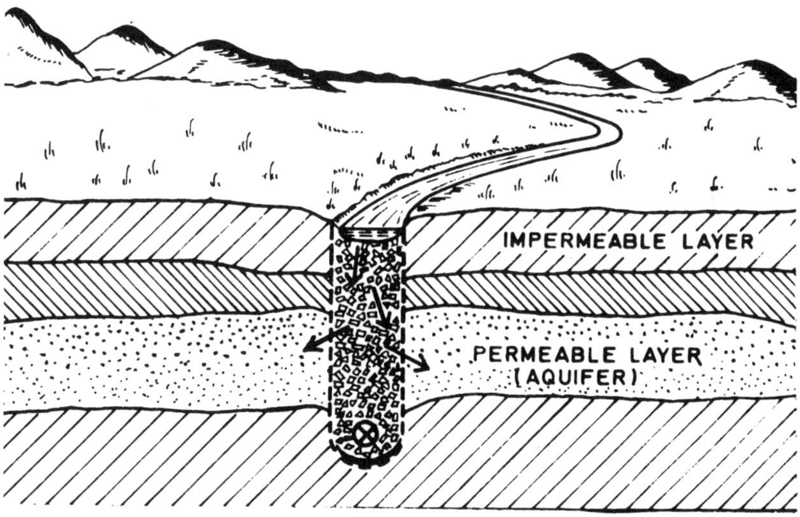

Fig. 23.9: Recharge of underground aquifiers using nuclear explosives.

regions. Compared with canal or waterway construction, the additional requirement imposed on highways and railways is the smoothness of the bottom of the cut. If the bottom is very uneven the cost of post detonation cleaning could nullify some of the economic benefits gained. A series of properly spaced nuclear explosives would produce a parabolic channel, reasonably free of discontinuities between the individual blasts. The upheaved lips of the cut so produced prevent the material from the surroundings from entering the channel.

d) Stimulation of Gas and Oil Reserves

Numerous gas fields and oil wells are not economical anymore because of the low permeability of the rock. It has been found that applications of heat and pressure help in stimulating the reluctant wells. When a well is not disturbed by applying high pressure and temperature for a very long time, the flow rate of the oil or gas starts decreasing and ultimately the flow ceases altogether. Various studies have indicated that reluctant oil and gas fields can be stimulated by using nuclear explosives for fracturing the surrounding rocks and hence increasing the permeability.

The development of techniques for the stimulation of oil and gas fields is a practical response to a recognized need and may result in a very significant contribution by making large quantities of natural gas and oil available at reasonable prices. Presently the stimulation of oil and gas reserves by nuclear explosives is very close to commercial exploitation. This estimate is based upon two field experiments, i.e. Gasbuggy & Rulison, conducted by the USAEC in collaboration with the private industry in 1967 and 1969 respectively. The aim of both experiments was to stimulate the flow of gas which was known to exist in substantial quantities in low permeability reservoirs. The conventional methods were found to be not feasible both economically and technically.

Gasbuggy employed a 30 kT device buried at a depth of about 1290 meters in a low permeability sandstone gas reservoir. The fracture zone extended to a radius of about 100 meters. Calculations of the porosity of the fracture zone before and after the detonation showed an increase in permeability by a factor of 5 and a stimulation factor of 5 to 8. Rulison used a 40 kT device at a depth of about 2570 meters. The response in this case was also found to be quite encouraging. Stimulation of reluctant oil reservoirs has been studied by the USSR using small nuclear explosives. The increase in oil production was estimated at 30 to 60 percent over a period of several years.

e) Mining

A very promising application of completely contained nuclear explosions is the in situ leaching of ores from low grade deposits. Throughout the world a large number of low grade ore deposits, containing large quantities of various minerals, exist. The distribution of these minerals, however, is such as to make their exploitation uneconomical by conventional techniques. It has been proposed to use nuclear explosives to create a chimney and the associated fracture zone in the ore deposits followed by the introduction of acid solutions at the top of the chimney. As the acid percolates through the broken rocks, the minerals would be leached from the new surfaces as well as from the pores and fissures. The acid with the dissolved mineral would then be recovered at the bottom and pumped to the surface where separation of the dissolved mineral would take place. Figure 23.10 illustrates in situ leaching of ore.

The mining of mineral deposits can also be made easier by using nuclear explosives for overburden removal when the minerals are buried not too deep underground. The technique, generally known as strip mining, is still in the initial stages of investigation and needs detailed studies as it involves problems of air-blast and radioactive fallout which is not present in the completely contained nuclear explosives.

Problems

23.1 For a fast critical assembly discuss the effect on criticality due to temperature variations.

23.2 Calculate the amount of pure Pu-239 which must be used in fission to

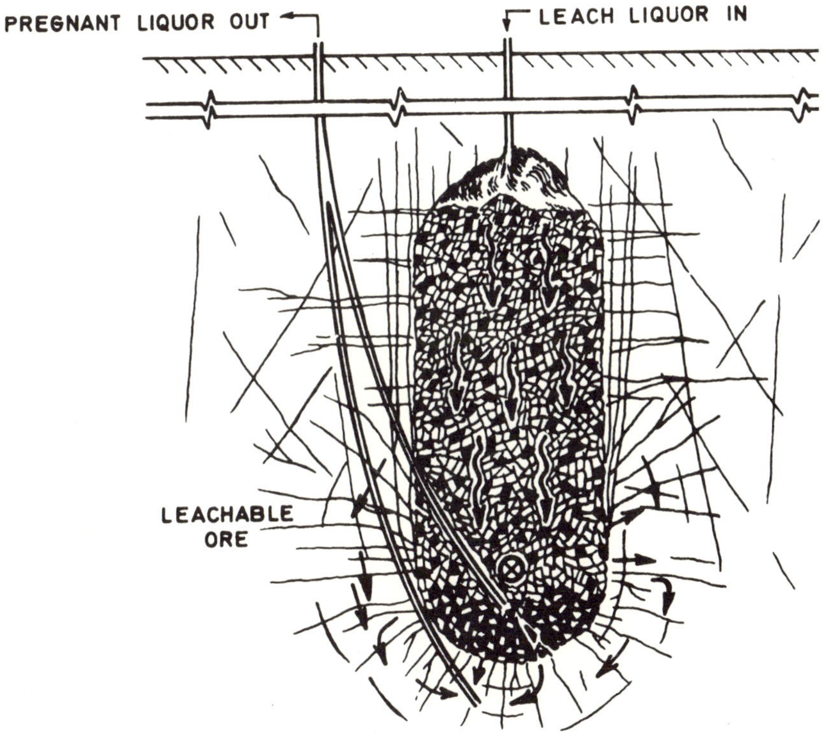

Fig. 23.10: In situ leaching of ore.

produce a power equivalent to one Megaton of TNT.

23.3 If the density of the fuel of a homogeneous critical system is doubled, discuss the effect on the critical dimension and the critical mass of the system.

23.4 Describe a detailed application in geographical engineering in the construction of a reservoir for flood control and irrigation.

23.5 A flood control reservoir is made by exploding a nuclear device equivalent to 10 kilotons of TNT. One third of the water in a river is expected to go through the reservoir and rejoin the downstream flow when the reservoir is filled to its maximum capacity. The mass flow rate of the river water is 5×10^7 kg/hr. If 1% of the radioactivity from fission is retained and mixed evenly with the water passing through the reservoir, what is the initial specific activity of the downstream water?

Assuming uniform mixing and decay, calculate the time required for the downstream water specific activity to be reduced to 5% of the initial value. The reservoir capacity is assumed to be 10^{10} kg of water.

23.6 What are the possible effects on a gun-type of nuclear explosive if Li-6 (or Li-7) is evenly distributed in the reflector?

23.7 Discuss the environmental effects due to the explosion of a 20 kT nuclear explosive for gas and oil stimulation at a depth of 1,000 meters.

23.8 Explain the concept of a "neutron bomb".

CHAPTER 24

RADIOGRAPHY

24.1 Introduction

A radiograph is a photographic record obtained by using radiation sources such as x-rays, gamma rays, or neutrons. It is a basic tool in nondesctructive testing which, among other techniques, includes penetrant, thermal, ultrasonic, and electric and magnetic methods. The most frequently used radiation sources for radiography are x-rays and gamma rays. Recently, neutron radiography has gained some importance because of the availability of neutron sources of various types and strengths. The basic principle of radiography is similar to that of photography, the only difference being that light sources are replaced by radiation sources. One major use of radiography is the inspection of mechanical parts and metal objects. Thickness measurements have a direct bearing on quality control in industry. X-rays have been used in medical diagnostics and checkups for a long time. It is quite certain that they will continue to be used as such in the future. Medical diagnostics and industrial inspection applications of neutron radiography are still in a state of infancy. The recent production of californium-252 in large quantities (milligram to gram) will further enhance the many applications of neutron radiography. Figure 24.1 shows the principle of radiography. A radiation source S emits radiation (x-ray, gamma ray, or neutron) that travels in straight lines to the specimen O. Depending on the nature of the speciman and the incoming radiation, the image I appearing on the screen shows not only the details of the object about its contour but also the structure of the object. If there is a flaw in the object, the contrast in the enlarged image is

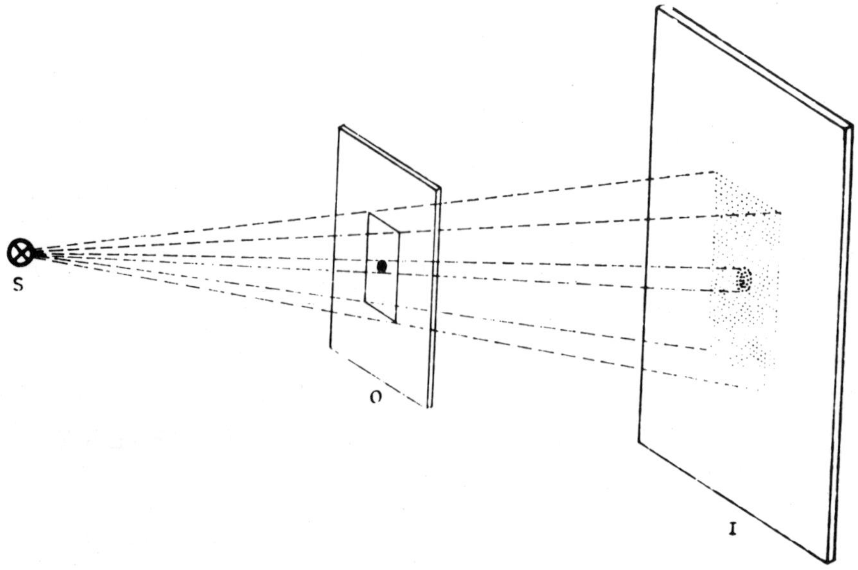

Fig. 24.1: Radiography

usually striking and easier to identify as compared with other inspection methods.

Figure 24.2 features a radiographic picture of a jeep from the cover of the booklet 'Nondestructive Testing' of the series: 'Understanding the atom' published by USAEC. The picture is a composite of several small x-ray radiographs. The penetrating views of various metal components and parts of the jeep are quite obvious.

The production of radiation sources for radiography consists of the usual radioisotope production and source encapsulation. The source strength usually lies in the medium range, i.e. a fraction of a curie up to a hundred curies. The detailed methods of radioisotope production are discussed in chapter 18. Introductory radiation engineering and source encapsulation (and some special discussions on radioisotope production) techniques are discussed later in Radioisotope Engineering.

The use of x-rays for the radiography of the human body is a very developed field in medical applications and is called radiology. Detailed discussions on this subject can be found in any standard radiology textbook. Neutron radiography is at the present time not widely used in industry. For a study of this subject, we refer the reader to an excellent introduction written by Harold Berger in 'Scientific American', November, 1962. In this chapter, discussion is concentrated on the most widely used radiography technique in industry and many other applications — gamma radiography.

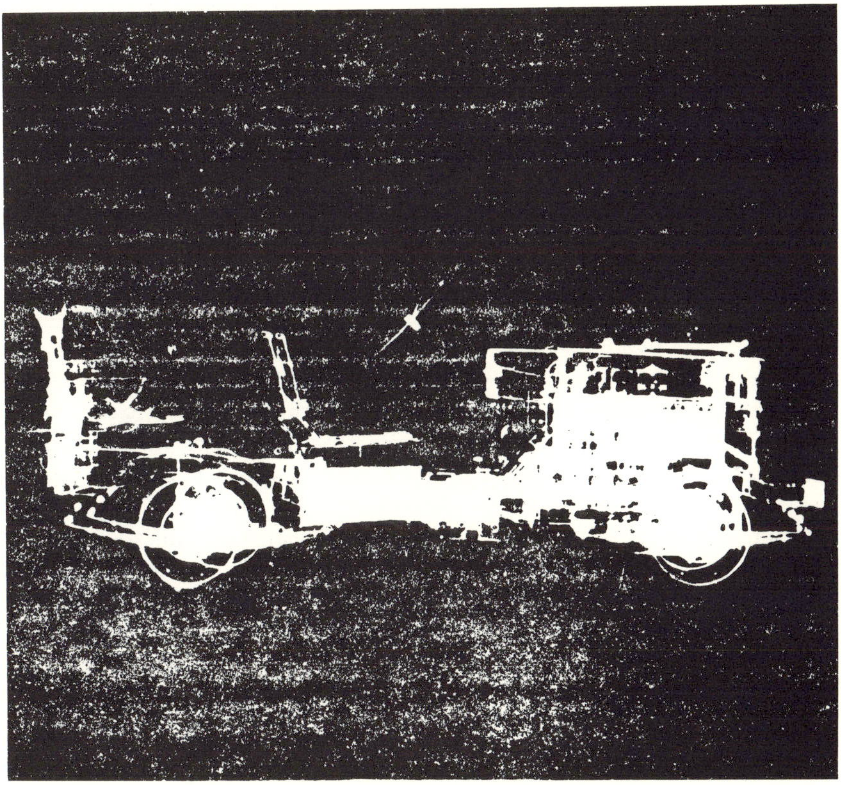

Fig. 24.2: An x-ray radiograph of a jeep.

24.2 Gamma Radiography Sources

There are certain criteria for selecting appropriate sources for radiography purposes. The penetrating power of the radiation under consideration must match the object to be radiographed. The half-life of the radioisotope selected must be long enough to permit repeated exposures of the object. The geometry of the source is also a very important factor which leads to the requirements of (a) standardization of the shapes and dimensions and (b) the high initial activity of the sources for radiography. Of course the availability and cost must also be evaluated. When these factors are considered, only a handful of radioisotopes remain as suitable radiographic sources. Table 24.1 lists some of the radioisotopes for gamma radiography sources.

Table 24.1: Radioisotopic source for radiography.

Source	Half-life	Normal activity (Ci)
Co-60	5.3 yr	0.5 - 2
Se-75	120 da	5 - 10
Cs-137	30 yr	0.5 - 5
Tm-170	127 da	0.2 - 1
Ir-192	74 da	0.2 - 5
Ra-226	1600 yr	0.2 - 1

The fabrication and encapsulation design and processing of radioisotope sources are discussed in chapter 25; in this chapter only the characteristics associated with radiation sources specifically used for radiography purposes are discussed. Such a source is usually sealed in a metal capsule, which in turn is inserted in one of a number of source containers. These containers serve both as storage containers and beam radiation sources. They are usually called gamma projectors or gamma cameras. Figure 24.3 shows a

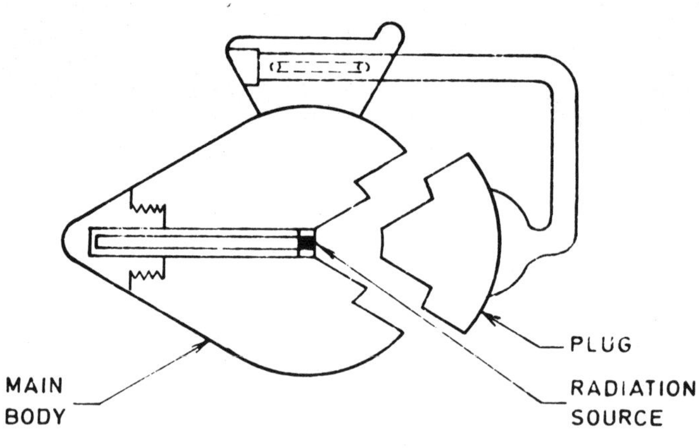

Fig. 24.3: Plug type camera.

typical stepped plug type camera that can be removed by the handle attached to it as shown in the figure. Another type of radiography camera is the D-type, Figure 24.4. In this case, the radiation source is attached to the rim of a rotating cylinder. When in use, the rotating cylinder is turned in such a way that the source is exposed to the flat surface of the D. After use, the cylinder is turned 180 degrees so that the source is shielded.

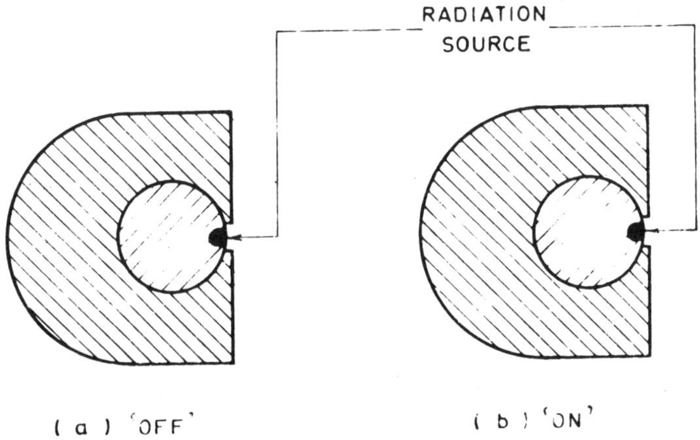

Fig. 24.4: D-type camera.

24.3 Optimal Features

An image of a flaw in a specimen or an object depends upon the size of the source, the object, the distance, and the exposure time. Each of these four factors has an optimum value that gives the best picture. The scattered radiation must also be considered. The opacity E_0/E of the image is usually indicated by the blackening density defined by

$$D \equiv \log \frac{E_0}{E}, \qquad (24.1)$$

where E_0 and E are the incident flux to the film with and without the object or the flaw, respectively. Note that the ratio E_0/E depends on the lumination Q in mCi/cm^2 defined by

$$Q = k \frac{A \cdot t}{d^2}, \qquad (24.2)$$

where A is the source activity, in mCi, t is the exposure time in hours and d is the distance between the source and the film in cm. When the blackening density D is plotted against the logarithm of the lumination Q, one obtains the sensitometric curve of a radiography system, Figure 24.5. The parameter k has the dimensions of h^{-1} and is a characteristic of the radiographic film which depends on the properties of the radiography system and the film development time. When the straight line portion of the curve is extrapolated

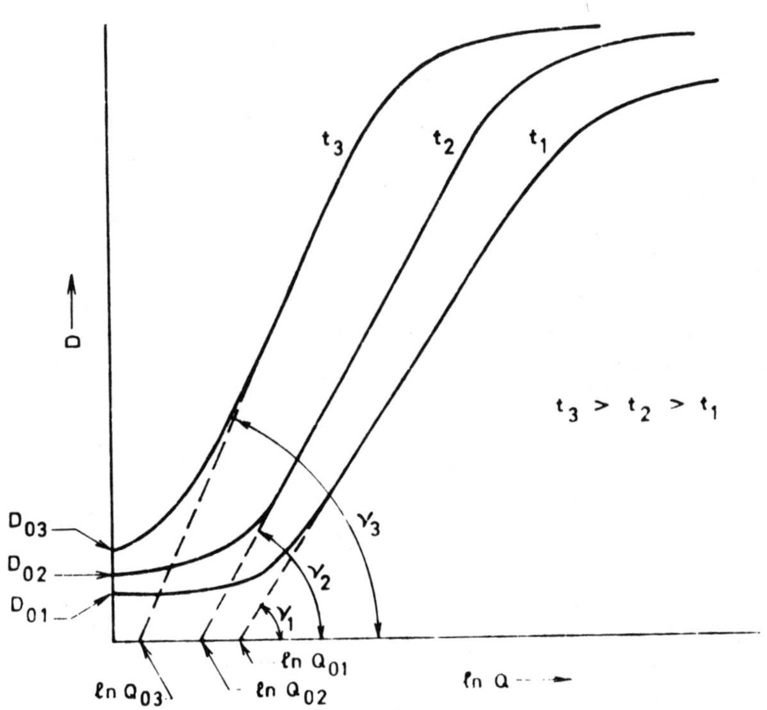

Fig. 24.5: Sensitometric curve.

to the horizontal axis and the angle γ is drawn, one can compare the contrast of the films under various exposure times. In general, the larger the angle γ, the more is the contrast achieved. For different systems and films these characteristic curves form the main criterion for performance comparisons.

The next important item in the operation of a radiography unit is the determination of the exposure time. This depends on the absorption characteristics of the film. In practice, detailed information on such characteristics is difficult to obtain. One must therefore rely on experimental methods. For this purpose, it is useful to define the radiographic sensitivity.

$$S \equiv \frac{\text{smallest detectable size}}{\text{specimen thickness}} \times 100\%. \qquad (24.3)$$

The experimental determination of this parameter can be done by using step wedges of increasing thicknesses of the same specimen material. The sensitivity depends on many factors. Some of these factors are listed below:

a) thickness differences in specimen;
b) source property such as size and radiation energy;

c) scattered radiation;
d) type of film and screen;
e) development and exposure time
f) source-film distance;
g) development temperature and other factors;
h) specimen-film distance;
i) activity of the developer;
j) abruptness of specimen thickness changes.

The blackening densities for different thicknesses and different times and other factors of exposure are obtained from a family of curves similar to the ones shown in Figure 24.6. When this figure is used together with the blackening density versus logarithm of Q curves (Figure 24.5) an exposure factor $Q \sim A \cdot t/d^2$ versus exposure time t can be plotted, Figure 24.7. This set of curves is plotted with A/d^2 as a parameter. From this curve the exposure time can be determined.

Sometimes the scattered radiation affects the quality of the film. In extreme cases, scattered radiation might have a higher intensity than the primary radiation. This effect is more pronounced with heavy specimens. To remedy this situation, one can a) arrange the entire system to be far from any flooring or walls, b) use collimators to limit the source radiation as a pencil or parallel radiation to impinge upon the specimen, c) use shielding materials with the least amount of buildup to mask off the unused portions of the film or the specimen, d) use lead intensifying screens in contact with the film to guard against the scattered radiation from the specimen and the backscatter,

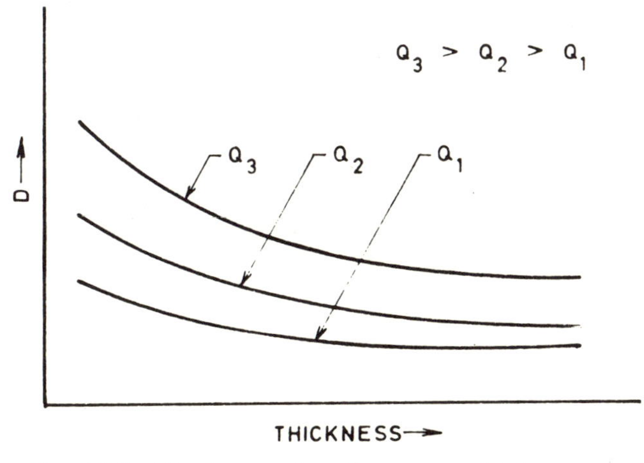

Fig. 24.6: Blackness density versus thickness.

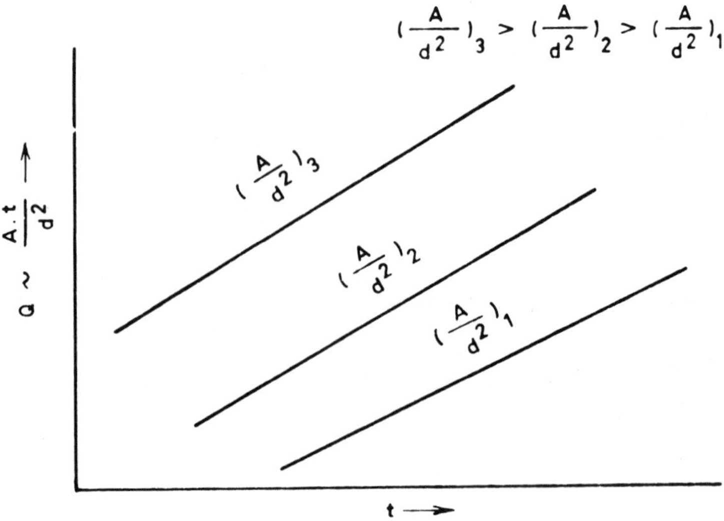

Fig. 24.7: Exposure factor versus exposure time.

respectively, or e) use filters to reduce scatter radiation due to low energy radiation. For gamma rays with high energy, the scattering radiation actually becomes less of a problem because it has a tendency to move in the forward direction, thereby reinforcing the forward moving primary radiation to yield a better quality picture of the object.

24.4 Applications

Compared with x-ray radiography, gamma radiography has the advantage of a) smaller equipment size, b) requiring no electric power supply, c) having no cooling system, and most important of all, d) lower cost. The only disadvantage is the quality of the film, which usually cannot match the x-ray film's definition. One must, however, emphasize the fact that a gamma radiographic unit can be made mobile because of its size. It is because of the first two factors that many portable models of radiographic units are being commercially made.

The most noted application of gamma radiography is in welding inspection. For a plane specimen the film is simply placed behind the weld to receive the radiation. For a welded circular pipe, the film may be placed inside the pipe if possible. However, for long pipes, this might prove to be impractical. In such a case, the film can be put outside the pipe next to the weld, with the source on the other side of the pipe.

The film badge dosimeter is a very important application of radiography. The blackening density of the film is proportional to the radiation dose. In a film badge dosimeter the radiographic film is enclosed in paper covers to shield it from light. Depending on the type of radiation the dosimeter is designed to detect, a suitable film material can be chosen. In some film dosimeters, metal filters might be used to correct the non-linear responses of a film.

A very useful application of radiography is the detection of material or component defects. Since most of the defects are essentially three dimensional, more than one film may have to be taken to qualitatively analyze the defect. One commonly used method to yield a three dimensional analysis of a defect is to use two radiographic cameras arranged in such a way that two perpendicular radiation beams pass through the suspected defect at the same time. The two films obtained would then collectively yield a three-dimensional description of the specimen.

In some cases radiographic methods can be used to measure the thickness or impurities of a specimen. Many such measurements using recording devices other than films have been discussed in chapter 22 — Radioisotope Gauging. In theory, films can be used in place of all other recording devices. This is particularly true for machines and parts in which bulky radiation detectors and other recording devices cannot be placed.

The quality control of castings is greatly facilitated if radiography is used. Presently, two basic methods are used for internal control of castings or ingots. They are the straight forward single thickness casting and the multi-thickness casting. Several techniques are used in the latter category. The addition of perfect specimen material of varying thicknesses enables the filming of a specimen with a regular shape but a finite number of thicknesses. A diagonal incidence of radiation can be applied to a specimen if its thickness is less than the value found in the exposure chart. Filters are sometimes used to harden the incident radiation so that the thicker specimens can be photographed.

In relation to nuclear power development, radiography is used to inspect rolled fuel elements for reactors. A typical radiograph may involve the use of beta radiation from Sr-90 with a source strength of 0.5 Ci for a total of five minutes exposure time.

Steel construction and steel wires such as cable car wires can be easily inspected by the radiographic method. The picture of a jeep, Figure 24.2 is an excellent example of this application. In a similar manner concrete structures can be inspected.

The following is a list of some other applications of gamma radiography:

a) examining submarine cables under water,
b) inspecting electric wirings,
c) checking vacuum tube elements,

d) locating impurities in process equipment,
e) detecting corrosion pits,
f) locating the plug in lead tubes.
g) testing solid rocket propellant,
h) testing transportation equipment.

In a technical report published by the IAEA in 1965, *Industrial Radioisotope Economics*, it was observed that the types of radioisotopes used as gamma radiography sources by many users throughout the world are almost identical (see table 24.1). This shows that the gamma radiography technique has reached a state of maturity. Its applications and operations can be standardized so that commercial units can be manufactured for wider usage.

Problems

24.1 Explain why Co-60 is not an ideal radiation source for radiography.

24.2 A point radiation source is suspended in air at a height of 90 cm above the floor. A worker is required to work in an area extending up to 2 meters from the source. How tall a plate shield must be placed at half a meter from the source to completely shield the worker? Assume the height of the worker to be 2 m.

Ans: 3.4 m.

24.3 If the opacity which is defined as the ratio of the intensities on the negative with and without the object is 0.1, what is the blackening density?

Ans: 1

24.4 A film negative is placed at 20 cm from a point source of 500 mCi. If the exposure factor $Q = 2\ mCi/cm^2$, what should be the exposure time?

Ans: 1.26 hr

24.5 What are the major advantages of using a gamma source such as Co-60 for radiography compared with x-ray?

24.6 Usually, in radiography, a point radiation source is used. This is so because the smaller the source size compared with the size of the object (or a fault in a piece of the material) the better the definition of the image. Illustrate the above statement graphically, letting the source-to-object and object-to-screen distances be constant.

CHAPTER 25

RADIOISOTOPE ENGINEERING

25.1 Introduction

During the last 20 years radioisotopes have been used in ever increasing applications in many fields, involving practically all aspects of human life. As a matter of fact, numerous applications of radioisotopes in the fields of agriculture, industry, medicine, etc. lead one to believe that in the future these applications may play a larger role in our everyday life than electric power generation from nuclear reactors. Radioisotope applications may be artificially divided into three categories according to the radiation source strength. Minute radiation quantities, in fractions of a microcurie to a few curies, are used in tracer applications, ionization, and gauging as described in chapters 21 and 22. Larger quantities, up to hundreds of curies, are used in radiography and radioisotope power plants, as described in chapters 19 and 24, respectively. In this chapter radioisotopes ranging from hundreds of curies up to several Megacuries or higher in strength will be dealt with.

RADIOISOTOPE ENGINEERING, according to G.G. Eichholz, IS THE BRANCH OF NUCLEAR ENGINEERING THAT CONCERNS ITSELF WITH THE PRODUCTION, APPLICATION, AND INSTALLATION OF INTENSE RADIOISOTOPE SOURCES. The emphasis is on the design of the irradiation facilities and the safe handling of sources and processing operations. Many unusual problems are encountered by design engineers in radioisotope engineering. First of all, the use of high intensity radioactive sources poses potential radiation hazards. Then there is the problem of utilizing the radioactive waste produced from nuclear power plant operations. As the disposal of radioactive waste is a major hindrance in the large scale development of nuclear power generation, utilization of large amounts of these wastes for intense radiation sources would considerably ease the problem. Another problem facing a radioisctope engineer is the challenge of developing feasible applications. Many useful and promising processes have been suggested that make use of intense radiation sources. A great majority of these have, however, failed to achieve fruition. This is often due to the hesitation to adopt new techniques. On the other hand, the lack of fundamental understanding and the cost and availability of radiation sources further deters the progress of such applications. Even under these unfavorable situations, the commercial radiation processing applications have grown steadily. The dollar value on these applications in the USA has grown from around \$25 million in 1963 to around \$300 million in 1975. With the inevitably increasing usage of high intensity radioactive sources, it is evident that more data and a greater understanding of radioisotope engineering must be obtained.

Another problem in the development of radioisotope engineering was the availability of high level radioactive sources. Up to the late nineteen fifties, large radiation sources suitable for commercial use were very expensive. Although accelerators were available to furnish these, their applications were limited. When high neutron flux density reactors, $\phi \equiv 10^{15}/cm^2s$ or higher, were made available in the early nineteen sixties, the production of high intensity radioisotopes in reactors by neutron activation was initiated. Some methods of radiosotope production are discussed in chapter 18 — "Introduction to Radiation Engineering". In the following section, intense radioisotope source types and production methods, mainly those for radioisotope engineering, will be reviewed.

25.2 Intense Radiation Sources

Compared with accelerators as radioactive sources for commercial applications, a radioisotope source has the following advantages:

a) It is simpler to operate and therefore costs less.

b) The source strength is fairly constant during the period of operation.

c) Small size and no power requirement means that it can be used in many circumstances where accelerator radiation sources cannot be installed.

d) Deep penetration can be achieved.

However, unlike accelerator sources, a radioisotope source cannot be turned off like an accelerator. Its energy and intensity is fixed except for the initial selection. Many of the factors for selecting the best material for radioisotope sources are discussed elsewhere. The most frequently used source material is Co-60. Table 25.1 lists some of the sources available or in use today.

Table 25.1: Reactor produced intense radioactive sources for commercial processing.

Radioisotope	Radiation emission	Half-life	Specific power (W/g)
Ac-227		22 yr	14
Cf-252		2.65 yr	40
Cm-242		163 da	120
Cm-244		18 yr	2.8
Co-60		5.3 yr	17.3
Cs-137		30 yr	0.42
Po-210		138 da	140
Pu-238		87 yr	0.57
Sr-90		28 yr	0.96
Tl-204		3.56 yr	0.73
Tm-170		127 da	13
Tm-171		1.92 yr	0.22

At the present time, the most practical methods of obtaining radioisotopes for large scale use are reactor production and fission products. Of the former, Co-60 is a common example. It is produced from Co-59 neutron activation, $^{59}Co(n, \gamma)^{60}Co$, in a reactor. In a reactor of a neutron flux density ϕ, the Co-60 production rate can be described by

$$\frac{dN_{60}}{dt} = N_{59}(\sigma_{act})_{59}\phi - \lambda N_{60} - N_{60}(\sigma_a)_{60}N_{60},$$

where N_{59} and N_{60} are Co-59 and Co-60 atom densities, $(\sigma_{act})_{59}$ and $(\sigma_a)_{60}$ are the microscopic activation cross section of Co-59 and the absorption cross section of Co-60, respectively, and λ is the disintegration constant of Co-60. From this equation the induced Co-60 activity can be obtained;

$$A(t) = \frac{\lambda\phi(\sigma_{act})_{59}N^{59}}{3.7 \times 10^{10}\,[\lambda + (\sigma_a)_{60} - (\sigma_a)_{59}]}\left\{e^{-(\sigma_{act})_{59}\phi t} - e^{-[\lambda + (\sigma_a)_{60}\phi]t}\right\},$$

assuming that $N_{59} = (N_{59})_o e^{-(\sigma_a)_{59}\phi t}$ and $N_{60} = (N_{60})_o$ when $t = 0$.

Depending on the activation time and neutron flux density of the reactor, the specific activity of the activated sample ranges from 200Ci/g up to 1200 Ci/g.

The lower figure is easily obtainable from many operational reactors. Californium-252 is another reactor produced radioisotope of importance although large quantities (in grams) of this isotope was produced in high flux reactors only recently. The detailed production chain is described in chapter 6. Cf-252 has some very attractive features for use as an irradiation source. It emits alphas and neutrons. The neutron specific emission rate is 2.3 $\times 10^{12}$ n/s g which makes it a very intense neutron source with a relatively long half-life, 2.65 yr. Of the latter class of radioisotopes, Cs-137 is one of the most important radiation sources. Because of its long half-life (30 yr), relative abundance, and ease of chemical separation from the fission product Cs-137 can be used as a large irradiation source. Cs-137 has a gamma energy of 0.662 MeV from 85% of its disintegrations. The average energy output of Cs-137 is about 0.56 MeV per disintegrations. Co-60 emits two gammas of energies of 1.17 MeV and 1.33 MeV, giving a total of 2.5 MeV per disintegration. From this point of view, Co-60 is 4.4 times more energetic than Cs-137. In other words, 4.4 curies of Cs-137 have the same energy output as one curie of Co-60. Furthermore, the specific activity of Cs-137 is around 25 Ci/g compared with at least several hundred Ci/g for Co-60. The handling of Co-60 is also simpler compared with the processing of Cs-137. Because of these reasons, Co-60 is a preferred radioactive source material than Cs-137.

The major tasks in radiation source preparation and production are a) radioisotope selection and production and b) radioactive source encapsulation. The engineering problems involved in radioisotope production using a reactor are many. Some of these are listed below:

a) selection of suitable radioisotopes and the target materials,
b) reactor criticality considerations,
c) efficiency of excess neutron utilization,
d) neutron flux effects,
e) reactor poison considerations, and
f) product separation.

In all of these considerations, computer aided programs and calculations are often used. A typical design calculation series may be illustrated by the following block diagram.

In order to carry out these calculations and obtain an accurate estimation of production, one must have the following information:

a) reactor power,
b) fuel composition and burnup,

c) temperature distribution,
d) control rod and coolant material properties, and
e) target exposure and self-shielding data.

Radioactive source encapsulation involves environmental and isotopic characteristic considerations. A typical design project consists of a) the capsule design, b) processing considerations, c) inner capsule assembly, d) outer capsule assembly, and e) quality control. In Figure 25.1, the flow chart illustrates the entire chain of processes from the irradiated sample to the final shipment of the encapsulated radioactive source.

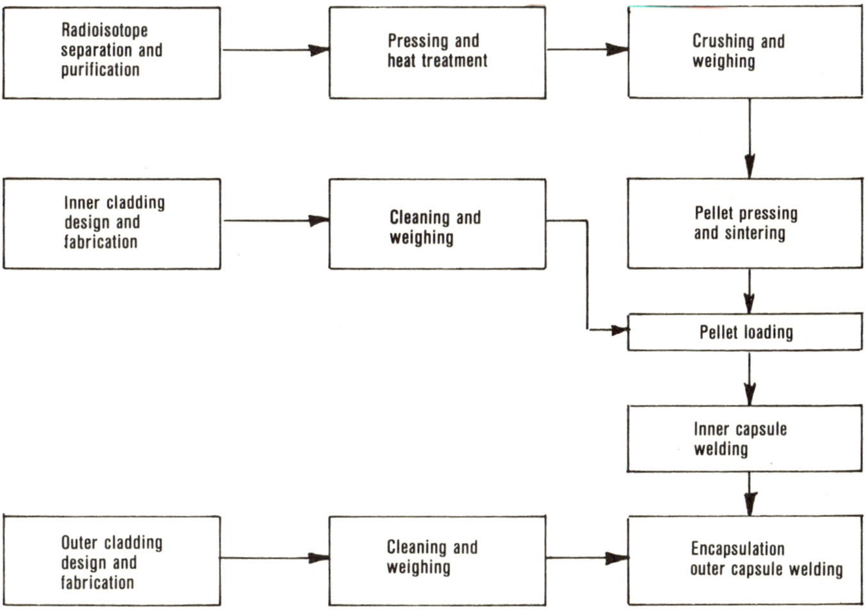

Fig. 25.1: Radioactive source encapsulation.

The fabrication of the source material starts with the extraction and purification of the selected radioisotope(s) from the neutron activated sample. Radiochemical techniques are used to test the purity, then oxide conversion forms the source pellet material. Pressing and heat treatment are used to increase the density of the powder, which is then compressed by a punch and die set. Usually the proper amount of source is weighed to comply with design specifications. These weighed source materials in pellet form are then ready for encapsulation.

After a series of material tests such as gauging, radiography, penetrant test, or chemical analysis, the inner capsule material is machined from bar. It is then loaded with source pellets. The preparation of an outer capsule follows similar stages. After the welded inner capsule has been loaded into the outer capsule, the outer capsule is welded. The finished product is then inspecpected and tested for shipment.

Throughout the entire chain of parallel or series stages of encapsulation, the quality control program must be carefully executed for the confirmation of design specifications and radiation safety. It is a usual practice to draw a flow chart with detailed stages in the beginning of the project. Inspections, tests and the progress of every stage of operation are recorded. For safety and hazard control, health physics precautions and monitoring must be observed throughout the entire operation.

25.3 Irradiation Facility Design

The major components of an irradiation facility are a) the radiation source assembly, b) the irradiation chamber, c) the conveyor system, and d) shielding. The design of these components is dependent upon the characteristics of the unprocessed and processed products. The source assembly consists of the radioactive sources in capsules and the assembled elements. Source element movement is controlled by motors. The design and operation of the source positioning control unit is an important task. It is stored in a storage tank when it is not being used. The irradiation chamber serves as the housing for both the source assembly and the products to be processed during irradiation. In normal operations, the source assembly is moved into position and the conveyor system fetches the material to be processed. After irradiation the source assembly is withdrawn back into its storage. Figure 25.2 shows the Hawaii development irradiator, which is used to irradiate fruits and other foods. The Co-60 source strength is 250,000 Ci. It has a capacity of 4000 lb/hr with a radiation dose of 75,000 rad. The figure shows the source assembly and its storage tank, the irradiation chamber, and the transfer conveyor system.

The first task of a radioisotope engineer in designing an irradiator is to collect data with respect to the use of the facility. This determines the source activity required and the irradiator geometric arrangement. For any irradiation application there are fixed limits on the maximum and minimum

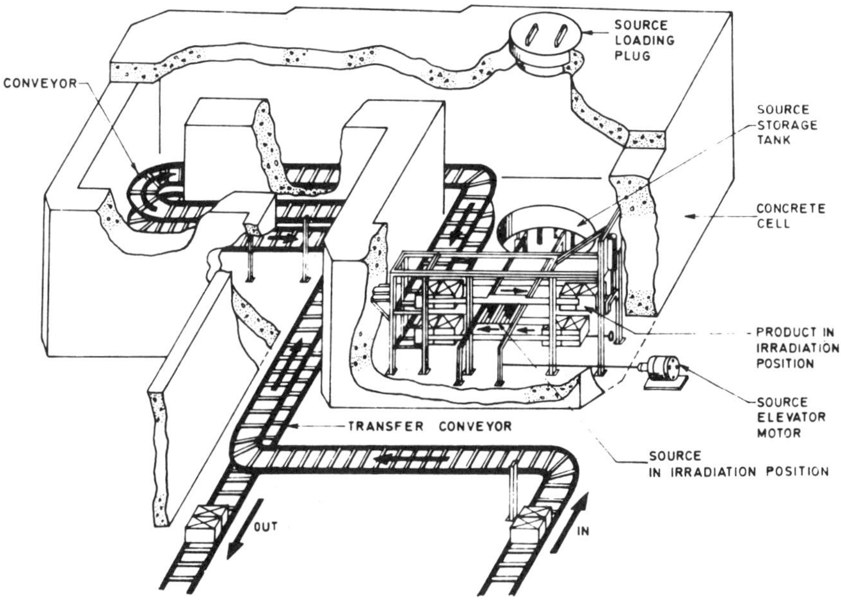

Fig. 25.2: Food preservation by irradiation.

radiation doses to the material, determined by the dose uniformity which is defined by

$$U \equiv \frac{\text{maximum allowable dose}}{\text{minimum allowable dose}} = \frac{D_{max}}{D_{min}}. \tag{25.1}$$

This parameter is a function of geometry, target properties, and the product residence time. Since U is a fixed quantity, the main task of the designers is to choose the optimum geometry and residence time within the prescribed value of U. The mass flow rate (W) affects the residence time directly. It is an important parameter which is frequently of fixed specification. Typically, the mass flow is in the range of 200 - 5000 kg/hr. Another parameter is the irradiator capacity Q defined by

$$Q = WD_{min}. \tag{25.2}$$

Various geometrical schemes and arrangements must be examined to satisfy this specified value. The target package dimensions can also be varied to satisfy the specifications.

Having examined these so-called target parameters, the designer then turns to the source assembly design. First, he must calculate the source strength required and choose the suitable source material. The radiation dose calculations then follow the selection of a set of source-target-irradiator geometry. Next, he has to calculate the irradiator efficiency defined by

$$\eta = \frac{\text{energy absorbed by target at } D_{min}}{\text{total source energy}}. \tag{25.3}$$

This is given by

$$\eta = \frac{k_1 Q_{min}}{k_2 C}, \tag{25.4}$$

where k_1 is the power absorption conversion factor which is approximately 0.58 W/(Megarad-kg/hr), Q_{min} is the irradiator capacity in Megarad-kg/hr, k_2 is the isotopic gamma power which is 0.015 W/Ci for Co-60 and 0.0032 W/Ci for Cs-137.

The geometric design of an irradiator is usually cylindrical or rectangular. In the first case the sources are fabricated into cylinders while for the second they are plates. For all source geometries, the designer strives for uniformity of irradiation by minimizing the lateral and depth-wise radiation distribution. The second objective can be obtained by the irradiation of the package on both sides simultaneously by two sources or through a multipass. To minimize the lateral dose distribution, one can introduce additional source material in the vicinity of low dose regions (augmentation) or adjust the conveyor speed such as in a stop-dwell mode of motion. In Figures 25.3, 25.4, 25.5, 25.6, and 25.7, four different source-target arrangements are shown. The first two figures show the stationary type irradiators: single-slab, two-position and two-slab, single-position. The last three figures illustrate three design concepts of irradiation operation.

A recent development in irradiation techniques is the indium-gallium (eutectic alloy) loop used in the USSR. An electromagnetic pump is used to circulate the liquid In-Ga alloy which is a liquid at room temperature (~20° C) and remains so at higher temperatures if the gallium mass percentage is around 75%. Indium has large thermal and resonance neutron absorption cross sections. Gallium has a low melting point, 30° C. After the In-Ga alloy is activated in a reactor, the induced activities from the sample are mainly the gammas from the 54 minute and thirteen second half-life decays of In-116. Therefore, the circulation of this alloy through a loop outside a reactor can be used for irradiation purposes. Due to the short half-life (54 minutes), such a device would pose no radiation hazard problem when not in use. Perhaps the best feature of this type of in-core activation, out-core irradiation arrange-

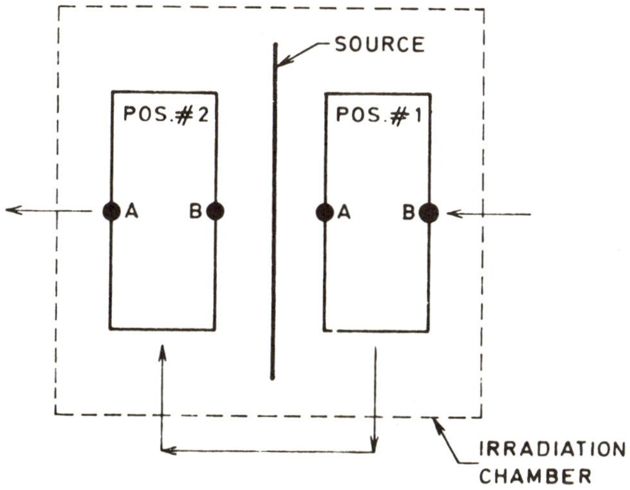

Fig. 25.3: Single-slab, two-position stationary irradiator.

ment, besides the fact that it can be 'turned' off, is the high irradiation power it can furnish. The only disadvantage is the possible leakage and subsequent safety problems.

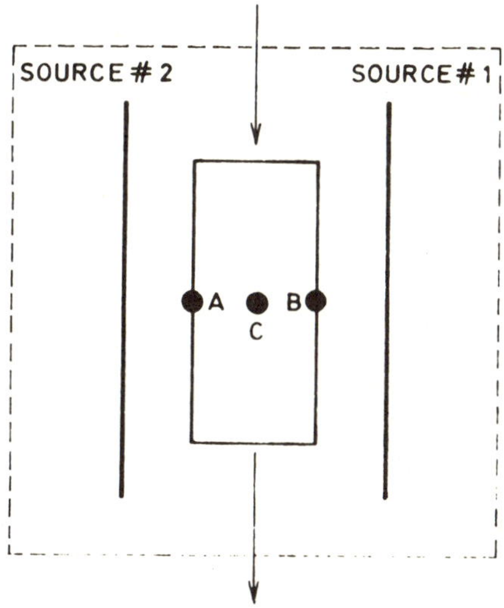

Fig. 25.4: Two-slab, single-position stationary irradiator.

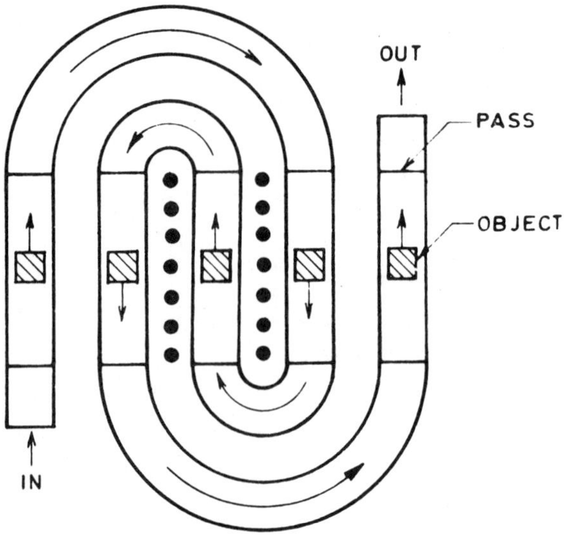

Fig. 25.5: Two-slab, five-pass mobile irradiator.

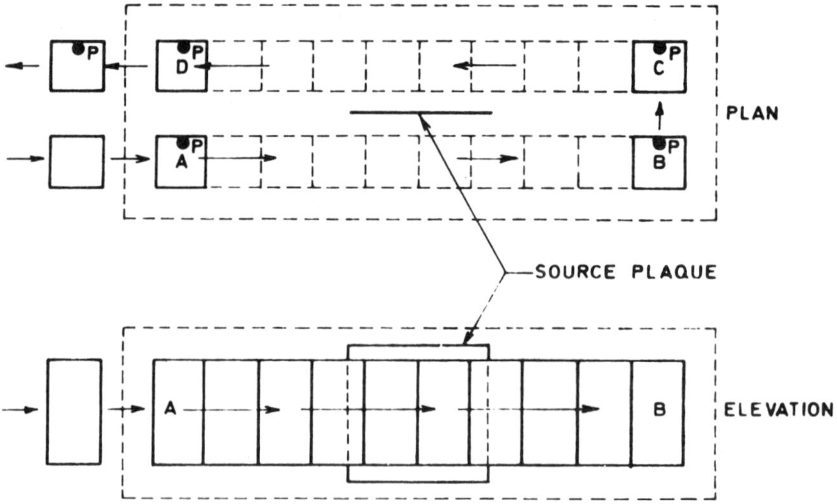

Fig. 25.6: Single-slab, two-pass single-direction irradiator.

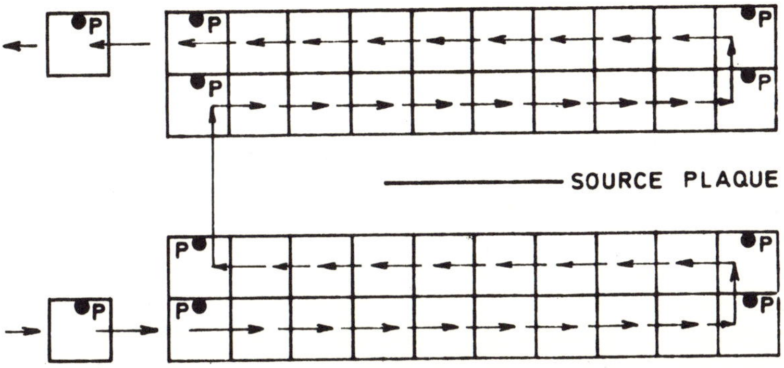

Fig. 25.7: Single-slab, multipass single-direction, stop-dwell irradiator.

25.4 Applications

X-rays have been widely used in medical diagnostic applications for about a century. Their main source is the radium radioisotope and x-ray tubes. Electron accelerators are also used in some applications. Due to the difficulties encountered in operation and cost considerations, these machines are being gradually replaced by radioisotopes. While x-rays remain a major diagnostic tool, Co-60 is now widely used for cancer treatment. The radioactivity of Co-60 in a radiotherapy unit ranges from several hundreds of curies up to a few thousand curies. Unlike the radioactive sources sold in terms of curies the commercially available Co-60 sources for radiotherapy are described and sold in terms of roentgens per minute at one meter (Rmm). The relationship between the source activity and the exposure rate E(Rmm) may be calculated from

$$E = KGS, \tag{25.5}$$

where K is the dose conversion constant, approximately 0.025 Rmm per curie for Co-60; G is a factor depending on the source geometry, source density, and other factors; S is the specific activity of the source. The G-factor varies according to system design. The specific activity of a source is difficult to determine accurately. Thus a Co-60 source reflects its usefulness when expressed in terms of Rmm instead of activity in curies or specific activity.

Nuclear radiations emitted by radioactive isotopes show great promise for the preservation of food for prolonged periods without introducing any significant changes in the food values and without inducing radioactivity in the foods. Preservation of food by nuclear radiations is accomplished by inhibiting or destroying bacteria and other micro-organisms. The destruction mechanism is complex and probably involves the production of extensive genetic damage, as well as direct effects on cell cytoplasm of the irradiated material. Different types of bacteria have different radiation sensitivities. Generally, very heavy radiation doses are necessary because micro-organisms are much less sensitive to radiation than higher plants and animals. The amount of radiation to be delivered depends upon the food itself and the results desired. In some species, practically all organisms in a culture are killed by 400,000 rads, while as many as 40,000,000 rads may be required to produce the same effects in other species. Since bacteria population decreases exponentially on irradiation, it is theoretically impossible to kill all the bacteria. For this reason experimental doses of nuclear radiation for different types of bacteria are determined under different environmental conditions which would prolong the life of various foods for given periods of time.

Radiation preservation of food may be accomplished either by pasteurization or sterilization. Pasteurization is accomplished with low dosages of radiation and may be called partial sterilization, while sterilization requires much higher levels. For the prolongation of shelf life or storage period, a pasteurization dose of 200,000 to 500,000 rads is generally sufficient. However, if the aim is to sterilize the food for long term storage without refrigeration, the required dose is between 2,000,000 to 4,500,000 rads. Different doses of radiation have been found to be effective for the preservation of different kinds of food under given environmental conditions. For example, a dose of 4,000 to 10,000 rads inhibits sprouting in potatoes or onions, while doses as high as 20,000 to 500,000 rads are needed in extending the refrigerated storage life of fresh fish and other marine products. For meats such as beef, chicken, bacon, and other pork products, a radiation dose of 4,500,000 rads results in the preservation of the food at room temperatures for as long as one year. Grains and cereals can be disinfested of insects at 20,000 to 50,000 rads, while radiation-pasteurization of fresh fruits and vegetables requires up to 100,000 rads. Some reports have been published on design studies of mobile gamma irradiators as well as packaging requirements for the radiation pasteurization of foods.

As a result of broad studies conducted in universities and industries on various aspects of food preservation by irradiation, it has been found that it is possible to effectively destroy all living organisms in any food by the use of radiation without significantly affecting the nutritive value of foods. Upon sterilization, some flavor changes occur at varying degrees depending upon the nature of the food irradiated. For example, milk products change markedly, meat and fish change moderately and cereals and breads are hardly changed at all. Pasteurization results in far less flavor changes than sterilization. Recent studies have indicated that most of the flavor is retained at sterilization doses if irradiation is performed at very low temperatures. At present there is no evidence of toxicity attributable to the radiation process in the irradiated food. However all the new developments of radiation preservation of food should be constantly watched.

The work done in the preservation of food by radiation indicates that this field will soon be commercialized. In the coming years it will result in substantial benefits to the human race by saving large quantities of foodstuffs which are presently spoiled during storage and distribution. Initial commercial applications of food preservation by radiation will occur gradually as determined by technical as well as economic developments. Large scale use of radiation in food preservation will ultimately affect many industries like food processing, packaging, refrigeration chemicals, marketing facilities and operations, and the world trade. The main benefits which may result, besides savings resulting from the reductions in spoilage loss, are a reduction in the incidence of foodborne diseases, increased variety of food in various areas due to the extension in shelf life, increased availability of a more nutritious and varied diet for the undernourished people of the world, expanded export of food from developing countries, and the stabilization of food markets due to reduction of food shortages.

Studies also indicate that through radiation-induced mutations, improved crop varieties can be developed. Although plant breeders are already using the conventional methods involving selection and hybridization, it is agreed that nuclear radiations applied either in the early flowering stages in order to affect the developing gametes, or to seeds, would result in more rapid mutations which might be transmitted to later generations. Although the vast majority of mutations are deleterious, desirable types occur in low proportions and have been selected in a wide variety of crops. Radiation-caused mutations have not only helped in obtaining improved types of crops including high-yield, disease-resistant species, but have also shown good prospects of creating new plant varieties. Crops with extended or reduced maturity periods as well as those with changed ecological requirements have also been produced. Various studies performed using Co-60 sources on growing plants and seeds indicate that the genetic effects produced would not only improve the quality of food but may also increase the quantity of food produced. Radioisotopes are also used for tracing various plant diseases and for the disinfection of crops and forests. It has been estimated that overall plant diseases result in food losses in billions of dollars every year. Although at one time the elimination of any epidemic spread of a plant disease was virtually impossible, today radioisotopes show great promise in this field by breeding resistant plant varieties as well as by facilitating the investigation of the behavior of various chemicals which are needed for controlling these diseases.

Radioisotopes could also bring about a significant increase in food production by providing methods for the study of crop growth and soil improvement. Studies on the causes of low yields of various crops due to factors in the roots and soil conditions, as well as on the protection of crops against frost and other deleterious effects, are now made possible with the use of radioisotopes. Table 25.2 summarizes some specific radioisotope engineering applications.

Table 25.2: Specific radioisotope engineering applications.

AGRICULTURE:	1. Food pasteurization and sterilization 2. Grain disinfection 3. Seed preservation and growth stimulation 4. Insect disinfection 5. Fruit preservation
INDUSTRY:	1. Wood polymerization 2. Coating curing 3. Sewage treatment 4. Synthesis of chemical products 5. Sea food processing and preservation 6. Crosslinking of polymers 7. Catalyzing chemical reactions (as ethyl bromide processing) 8. Wood hardening
MEDICAL:	1. Radiotherapy 2. Sterilization of medical supplies 3. Virus inactivation 4. Bacteria pasteurization and sterilization.

Problems

25.1 A 5000 curie point Co-60 source is to be stored in water. At what depth in water should this source be kept in order to reduce the radiation dose rate at the surface to 0.2 mrem/hr? The water storage tank is assumed to be large in cross sectional area.

25.2 A spherical radioisotope container has an outer radius of 15 cm and an inner radius of 5 cm. For 1.5 MeV gamma rays, lead has a linear absorption coefficient of 0.55 cm^{-1}. The buildup factor is given by $B(\mu t) = 1 + \mu t$, where t is the wall thickness. What is the fractional reduction of the radiation intensity due to the container wall?

Ans: 99.3%

25.3 Find the thickness of water required to reduce the intensity of a beam of gamma rays of 2 MeV energy to 1%.

25.4 Calculate the maximum specific activity of Co-60 produced from Co-59 neutron irradiation in a reactor with a thermal neutron flux density $\phi = 10^{13}$ n/cm^{2}s.

25.5 Discuss the concept of a fission product power plant with a capacity of one megawatt.

25.6 What is the power output of a two MCi Co-60 radiation source if there is no energy loss?

Ans: 31 kW.

25.7 Twelve cylindrical Co-60 sources of one MCi each are to be used as the source of an irradiator. In order to receive an even exposure, the targets are packaged in cylindrical containers of the same dimensions as the sources. Describe an irradiation concept for achieving the maximum radiation dose in a fixed time interval and a uniform dose throughout the entire period.

25.8 Discuss the cost of the four factors of a process irradiation system: a) the source, b) the target, c) the shield, and d) the buildings.

25.9 An important factor for the cost analysis of a processing irradiator is the source efficiency defined as

$$\epsilon = \frac{\text{useful radiation absorption in product}}{\text{total radiation emission from source}} .$$

An irradiation arrangement uses a cylindrical source with an outer radius of 2 cm. All the products are packaged in cylindrical containers of the same size as the source. During irradiation there are eight bundles of the cylindrical product spaced uniformly from the source with a center to center distance of 10 cm. Assuming that all radiations incident upon the product are absorbed and neglecting the end effect, calculate the source efficiency of the arrangement.

Ans: 50%

25.10 What is the irradiator capacity of an irradiator through which the product mass flow rate is 1000 kg/hr. The minimum dose required is 2×10^6 rad.

APPENDIX 1: *TABLE OF ELEMENTS (with some of their properties)*

Element	**Chemical symbol**	**Atomic Number Z**	**Atomic Mass (C-12) A**	**Density ρ (g/cm³)**	**Th.n. cross† sections** σ_a (b)	σ_s (b)
Actinium	Ac*	89	227	—	510	—
Alumirium	Al	13	26.9815	2.7	0.24	1.4
Americum	Am*	95	243	13.67	—	—
Antimony	Sb	51	121.75	6.69	5.7	4.3
Argon	Ar	18	39.948	0.0018	0.63	1.5
Arsenic	As	33	74.9216	5.73	4.5	6
Astatine	At*	85	~ 210	—	—	—
Barium	Ba	56	137.34	3.5	1.2	8
Berkleium	Bk*	97	247	14	—	—
Beryllium	Be	4	9.0122	1.85	0.01	7.0
Bismuth	Bi	83	208.9806	9.75	0.034	9
Boron	B	5	10.811	2.45	759	4
Bromine	Br	35	79.909	3.12	6.7	6
Cadmium	Cd	48	112.40	8.65	2450	7
Calcium	Ca	20	40.08	1.55	0.44	3.0
Californium	Cf*	98	251	—	—	—
Carbon	C	6	12.0115	1.6	0.0037	4.8
Cerium	Ce	58	140.12	6.78	0.73	9
Cesium	Cs	55	132.9055	1.873	28	20
Chlorine	Cl	17	35.453	0.0032	33.8	16
Chromium	Cr	24	51.996	7.20	3.1	3
Cobalt	Co	27	58.9332	8.9	37	7
Copper	Cu	29	63.54	8.96	3.8	7.2
Curium	Cm*	96	247	13.51 (Calc.)	—	—
Dysprosium	Dy	66	162.50	8.56	950	100

* No stable isotope exists

†At 2200 m/s.

Element	Chemical symbol	Z	A	ρ	σ_a	σ_s
Einsteinium	Es*	99	254	—	—	—
Erbium	Er	68	167.26	9.16	173	15
Europium	Eu	63	151.96	5.22	4300	8
Fermium	Fm	100	257	—	—	—
Fluorine	F	9	18.9984	0.0017	0.010	3.9
Francium	Fr*	87	223	—	—	—
Gandolinium	Gd	64	157.25	7.95	46000	4
Gallium	Ga	31	69.72	5.91	3.0	—
Germanium	Ge	32	72.59	5.323	2.4	3
Gold	Au	79	196.9665	19.32	98.8	9.3
Hafnium	Hf	72	178.49	13.3	105	8
Hahnium	Ha*	105	—	—	—	—
Helium	He	2	4.0026	1.77×10^{-4}	0.007	0.8
Holmium	Ho	67	164.9303	8.76	65	—
Hydrogen	H	1	1.00797	5.988×10^{-5}	0.33	38
Indium	In	49	114.82	7.28	191	2.2
Iodine	I(0)	53	126.9044	4.93	7.0	3.6
Iridium	Ir	77	192.22	22.42	440	—
Iron	Fe	26	55.8470	7.87	2.53	11
Krypton	Kr	36	83.80	0.0037	31	7.2
Kurchatovium	Ku*	104	—	—	—	—
(Rutherfordium	Rf)					
Lanthanum	La	57	138.9055	6.19	8.9	15
Lead	Pb	82	207.19	11.35	0.170	11
Lawrencium	Lw*	103	257	—	—	—
Lithium	Li	3	6.939	0.534	71	1.4
Lutetium	Lu	71	174.97	9.74	112	—
Magnesium	Mg	12	24.312	1.74	0.069	3.6
Manganese	Mn	25	54.9380	7.2	13.3	2.3
Mendelevium	Md*	101	256	—	—	—
Mercury	Hg	80	200.59	13.55	380	20
Molybdenum	Mo	42	95.94	10.2	2.7	7
Neodymium	Nd	60	144.24	6.95	46	16
Neon	Ne	10	20.183	0.0009	0.26	2.4
Neptunium	Np*	93	237	20.25	170	—
Nickel	Ni	28	58.71	8.9	4.6	17.5

Element	Chemical symbol	Z	A	ρ	σ_a	σ_s
Niobium	Nb	41	92.9064	8.57	1.16	5
Nitrogen	N	7	14.0067	0.0013	1.88	10
Nobelium	No*	102	—	—	—	—
Osmium	Os	76	190.2	22.48	15.3	11
Oxygen	O	8	15.9994	0.0014	0.0002	4.2
Palladium	Pd	46	106.4	12.16	8	3.6
Phosphorus	P	15	30.9738	1.82	0.20	5
Platinum	Pt	78	195.09	21.37	8.8	10
Plutonium	Pu*	94	239	19.7	1029	9.6
Polonium	Po*	84	210	9.24	—	—
Potassium	K	19	39.102	0.862	2.1	1.5
Praseodymium	Pr	59	140.9077	6.78	11.3	4
Promethium	Pm*	61	147	—	60	—
Protactinium	Pa	91	231.0359	15.4	200	—
Radium	Ra*	88	226.0254	5	20	—
Radon	Rn*	86	222	0.0097	0.7	—
Rhenium	Re	75	186.2	20.53	85	14
Rhodium	Rh	45	102.9055	12.5	149	5
Rubidium	Rb	37	85.47	1.53	0.73	12
Ruthenium	Ru	44	101.07	12.2	2.56	6
Samarium	Sm	62	150.35	7.7	5600	5
Scandium	Sc	21	44.956	2.99	23	24
Selenium	Se	34	78.96	4.8	12	11
Silicon	Si	14	28.086	2.42	0.16	1.7
Silver	Ag	47	107.868	10.50	63	6
Sodium	Na	11	22.9898	0.971	0.525	4.0
Strontium	Sr	38	87.62	2.54	1.21	10
Sulfur	S	16	32.064	~ 2	0.52	1.1
Tantalum	Ta	73	180.9479	16.6	21	5
Techretium	Tc*	43	98.9062	11.50	22	—
Tellurium	Te	52	127.60	6.24	4.7	5
Terbium	Tb	65	158.9254	8.33	46	—
Thallium	Tl	84	204.37	11.85	3.4	14
Thorium	Th	90	232.0381	11.3	7.56	12.6
Thulium	Tm	69	168.9342	9.35	128	7
Tin	Sn	50	118.69	6.5	0.625	4

Element	Chemical symbol	Z	A	ρ	σ_a	σ_s
Titanium	Ti	22	47.90	4.5	6.1	4
Tungsten	W	74	183.85	19.3	19.2	5
Uranium	U	92	238.329	18.9	7.68	8.3
Vanadium	V	23	50.942	6.11	4.9	5
Xenon	Xe	54	131.30	0.0059	35	4.3
Ytterbium	Yb	70	173.04	7.01	37	12
Yttrium	Y	39	88.906	5.51	1.31	3
Zinc	Zn	30	65.37	7.133	1.10	3.6
Zirconium	Zr	40	91.22	6.506	0.185	8

APPENDIX 2: *RADIATION DOSE AND HAZARD*

A.2.1: *Exposed Dose to Common Radiations*

Type of Exposure	**Dose**
External exposure to naturally radioactive	0.07 mrem/d
Internal exposure to radium and its decomposition products	0.13 mrem/d
Internal exposure to K-40	0.15 mrem/d
Total background exposure	0.48
One Ci Co-60 at 1 m	1.3 R/hr
Chest X-ray, best conditions	0.040 R
Average chest X-ray	0.100-2 R
Dental X-ray	1.5 - 15 R
Fluoroscopic examinations	5 - 50 R
Pregnancy X-ray examination	20 - 65 R
Treatment of malignant tumors (local dose)	3000 - 7000 R
Fatal dose applied to the entire human body in a short period of time	200 - 800 R

A.2.2: *Expected Effects of Radiation Doses (due to acute doses)*

Dose (rem)	**Probable Effect**
0 - 100	No obvious effect
100 - 200	Vomiting in 5% - 50% of the cases for about one day, followed by other symptoms of radiation sickness. Require several weeks of convalescence. No death.
200 - 600	Vomiting and nausea in nearly all cases on first day, followed by other radiation sickness. Up to 50% deaths within one month.
600 - 1000	Vomiting and nausea in 100% of the cases. Probably no survivors.
Above 1000	No survivors.

Data extracted from Glasstone, S., Ed., *The Effects of Nuclear Weapons,* USAEC, 1964.

APPENDIX 3: *CONSTANTS, CONVERSION FACTORS, AND UNITS*

Three significant figures are used for most of the items in this compilation. If more accurate data is required one should consult either the *Handbook of Chemistry and Physics,* the Chemical Rubber Company or Reactor Physics Constants, ANL. Note that we are using the superscript to indicate the power of 10. As Examples, the Avogadro's number $A_0 = 6.02 \times 10^{23}$ atom/g mole is written as $A_0 = 6.02^{23}$ atom/g mole and $10.5 = 1.05 \times 10^{1} = 1.05^{1}$.

A. *CONSTANTS:*

Absolute zero:	$^\circ K = - 2.73^{2}\ ^\circ C$
Atomic mass unit:	amu = 1.66^{-27} kg
Avogadro's number:	$A_0 = 6.02^{23}$ atom/g mole
Becquerel:	Bq = 1.00 dis/s
Boltzmann constant:	k = 1.38^{-16} erg/°K
Bohr magnetron:	B = 9.27^{-24} A m^2
Bohr first radius:	$a_0 = 5.26^{-11}$ m
Compton electron wavelength:	$\lambda c = 2.43^{-12}$ m
Coulomb's law constant:	K = 9.00^{9} N m^2/C^2
Curie:	Ci = 3.70^{10} disintegrations/s.
Density of Air at STP:	$\rho = 1.29^{-3}$ g/cm^3
Density of water at 3.98°C:	$\rho = 1$ g/cm^3
Electron magnetic moment:	$\mu_e = 9.28^{-24}$ A m^2
Electron (classical) radius	$r_e = 2.82^{-15}$ m
Electronic charge:	e = 1.60^{-19} C
Electronic charge to mass ratio:	e/m = 1.76^{11} C/kg
Faraday constant:	F = Ae = 9.65^{4} C/kg
Fine structure constant:	$\alpha = 7.30^{-3}$
Gas constant:	R = 8.31 J/°K mole
Gravitational acceleration:	g = 9.80 m/s^2
Gravitational constant:	G = 6.67^{-11} N m^2/kg^2
Gray (of absorbed dose):	Gy = 1.00^{2} J/kg
Heat of fusion of water:	Q = 8.00^{1} kcal/kg
Heat of vaporization of water:	Q = 5.40^{2} kcal/kg
Impedance of free space:	$Z_0 = (\mu_0/\epsilon_0)^{1/2} = 3.77^{2}\ \Omega$
Ionization energy of normal hydrogen atom:	E = 1.35^{1} eV
Mechanical equivalent of heat:	J = 4.19 J/cal
Nuclear magneton:	$\mu_N = 5.05^{-27}$ A m^2
Permeability of vacuum:	$\mu_0 = 1.26^{-6}$ Wb/A m
Permitivity of vacuum:	$\epsilon_0 = 8.85^{-12}$ C^2/N m^2
Planck's constant:	h = 6.63^{-34} J s
Proton magnetic moment:	$\mu_p = 1.41^{-26}$ A m^2
Rad:	R = 1.00^{2} erg/g

Ratio of O-16/C-12 mass scale: M(O-16)/M(C-12)=1.000043

Rest masses — Electron: $m_e = 9.11^{-28}$ g = 0.511 MeV

Neutron: $m_n = 1.675^{-24}$ g

Proton: $m_p = 1.673^{-24}$ g

α-particle $m_\alpha = 6.60^{-24}$ g

Roentgen: R = radiation energy to create 1 esu of charge in 1 cm^3 of air at STP

Rutherford: rd = 1.00^6 dis/s

Rydberg constant: R = 1.10^7 /m

Solar system data:

Planet	**Mass (kg)**	**Diameter (km)**	**Distance to sun (km)**
Sun	1.97^{30}	1.39^6	—
Mercury	3.24^{23}	5.14^3	5.80^7
Venus	4.83^{24}	1.26^4	1.08^8
Earth	5.98^{24}	1.28^4	1.50^8
Mars	6.37^{23}	6.86^3	2.28^8
Jupiter	1.88^{27}	1.44^5	7.78^8
Saturn	5.63^{26}	1.21^5	1.43^9
Uranus	8.62^{25}	5.34^4	2.87^9
Neptune	1.00^{26}	4.97^5	4.50^9

Density of the earth: D = 5.52^3 kg/m^3

Earth's distance to the moon: S = 3.84^5 km

Mass of the moon: m = 7.35^{22} kg

Ocean area of the earth: A = 3.61^6 km^2. Ocean depth: ~ 5 km.

Total thickness of the earth's atmosphere: T = 6.30 km

Total mass of the earth's atmosphere: M = 5.20^{18} kg

Speed of light in vacuum: C = 3.00^8 m/s

Speed of sound in air at 0°C: v = 3.31^2 m/s

Speed of sound in water at 20°C: v = 1.47 m/s

Standard volume of a perfect gas: v = 2.24^4 cm^3/mole

Stefan-Boltzmann constant: $\sigma = 5.67^{-8}$ w/m^2 $°K^4$

Wien displacement constant: b = 2.90^{-3} m °K

B. *UNITS:* The SI basic units are: *meter* (m), *kilogram* (kg), *second* (s), *Ampere* (A), *Kelvin* (°K), *candela* (cd), and *mole*.

The decimal prefixes are:

exa (E) = 10^{18}

deci (d) = 10^{-1}

peta (P) = 10^{15}	centi (c) = 10^{-2}
tera (T) = 10^{12} (trillion)	milli (m) = 10^{-3}
giga (G) = beva (B) = 10^{9} (billion)	micro (μ) = 10^{-6}
mega (M) = 10^{6} (million)	nano (n) = 10^{-9}
kilo (K) = 10^{3}	pico (p) = micromicro ($\mu\mu$) = 10^{-12}
hecto (h) = 10^{2}	femto (f) = 10^{-15}
deca (da) = 10	atto (a) = 10^{-18}

UNIT-OPERATION RULES:

Ampere = Coulomb/s	($\because$ i = dq/dt)
Coulomb = Faraday · volt	($\because$ q = cV)
Henry = volt/Ampere	($\because$ M = E/(di/dt))
Joule = Newton · meter	($\because$ w = FS)
= kg·m²/s² → $= kg \cdot m^2/s^2$	($\because$ $E = mc^2$)
$Joule/m^3 = ({}^{wb}/{}_{m^2})({}^{Ampere}/{}_{m})$	($\because$ E/Vol. = 1/2 BH)
Lumen = candels/steradian	
Lux = $Lumen/m^2$	
Newton = $kg \cdot m/s^2$	($\because$ $\vec{F} = m\vec{a}$)
= Coulomb · volt/m	($\because$ $\vec{F} = q\vec{E}$)
= Coulomb · m/s · wb/m^2	($\because$ $\vec{F} = q\vec{v} \times \vec{B}$)
Tesla = wb/m^2	
Volt = Ampere-Ohm	($\because$ V = IR)
Watt = Joule/s	($\because$ P = dw/dt)
= Ampere · Volt	($\because$ P = IV)

From these relationships we see that

a. wb = (Newton-s) m/Coulomb = volt · s
b. m = Joule/Newton = Newton s^2/kg = $(wb/Tesla)^{1/2}$ = $(lumen/lux)^{1/2}$
c. $Coulomb^2/Newton\text{-}meter^2$ = Faraday/meter
d. $A \cdot m^2 = J/(wb/m^2)$ because J = A · wb

C. CONVERSION FACTORS:

Acceleration: $1\ m/s^2 = 3.28\ ft/s^2 = 1.0^2$ gal $= 3.94^1\ in/s^2$

Angle: 1 radian = 1.0^3 angular mil = 5.73^1degree = 3.44^2 min = 2.06^4s. 1 steradian = 1.59^{-1} hemisphere = 7.96^{-2} solid angle = 3.28 solid degree.

Area: $1\ m^2 = 2.47^{-4}$ acre $= 1.0^{-2}$ are $= 1.0^{28}$ barn (b) =

4.00^{-4} Chinese square li = 1.97^{9} circular mil = 1.0^{28} Fermi = 1.08^{1} ft^2 = 1.0^{-4} hectare = 1.55^{3} in^2 = 3.86^{-7} mi^2 = 3.86^{-7} section = 2.47^{-3} square chain = 1.07^{-8} township = 1.20 yd^2.

Density: 1 kg/m^3 = 1.00^{-3} g/cm^3 = 3.61^{-5} lb/in^3 = 6.24^{-2} lb/ft^3 = 1.94^{-3} $slug/ft^3$.

Electric Charge: 1 coulomb (C) = 1.00^{-1} abcoulomb (emu) = 2.78^{-4} amp. hr = 1.00 amp. s = 3.00^{9} statcoulomb (esu of charge) = 1.04^{-5} Faraday of charge.

Electric Current: 1 ampere (A) = the constant current which, if maintained in two straight parallel conductors of infinite length, of negligible circular cross section, and placed one meter apart in vacuum, would produce between these conductors a force equal to 2.0^{7} N per meter of length = 1.0^{-1} abampere (emu) = 1.0 coulomb/s = 3.00^{9} statampere (esu).

Electric capacity: 1 Farad = 1.0^{-9} emu of capacitance = 8.99^{11} esu of capacitance.

Electric inductance: 1 Henry = 1.0^{9} emu of inductance = 1.11^{-12} esu of inductance.

Electric resistance: 1 Ohm (Ω) = 1.0^{9} emu of resistance = 1.11^{-12} esu of resistance.

Electric voltage: 1 volt (V) = 1.00^{8} abvolt (emu) = 3.34^{-3} statvolt (esu).

Energy: 1 Joule (J) = 9.48^{-4} British Thermal Unit (BTU) = 2.39^{-1} calorie = 6.24^{18} electron volt (eV) = 1.0^{7} erg = 7.38^{-1} ft.lb. = 2.37^{1} ft poundal = 2.39^{-4} kilocalorie = 2.78^{-7} kilowatt hour (kWH) = 2.38^{-16} Megaton TNT = 6.24^{12} MeV = 3.17^{-14} MW yr = 2.38^{-10} ton TNT = 2.78^{-4} watt·hour = 8.17^{-8} watt·year = 3.2^{-11} metric ton of coal = 1.65^{-10} barrel of oil = 1.00^{-6} ft^3 of natural gas = 1.2^{-11} g of uranium-235 = 4.20^{-12} g of deuterium.

The following is a compilation of some frequently used energy conversion factors some of which are based on heat content

1 amu = 9.31^{2}MeV; 1 Quad = 1 mQ = 1.00^{15}BTU.

1 C = 1.00^{16} BTU; 1Q = 1.00^{18} BTU.
1 eV = 1.60^{-19} J; 1 Quadrillion = 1 Qaud.
1 eV/molecule = 2.30^{1} kcal/mole.
1 gallon of gasoline = 1.25^{5} BTU.
1 metric ton of U ore (U_3O_8) = 1.351 MW-yr = 4.01^{11} BTU.
1 Mtoe = 1.00^{13} kcal.
1 Metric ton of coal = 5.00 bbl (42 gallons) of oil = 3.00^{4} ft^3 of natural gas = 7.5^{-5} metric ton of U_3O_8 = 7.50^{-1} gallon of sea water.
1 therm = 1.00^{5} BTU.
Daily energy needs per man = 2.50^{2} kWH.

Force: 1 Newton (N) = 5.76^{1} drams = 1.00^{5} dyne = 1.02^{-1} kilograms of force = 2.25^{-4} kip (kilopound of force) = 2.25^{-1} lb = 7.23 poundal.

Heat Transfer coefficient: 1 w/cm^2-°C = 2.39^{-1} cal/s·cm^2·°C = 1.76^{3} BTU/hr·ft^2·°F.

Length: 1 meter (m) = 1.65076373^{6} wave length of Kr-86 between the levels of $2P_{10}$ and $5d_5$.
= 1.00^{10} angstrom (A) = 5.67^{-12} astronomical unit = 4.55^{-3} cable = 3.94^{3} caliber = 2.00^{-3} Chinese li = 3.00 Chinese tze = 3.28^{-2} engineer chain = 2.19 cubit = 5.46^{-1} fathom = 1.00^{15} fermi (fm) = 3.28 feet (ft) = 4.97^{-3} furlong = 9.84 hand = 3.94^{1} inch (in) = 1.06^{-16} light year = 1.00^{6} micron = 3.94^{9} mil = 6.21^{-4} mile (mi) = 5.40^{-4} nautical mile = 1.31 pace = 3.14^{-17} parsec = 1.99^{-1} rode = 9.09^{-3} skein = 4.47 span = 4.98^{-2} surveyor chain = 1.00^{13} x-ray unit (XU) = 1.09 yard (yd).
1 inch = 2.54 cm (exactly).

Luminance: 1 candela/m^2 = 1.00^{-4} stilb = 3.14^{-4} Lambert.

Luminous Intensity: 1 candela = the luminous intensity, in the perpendicular direction of a surface of 1/600000 square meter of black body at the temperature of freezing platinum under a pressure of 101325 N/m^2 = 1.00 candle = 1.11 Hefner unit = 1.00 Lumen/steradian.

Magnetic Field Intensity: 1 ampere-turn/m = 1.26^{-2} oersted (emu) = 1.26^{-2} dyne/pole = 1.26^{-2} emu unit = 3.78^{8} esu unit = 1.26^{-2} gilbert/cm.

Magnetic Flux Density: 1 weber/m^2 = 1.0^9 Gammas = 10^4 gauss (emu) = 1.0^4 lines/cm^2 = 1.0^4 Maxwell/cm^2 = 1.0 tesla.

Magnetic Potential: 1 ampere-turn = 1.26 gilbert = 1.26 magnetomotive force (umf) (amu) = 3.78^{10} (esu) (mmf).

Mass: 1 kilogram (kg) = the mass of the international prototype of the kilogram preserved by International Bureau of Weights and Measures at Sevres, France = 6.02^{26} atomic mass unit (amu) = 5.0^3 carat = 1.0^9 gamma = 1.54^4 grain = 2.52^1 ounce mass = 2.21 pound mass = 6.41^2 pennyweight = 6.85^{-2} slug = 1.58^{-1} stone = 1.0^{-3} ton (long) = 1.12^{-3} ton (short, 2.0^3 lb.) = 1.0^{-3} tonne (metric ton)
1 amu (C^{12} = 12) = 1.00032 amu (physical scale) = 1.000043 amu (chemical scale)

Power: 1 watt (w) = 9.48^{-4} BTU/s = 5.69^{-2} BTU/m = 3.41 BTU/hr = 2.39^{-1} calories/s = 7.38^{-1} ft. lb/s = 1.34^{-3} horsepower (hp) = 6.24^{15} MeV/s.
1 MAF/ft/Y = 1.00^{-1} MW.

Pressure: 1 Newton (N)·m^{-2} = 9.87^{-6} atmospheric pressure at standard temperature and pressure = 1.0^{-5} bar = 1.0^1 barge = 7.50^{-4} cm of mercury at 0°C = 1.02^{-2} cm of water at 4°C = 1.0^1 dyne/cm^2 = 2.09^{-2} lb/ft^2 = 1.45^{-4} lb/in^2 = 1.0^{-2} milibar = 1.0 pascal = 1.45^{-4} psi = 7.50^{-3} torr (0°C).

Radioactivity 1 Bq = 2.7×10^{-11} C = 1 dis/s = 1.00^{-6} rd.

Specific Heat: 1 kcal/kg·°C = 1.00 BTU/lb·°F.

Speed: 1 m/s = 1.18^4 ft/h = 1.97^2 ft/m = 3/28 ft/s = 3.94^1 in/s = 3/50 km/hr = 1.94 knot = 2.24 mi/hr = 3.73^{-2} mi/m = 6.21^{-4} mi/s.

Thermal Conductivity: 1 w/cm·°C = 5.78^1 BTU/h·°F = 2.39^{-1} cal/s cm °C.

Time: 1 second (s) = the duration of 9.192631770^9 periods of the radiation transition between the two hyperfine levels of the ground state of the cesium-133 atom = 1.16^{-5} day = 2.78^{-4} hour (h) = 1.67^{-2} minute (min) = 3.80^{-7} month = 3.17^{-8} year (yr).

Viscosity: $1 \text{ N s/m}^2 = 1.0^3$ centipoise $= 1.0^1$ dyne s $\text{cm}^2 = 1.0^1$ poise $= 1.0^1$ rhe.

Volume: $1 \text{ m}^3 = 8.11^{-4}$ acre foot $= 6.29$ barrel $= 4.24^2$ board foot $= 2.84^1$ bushel $= 2.70^1$ Chinese cubic tz $= 2.76^{-1}$ cord $= 1.0^6$ cubic centimeter (cc) $= 4.23^3$ cup $= 2.71^5$ dram $= 1.82^3$ dry pint $= 9.12^2$ dry quart $= 3.33^4$ fluid ounce $= 3.53^1 \text{ ft}^3 = 2.64^2$ gallon $= 8.45^3$ gill $= 4.20$ hogshead $= 6.10^4 \text{ in}^3 = 1.0^9$ lambda $= 1.0^3$ liter (l) $= 1.14^2$ peck $= 2.11^3$ pint $= 1.06^3$ quart (qt) $= 1.0$ stere $= 6.76^4$ tablespoon $= 2.03^5$ teaspoon $= 3.53$ ton (register) $= 1.31 \text{ yd}^3$.

1 MAF $= 1.00^6$ acre foot (mega-acre foot).

Others: $1°$ Kelvin $= 1°\text{C} + 273°$; $1°$ Rankine $= 1°\text{F} + 460°$;

1 stoke $= 1.0 \text{ cm}^2/\text{s} = 1.0$ poise·cm^3/g;

$\pi = 3.14$; $e = 2.72$; $^1/e = 3.67^{-1}$; $^1/e^2 = 1.35^{-1}$; $^1/e^3 = 4.93^{-2}$; $\sqrt{2} = 1.41$; $\sqrt{3} = 1.73$; $\sqrt{5} = 2.24$;

$\ln 2 = 6.93^{-1}$; $\log \pi = 5.00^{-1}$; $\log e = 4.30^{-1}$;

$\ln 10 = 2.30$.

1 candle power $= 1.50^{-7} \text{ J/cm}^2 \text{ s}$;

1 cm^3 of water at $4°\text{C} = 1.00$ g of water.

1 gal of water at $62°\text{F} = 8.33$ lb of water.

BIBLIOGRAPHY

1. Anderson, H.L., et. al., *American Journal of Physics, 20,* 541-550, 1952.

2. Benedict, M. and T.H. Pigford, *Nuclear Chemical Engineering,* McGraw Hill Book Co., 1957.

3. Berger, H., *Nondestructive Testing,* Understanding the Atom Series, U. S. AEC, 1968.

4. Bishop, A.S., *Project Sherwood,* Addison-Wesley Co., Inc. 1958.

5. Bishop, A.S., Nuclear Fusion, *10*, 85, 1970.

6. Boley, F.I., *Plasmas —Laboratory and Cosmic,* D. Van Nostrand co., Inc. 1966.

7. Bonilla, F., *Nuclear Engineering,* McGraw-Hill Book Co., 1957.

8. Boorse, H.A. and L. Motz, *The World of Atom,* Basic Books, Inc. 1966.

9. Booth, L.A., *U.S. AEC Report,* LA-4858-MS, 1, 1972.

10. Booth, L.A. Nuclear Engineering and Design, 24, 296, 1973.

11. Brinkworth, J.B., Solar Energy, Nature, *249*, 5459, 1974.

12. Chadwick, *Proceedings of the Royal Society,* London, Series A, 136, 692-707, 1932.

13. *Chart of the Nuclides,* Knolls Atomic Power Laboratory, Ninth Edition, July 1966.

14. Cork, J.M., *Radioactivity and Nuclear Physics,* Third Edition, D. Van Nostrand Company Inc., 1958.

15. Davis, H.L., *Nuclear Energy in Space,* Nucleonics, *19,* 53-100, 1961.

16. Eichholz, G.G. *Radioisotope Engineering,* Marcel Dekker, Inc., 1972.

17. Ellis, R.H., *Nuclear Technology for Engineers,* McGraw-Hill Book Co.

18. *Energy Digest, 3,* 19, Oct. 17, 1973.

19. Energy Digest, *4*, 3, Feb. 19, 1974.

20. Etherington, H., *Nuclear Engineering Handbook,* McGraw Hill Book Co., 1958.

21. Evans, R.D., *The Atomic Nucleus,* McGraw-Hill Co., 1955.

22. Fermi, E., *American Journal of Physics, 20,* 536-541, 1952.

23. Gibson, A.S., Ecological Considerations and the Fast Breeder Reactor, paper presented at IEEE Region Six Conference, May 11-13, 1971.

24. Glaser, P.E., "Power from the Sun", *Science, 162,* 3856, Nov. 22, 1968.

25. Glasstone, S., *Controlled Nuclear Fusion,* Understanding the Atom Series, US, AEC, 1968.

26. Glasstone, S. and A. Sesonske, *Nuclear Reactor Engineering,* D. Van Nostrand Co., Inc. 1967.

27. Gough, W.C. and B.J. Eastland, "The Prospects of Fusion Power", *Scientific American, 224,* 21-33, Feb. 1971.

28. Grubber, J.E., *Nuclear Fusion, 13,* 451, 1973.

29. Hahn, O. and F. Strassman, *Naturwiss, 27,* 11, 1939.

30. Hammond, A.L., *Science,* 177, 4053, Sept. 15., 1972.

31. Hammond, A.L., *Science,* 178, 4062, Nov. 17, 1972.

32. *Handbook of Chemistry and Physics,* The Chemical Rubber Co., 54th Edition, 1973-1974.

33. Hellman, H., *Lasers,* US, AEC, Division of Technical Information, 1968.

34. Hoag, J.B., *Nuclear Reactor Experiments,* D. Van Nostrand Co., 1958.

35. Hubbert, M.K., *Energy Resources for Power Production, Proceedings of IAEA Conference,* Vienna, IAEA-SM-146/1, 1970.

36. Hubbert, M.K., *Scientific American, 224,* 3, Sept. 1971.

37. Kaye, G.W.C. and T.H. Laby, *Tables of Physical and Chemical Constants,* Thirteenth Edition, Longman Group Limited, 1971.

38. Keepin, G.R., *Physics of Nuclear Kinetics,* Addison-Wesley Publishing Co., 1965.

39. Koenig, J.B., *Trans. Am. Nucl. Soc., 15,* 1, 1972.

40. Kohl, J., Zentner, R.D., and H.R. Lukens, *Radioisotope Applications Engineering,* D. Van Nostrand Co., Inc., 1961.

41. Lamarsh, J.R., *Introduction to Nuclear Reactor Theory,* Addison-Wesley Publishing Co., 1966.

42. Lawson, J.D., *Proc. of Phys. Soc., 70,* 6, 1957.

43. Lenihan, J.M.A., "Advances in Activation Analysis", *Nucleonics, 13,* 65, Feb. 1965.

44. Lubin, M.J. and A.P. Fraas, "Fusion by Laser", *Scientific American, 224,* 50-64, June 1971.

45. Meinel, A.B. and H.P. Meinel, Physics Today, 44, Feb. 1972.

46. Meitner, L. and C.R. Frisch, *Nature, 143,* 239, 1939.

47. Metz, W.D., *Science, 180,* 4092, June 22, 1973.

48. Mills, R.A., *IEEE Trans. Nucl. Sci.,* MS-*18,* 205, 1971.

49. Mukaibo, T., "Japan and the Energy Crisis", *Nature, 249,* 5459, June 21, 1974.

50. Murray, R.L., *Introduction to Nuclear Engineering,* Prentice-Hall, Inc., 1961.

51. Mustafa, N., Chauri, G.A., Ahmad, M. and A.A. Khan, *Notes on Industrial Radiography,* Pakistan Institute of Nuclear Science and Technology, 1975.

52. Norris, J.A., "The Guide to Siting of Nuclear Power Plants", Supplement No. 1 to Vol. 19, *Transactions of Am. Nucl. Soc.,* 1974.

53. Onega, R.J., *An Introduction to Fission Reactor Theory,* Univ. Publications, P.O. Box 47, Blacksburg, VA 24060 ,1975.

54. Piraux, H., *Radioisotopes and their Industrial Applications,* Phillips Technical Library, 1964.

55. Post, R.F., *Physics Today, 26,* 31, Apr. 1973.

56. *Reactor Physics Constants,* ANL-5800, 2nd Ed., US, AEC,1963.

57. Roentgen, W.C., *Harpers Scientific Memoir,* 1898.

58. Rose, D.J. and M. Clark, Jr., *Plasmas and Controlled Fusion,* John Wiley and Sons, 1961.

59. Rutherfor, E., *Philosophical Magazine, 47,* 109-116, 1899.

60. Schultz, M.A., *Control of Nuclear Reactors and Power Plants,* McGraw-Hill Book Co.,1961.

61. Seaborg, G.T., *Nuclear Power Two Years after Geneva,* Press Release, S-38-66, U.S. AEC, 1966.

62. Searl, M.F., *Fossil Fuels in Future,* TID-8209, U.S. AEC, 1960.

63. Seldernath, T.R., *World Energy Requirements and Resources in the*

Year 2000, Proc. 4th UN Conference on Peaceful Uses of Atomic Energy, Sept. 1971.

64. Simon, A., *An Introduction to Thermonuclear Research,* Pergamon Press, Inc., 1959.

65. Skrotzki, G.A. and W.A. Vopat, *Power Station Engineering and Economy,* McGraw-Hill Book Co. 1960.

66. Stephenson, R., *Introduction to Nuclear Engineering,* McGraw-Hill Book Co.

67. *Standard Format and Content of Safety Analysis Reports for Nuclear Power Plants,* Revision 1, U.S. AEC, Oct. 1972.

68. Strauss, S.D., "Starting Up Nuclear Power Plants", *Nucleonics, 23,* 45, Mar. 1965.

69. Teller, E., *The Constructive Uses of Nuclear Explosives,* McGraw-Hill Book Co.

70. Thomson, J.J., *Philosophical Magazine, 44,* 293-311, 1897.

71. United Nations, *Science and Technology for Development, 2,* Report No. E/Conf. 31/1, 1963.

72. Understanding the Atom Series, US AEC, 1969.

73. Valente, F.A., *A Manual of Experiments in Reactor Physics,* MacMillan Book Co. 1963.

74. Villard, P., *Comptes Rendus, 130,* 1010-1012, 1900.

75. Wang, C.H. et.al., *Radiotracer Methodology in the Biological, Environmental, and Physical Sciences,* Prentice-Hall, Inc. 1975.

76. WASH-1248, *Environmental Survey of the Uranium Fuel Cycle,* USAEC, 1974.

77. Zweifel, P.F., *Reactor Physics,* McGraw-Hill Book Co. 1973.

SUBJECT INDEX